Medizinische Länderkunde
Geomedical Monograph Series

6

KOREA

Springer-Verlag Berlin Heidelberg 1980

Medizinische Länderkunde

Beiträge zur geographischen Medizin

Geomedical Monograph Series

Regional Studies in Geographical Medicine

Schriftenreihe der / Series of Monographs of the
Heidelberger Akademie der Wissenschaften · Mathematisch-naturwissenschaftliche Klasse
Begründet von / Founded by

Ernst Rodenwaldt †

Herausgegeben von / Edited by

Helmut J. Jusatz

em. ord. Professor Dr. med., Direktor des Instituts
für Tropenhygiene und öffentliches Gesundheitswesen am
Südasien-Institut der Universität Heidelberg i. R.
Leiter der Geomedizinischen Forschungsstelle
der Heidelberger Akademie der Wissenschaften

Unter Mitarbeit von / In collaboration with
Prof. Dr. med. Hans Jochen Diesfeld, Direktor des Instituts für Tropenhygiene und öffentliches Gesundheitswesen
am Südasien-Institut der Universität Heidelberg · Dr. rer. nat. Heinz Felten, Säugetierabteilung des Forschungsinstituts
Senckenberg, Frankfurt/Main · em. Prof. Dr. phil. Hermann Flohn, Direktor des Meteorologischen Instituts der Universität Bonn i. R. · Professor Dr. rer. nat. Ulrich Freitag, Institut für Anthropogeographie, Angewandte Geographie
und Kartographie der Freien Universität Berlin · em. Prof. Dr. phil. Gerhard Piekarski, Direktor des Instituts für medizinische Parasitologie der Universität Bonn · Prof. Dr. rer. nat. Ulrich Schweinfurth, Direktor des Instituts für Geographie am Südasien-Institut der Universität Heidelberg

KOREA

A Geomedical Monograph of the

REPUBLIC OF KOREA

by

Chin Thack Soh

M. D., Dr. M. S.
Professor of Parasitology
Director of the Institute of Tropical Medicine
Yonsei University Seoul

with Cartographical Contributions of

Dr. Eckart Dege

Geographisches Institut
Universität Kiel

With 45 Photos, 18 Figures, and 5 Map-Plates

Prof. Chin Thack Soh
Institute of Tropical Medicine, Yonsei University
P. O. Box 71, Seoul, Korea

Additional material to this book can be downloaded from http://extras.springer.com.

ISBN 978-3-642-67137-1   ISBN 978-3-642-67135-7 (eBook)
DOI 10.1007/978-3-642-67135-7

Library of Congress Cataloging in Publication Data. So, Chin-t'ak. Korea : a geomedical monograph of the Republic of Korea. (Geomedical monograph series ; 6) Bibliography: p. Includes index. 1. Medical geography — Korea. I. Title. II. Series : Medizinische Länderkunde ; 6. RA934.K6S6   614.4′22519   79-23434   ISBN 978-3-642-67137-1

Production of the Maps 1 – 5: Geomedizinische Forschungsstelle der Heidelberger Akademie der Wissenschaften.

Vorwort des Herausgebers

Mit diesem Band 6 erscheint in der Reihe der Medizinischen Länderkunden die Darstellung eines Landes aus der Sicht eines Angehörigen dieses Landes, dessen Wunsch es ist, der Entwicklung seines Landes, der Republik KOREA, durch eine geographische Betrachtung der Probleme der Gesundheits- und Krankheitsverhältnisse nützlich zu sein. Wir haben diese Anregung des Verfassers, Professor Chin Thack Soh, M. D., einen umfassenden Beitrag aus dem ostasiatischen Raum über medizinisch-geographische Probleme seines Landes zu geben, dankbar aufgegriffen und verbinden mit der Herausgabe dieses Bandes zugleich die Hoffnung, daß auch in anderen Ländern der Dritten Welt sich Wissenschaftler finden werden, eine geomedizinische Betrachtungsweise ihrer Länder folgen zu lassen.

Die Schwierigkeiten bei der Abfassung einer Medizinischen Landeskunde sind dem Geographen wahrscheinlich sehr viel bewußter als dem Mediziner, noch dazu, wenn der Verfasser nicht in europäischer wissenschaftlicher Tradition wurzelt und eine Fülle von Material gesammelt hat. Es ist dann die Medizinische Landeskunde in Gefahr, aus dem besonders großen Umfang der Fragestellung und aus der Fülle des Materials zur Enzyklopädie auszuufern. Wenn der Verfasser der vorliegenden Medizinischen Länderkunde von KOREA stärker zu einer enzyklopädischen Behandlung seines Themas neigt, als die Verfasser der bisher veröffentlichten Bände Libyen, Afghanistan, Kuwait, Äthiopien und Kenia, so halten wir es nicht allein wegen der Bedeutung des Informationswertes, sondern ganz besonders auch für den Vergleich geomedizinischer Probleme seines Landes mit jenen anderer Länder und Regionen für vertretbar. Es bedeutet eine außergewöhnliche Leistung des Verfassers, wenn er ein umfangreiches Material geordnet zugänglich und greifbar für den Benutzer vorbereitet hat. Durch die Beschreibung von Einzelheiten, deren Bedeutung für das öffentliche Gesundheitswesen des Landes wahrscheinlich nur von einem Einheimischen so in das Blickfeld gerückt werden kann, darf auch diese Art enzyklopädischer Behandlung des Themas als wertvoller Beitrag für die Reihe der Medizinischen Länderkunden aufgefaßt werden. Die Fülle wissenschaftlicher Angaben über KOREA ist, soweit hier bekannt, noch nie in einem derartigen Umfang zusammengestellt worden.

Wenn es als vordringlichste Aufgabe der Medizinischen Geographie erscheint, brauchbare Medizinische Länderkunden abzufassen, so ist dabei die Behandlung politisch begrenzter Gebiete nur ein Kompromiß, der aus praktischen Gründen akzeptiert werden muß. Dem geomedizinischen Gedanken würde eine Darstellung natürlicher Räume besser entsprechen.

Im vorliegenden Falle bereitet die Gebirgsnatur des Landes der medizinisch-kartographischen Behandlung gewisse zusätzliche Probleme, weil sie zu bedeutenden Unterschieden in Landnutzung und Besiedelung geführt hat.

Zur Überwindung dieser Schwierigkeiten erhielten wir Unterstützung durch Herrn Dr. rer. nat. Eckart Dege vom Geographischen Institut der Universität Kiel. Durch einen Aufenthalt von mehreren Jahren als Visiting Professor an der Kyung-Hee University in Seoul hat er große Kenntnisse des Landes erworben und die von ihm entworfenen Karten über Orographie, Verwaltungsgliederung, Bevölkerungsverteilung und agrargeographische Verhältnisse zur Verfügung gestellt. Unter seiner Anleitung hat außerdem Herr Woo-ik Yu, Department of Geography, Seoul National University, zur Zeit am Geographischen Institut der Universität Kiel tätig, nach unserer Anregung eine Karte der Verteilung der Ärzte im Verhältnis zur Einwohnerzahl und der Krankenhäuser mit Angabe der Bettenzahlen entworfen.

Die beigegebenen Karten stellen die gegenwärtigen Verhältnisse der Republik Korea dar und sollten von Zeit zu Zeit durch Eintragungen neuer Daten eine Ergänzung finden, wie überhaupt die vorliegende medizinisch-geographische Übersicht über die Krankheitsverhältnisse und das Gesundheitswesen der Republik Korea Anregungen geben soll, die aufgezeigten Probleme im geomedizinischen Sinne weiter zu untersuchen.

Bei der weiteren Herausgabe von Medizinischen Länderkunden haben nach dem Tod der bisherigen Beiratsmitglieder, der Herren Berthold Carlberg, Ludolph Fischer und Carl Troll, die Herren Professor Dr. Ulrich Freitag, Institut für Anthropogeographie, Angewandte Geographie und Kartographie der freien Universität Berlin, und Professor Dr. Hans Jochen Diesfeld, Direktor des Instituts für Tropenhygiene und öffentliches Gesundheitswesen am Südasien-Institut der Universität Heidelberg, für die wissenschaftliche Beratung über Fragen ihrer Fachgebiete ihre Mitwirkung zugesagt.

Helmut J. Jusatz

Foreword of the Editor

With Volume 6 Korea of the Geomedical Monograph Series written by a scientist born in Korea, the health conditions and the diseases of his country, the Republic of Korea, are published as a contribution to the geomedical problems of East Asia. We have gratefully welcomed this suggestion of the author, Professor Chin Thack Soh, Director of the Institute of Tropical Medicine, Yonsei University, Seoul, who wishes to promote the development of his country, in presenting a comprehensive survey on health and disease in the Republic of Korea. With the publication of this volume we at the same time hope that in other third-world countries scientists will follow in his steps with a geomedical study of their countries.

The difficulties in composing a geomedical monograph are probably much clearer to the geographer than to the physician, especially if the author does not have roots in the European scientific tradition and has collected a wealth of information. In such a case medical geography, due to the especially broad scope of the questions posed and the profuse information, runs the risk of overflowing its borders and becoming an encyclopedia. Even if the author of this geomedical monograph of Korea does tend to an encyclopedic treatment of his theme to a greater degree than is the case in previously published volumes on Libya, Afghanistan, Kuwait, Ethiopia and Kenya, we consider it justified not only because of the importance of its informational value but also and in particular for the comparison of geomedical problems in his country with those in other countries and regions that it affords. It is an exceptional accomplishment of the author to have organized such extensive information in an accessible and handy form for the user. With the description of details, whose relevance for public health in Korea can probably only be put into perspective by a scientist born in Korea, this type of encyclopedic treatment of the theme can also be interpreted as a valuable contribution to this series of geomedical monographs. This wealth of scientific data on Korea has to our knowledge never before been collected in such scope.

If the most important task of medical geography is the writing of useable regional studies in geographical medicine, the treatment of politically defined areas is only a compromise that must be accepted for practical reasons. A depiction of areas delimited by natural borders would better suit the geomedical concept.

In the case of Korea, certain additional problems as regards medical cartographic treatment have been posed by the mountainous nature of the country which has led to significant differences in land usage and settlement.

In overcoming these difficulties, we received support from Dr. Eckart Dege of the Geographic Institute, University of Kiel. During a stay of several years at the Kyung-Hee University in Seoul as visiting professor, he acquired a broad knowledge of the country. He has made available his designs of the maps on orography, administrative organization, population distribution, and geographic agrarian conditions. Under his guidance, moreover, and at our suggestion, Mr. Woo-ik Yu of the Department of Geography, Seoul National University and presently at the Geographic Institute, University of Kiel, created a map of the distribution of doctors relative to population and hospital figures, including the number of beds.

The added maps depict the current conditions in the Republic of Korea and should be brought up to date with entries of new data. Indeed, in general, this geomedical survey on health conditions and diseases in the Republic of Korea should encourage further geomedical investigation into the problems discussed.

Following the deaths of the former members of the advisory council, Mr. Berthold Carlberg, Mr. Ludolph Fischer, and Mr. Carl Troll, Professor Dr. Ulrich Freitag, Institut für Anthropogeographie, Angewandte Geographie und Kartographie der Freien Universität Berlin, and Professor Dr. Hans Jochen Diesfeld, Direktor des Instituts für Tropenhygiene und öffentliches Gesundheitswesen am Südasien-Institut der Universität Heidelberg, have consented to cooperate on scientific questions involving their specializations in further editions of the Geomedical Monograph Series.

Helmut J. Jusatz

Preface

For over 30 years I have been engaged as a parasitologist in research on endemic diseases in our land. However, I have been somewhat dissatisfied within my heart of hearts from the point of view of a medical person. Most of us, myself included, handle disease sectionally, not comprehensively. Clinicians pay more attention to finding effective drugs, medical scientists concentrate more effort on clarifying pathologic etiology, and public health workers are more concerned with environmental sanitation. Thus, most of us generally neglect to search out the causal relation of a certain disease. For a disease to be established various factors must be involved: agent, ecology, hosts, carriers, transmitters, habits, geographic and geologic conditions. It is true that the prevalence of respective diseases differs by area, and some disease show endemicity. For this reason, a comprehensive understanding is needed to solve the problems of individual diseases.

Going one step further, it seemed to me that medicine from the geographic viewpoint (geomedicine) would be the primary step toward understanding and solving nosologic problems in individual regions.

During my visit to Athens for the Ninth International Congress on Tropical Medicine and Malaria in 1973, Dr. Boris Velimirovic, Chief of the Field Office of Pan American Sanitary Bureau, El Paso, U.S.A., introduced me to Professor Dr. Helmut J. Jusatz, Director of the Institute of Tropical Hygiene at the South Asia Institute of the University of Heidelberg and to the team of the Geomedical Research Unit of the Heidelberg Academy of Sciences. While enjoying a magnificent view of the "Acropolis" sanctuary from the roofgarden of the eight-storied Astor Hotel, our topics gravitated to geographic medicine. Being very interested in this subject, I was pleased to learn that several volumes have already been published on the subject of geomedicine. Therefore, it was perhaps inevitable that the suggestion be made to add one more monograph to the series, this time about Korea. Since the proposal coincided with a long-cherished wish, I fully accepted the advice of my sincere friend, Boris Velimirovic. However, the preparation began only with my visit to Springer-Verlag, Heidelberg, on 15 January 1976. The publishing agreement with Professor Dr. W. Gei-

nitz, Springer-Verlag, Heidelberg, was arranged in the presence of Professor Jusatz. It was indeed a great honor for me, but I also felt a very heavy responsibility; how to set up the scheme, how to collect the materials, how to digest, analyze and compile them in accordance with the original goals. I frankly confess to realizing how limited my knowledge was at the same time. Poor editing would result in dishonor not only to myself, but to my country as well.

Now at this juncture, the point of editing, I am filled with deep emotion. In spite of limited time and knowledge, this manuscript will be prepared for edition at all events. At this moment, I have to extend my heartful thanks to all concerned: Professor Dr. Helmut J. Jusatz for his encouragement and his staff of the Geomedical Unit Heidelberg, Dr. Boris Velimirovic for his instructions on how to set up the compilation scheme. Dr. Eckart Dege, Geographisches Institut der Universität Kiel, who generously shared me valuable cartographs: Professor Bo-ki Sohn, Director of Yonsei University Museum who greatly assisted me to describe the main portion of historical background of Korea; Professors Sa-Suk Hong, Il-Soon Kim, Dong Woo Lee, Ki-Ho Kim, Samuel Lee, and Chung Yong of Yonsei University Medical College for supplying me with valuable literature, Professor Chong Hwi Chun of the Catholic Medical College for his life-long collection of literature on acute communicable diseases and his kind corrections, Professor Kwang-Il Kim of Hanyang University for his valuable references on shamanism and folk medicine, the National Institute for Agricultural Technology for providing me with the soil maps, Professor Yong No Lee of Ehwa Woman's University for pictures of flowering plants, Mr. Chang Chull Chung and Mr. Si-Heun Lee, my sincere friends, for their generous suggestions on the preparation of the manuscript and photographs, Dr. Joo-Su Lee of the Ministry of Health and Social Affairs, Mr. Ik-Soon Park of the Ministry of Public Information, Mr. Jeung-Man Seo of the Ministry of Science and Technology, and Director of Central Meteorological Office in Seoul for their kindness in providing me with governmental publications. My sincere thanks is also extended to Miss Marion Current in the

Physiotherapy Clinic, Severance Hospital, for preparing the English manuscript. Miss Hyun-Sook Cheon, secretary in my laboratory, typed the complete manuscript. All staffs, especially Mr. Sang-Joon Kim, in the Department of Parasitology, Yonsei University Medical College, helped in preparing photos and manuscripts. Unfortunately, space does not allow me to add all the names of those who have helped me in various ways. Nevertheless, I have to express my apology to and appreciation of all the authors whose valuable references are cited in this description. My last thanks is to the publisher, Springer-Verlag, Berlin – Heidelberg – New York for its willingness to cooperate in the layout of this monograph and to the printer of the maps, Henning Wocke, Karlsruhe.

Seoul, Korea
January 1980

Chin Thack Soh

Contents

XIV

Contents

Figures

Maps

Introduction

The belief that diseases are closely related to "Wind and Soil" (Pung-tuo) has prevailed in Korea since ancient times. When we give this term a more modern translation, "wind and soil" relates to environment, which consists mainly of two factors — nonliving and living. Among the former, we may count soil, water, and air; and in the latter, plants and animals. The interrelationship between these two factors comprises the environment.

A harmonious environment offers good living conditions, but if the balance is disturbed, pathologic phenomena may be the consequence, i.e., geomedical problems. Diseases in human society are thought to be caused by these pathologic phenomena of the environment. For this reason, abnormal conditions in our society cannot be overlooked, nor should they be treated separately. A medical problem on the level of an individual country may have its own historical background for many varied reasons, and Korea is in this regard by no means an exception. Medical problems must first be analyzed from the viewpoint of historical, sociocultural, and environmental factors to find a clue to their cause, thus laying the groundwork for establishing a principle of effective prevention.

The contents of this monograph will be organized in three phases: environmental background, health problems, and analysis. In the course of data collection, however, some difficulties were encountered, e.g., the political barrier between the north and the south, and crude statistics collected in the health field. In the past, the Korean peninsula has experienced several foreign invasions by China and Japan. However, King Sejong (1449 AD) during the Yi dynasty expelled all the invaders into the peripheral region of the land and fixed the borders of this land [279]: from Manchuria and Siberia in the north, mainland China to the west up to the Yellow Sea (*Hwanghae*) and from Japan to the east up to the Sea of Japan (*Donghae*). Unfortunately, this land has been separated into north and south by two politically different regimes, Communism and Democracy, since World War II in 1945 [5, 21].

By the decision of the United Nations' General Assembly in 1948, the Republic of Korea was established and authorized to administrate all the land. But unfortunately the administrative power of the Republic of Korea is limited to the area south of the 38th parallel. Even though the land is inhabited by people of the same blood with similar environmental conditions, language, traditions, and customs, differences between the people of the north and the south have probably arisen during the past 30 years of separation.

Among these differences, the cultural gap of the people in the two parts is probably more serious than the environmental changes, but of this we cannot be sure because no information is as yet exchanged officially between the two parts. For this reason no reliable data on the morbidity and its related problems in North Korea are now available. Most descriptions relating to the north must remain only in the realm of conjecture; nevertheless, some facts can be gleaned since the 38th parallel is merely a symbolic line in the political sense and not a physical wall. Whatever the cultural and political differences might be, the land, the natural environment, genetic factors, and the geographic similarities will be the same. This endorses the view that the geomedical environment is similar in spite of the artificial segregation. The disease of today in the north will become the same problem in the south tomorrow, and vice versa. This is the reason why I refer to and introduce available data about North Korea in the respective chapters.

In South Korea (the administrative region of the Republic of Korea), highly valued development has been achieved in various fields during the past two decades. There is a saying in our land: "After ten years of time even the rivers elapse and mountains change." In past times the mortality rate was so high and the life span so short that when a person reached 60 years of age, he was honored and cherished by the people.

But now the situation has changed. Since the first "Five-year Economic Development Plan" was established in 1962, the shape of the land has been changing morning and evening. The New Village (*Saema-ul*) Movement in the third "Five-year Economic Plan" played an active role in achieving this cultural improvement. The annual per capita income was about U.S. $ 100 in 1966; it increased about tenfold in 1979. Along with the economic development, physical, behavioral, and cultural changes have been taking place. Oriental philosophy and old feudal practices are being converted to a more western style with moderate success. But other problems have arisen in the course of this development, namely, ignorance of our traditional heritage and increase of mental diseases, land erosion, public nuisances, and various other side-effects of modern life. Such factors influence the pattern of geomedical diseases. It is true that control of infectious diseases has met with success due to the continuous efforts and concern of the government; however, ecological changes due to dam construction, massive mobilization of the inhabitants, and rapid urbanization have resulted in other disease problems. In the past, treatment of disease was everything in medicine, but nowadays the emphasis is changing to preventive medicine. More importance is being attributed to social medicine than to individual diagnosis and treatment, and the reason should not be difficult to understand.

The matters constituting public nuisances are air, water, and soil, and these items are not owned by a single individual, but by the community. Increasing transportation may spur the spread of diseases. Dam construction along big rivers may modify the ecology, thus inviting new epidemics of various snail-borne or vector-borne diseases. In a way social changes are intermingled with public health problems. Unless the contradictions are resolved, the problems will be multiplied.

The primary concern of this monograph is to review the disease problems in the country by examining the backgrounds of culture, tradition, climate, geography, geophysics, and biologic relationships. The secondary concern is to analyze the factors of all health problems so as to resolve them quickly and efficiently. During the collection of information some difficulties inevitably arose in compiling reliable statistics on morbidities. For example, it is highly probable that reports of localized outbreaks of certain diseases never reached the health authorities in the capital. The shortage of well-trained statisticians in various levels of clinics might be another reason. Nevertheless, the author has put a lot of effort into ascertaining the latest data on each respective item. My last concern is to prepare this monograph for the benefit of our people and to introduce Korea's health situation to neighboring nations with the aim of improving mutual understanding.

Korean Words

In the text, some Korean words may be unfamiliar to the reader:

Honam,	an alias of Jeonra Do (Jeonra Nam Do and Jeonra Bug Do) and a part of Chungcheong Nam Do;
Yeugnam,	an alias of Gyeongsang Do (Gyeongsang Nam Do and Gyeongsang Bug Do);
Gang,	river, e.g., Nagdong Gang (Nagdong river);
San,	mountain, e.g., Jiri San (Jiri mountain);
Do,	province, e.g., Gangweon Do (Gangweon province);
Nam,	south, e.g., Jeonra Nam Do (south Jeonra province);
Bug,	north, e.g., Jeonra Bug Do (north Jeonra province);
Shi or *Si,*	city;
Gun,	county;
Gu,	administrative district in a city;
Bu,	old name of Shi;
Hyeon,	old name of Gun;
Myeon or *Eub,*	administrative unit (town level) in a Gun;
Ri or *Dong,*	village;
Pyeong,	3,24 m²;
Jeongbo,	3000 pyeong;
South Korea,	administrative area of the Republic of Korea for convenience sake;
Won,	money unit, e.g., U.S. $ 1 equals 485 won in 1976 currency;
Seog or *Seom,*	1 *seok* equals 47.6 U.S. gallons or 5.1 bushels.

Footnote by the Editor

A publication of geomedical data from Korea necessitates the transliteration of geographical names from the non-Latin Korean script into an internationally understandable Romanized script. As an expression of international respect for the national decision on the form of Romanization of its own script, international usage in production of maps, etc., should follow the form of script fixed by the government of the respective country with non-Latin script at the Third Conference of the United Nations in Athens (8 August – 7 September 1977) as the official spelling of their geographical names. We thank the Korean Embassy in Bonn, in particular the attaché for Education and Science, Mr. Chung-Gil Kim, for the information that the Ministry of Education had already published in 1959 a system, called M.O.E. system, for the international Romanization of Korean geographical names. We have complied with the request and Romanized according to the M.O.E. system in both text and map sections of this volume.

A. The Land and Its People

I. Historical Background

1. Race

Koreans form a race with a single language and a relatively homogeneous gene pool. Anthropologically the cephalic index of the Korean show considerable difference from that of the Chinese and the Japanese. The average cranial capacity of a modern Korean man, about 1490 cm^3, is classified as aristocephalic [23, 25, 26, 27, 28, 29, 30, 31, 32, 33, 34].

The characteristic physical features of Koreans, straight dark hair, slit eyes, straight nose, high cheek bones, and Mongolian eyelid fold, can be interpreted to reflect links with the people of central Asia, the Baikal region, Mongolia, Manchuria, and the coastal areas of the Yellow Sea.

2. Blood Type

As for blood types Koreans are akin to Mongoloids. According to the AB0 system they are as follows: Type A=30.3% (A$_1$ 30.03 and A$_2$ 0.70), B=25.8%, AB=9.7% (A$_1$B 9.7% and A$_2$B 0%). The MN system shows M= 29.8%, N=30.7%, and NM=39.4%. The gene frequencies are m=0.4952, n=0.5048, p=0.226, q=0.197, and r=0.585 [35, 36, 37]. By subgroup, the A$_2$ is very rare among Koreans, 0% – 0.27%, compared with the English, 9.7%. However, B-gene (q) is very common compared with other races; indeed this is one of the characteristics of Mongolians: Korean, 0.197 – 0.2315 and English, 0.06. Its frequency in Koreans is higher than that in Chinese and Japanese (Table 1), thus indicating that Koreans bear the closest resemblance to Mongoloids of all the nations in Asia. Even among Koreans, the distribution of genes differs by area; q (gene B) is more frequent in Kiho (Chungcheong Nam Do and Chungcheong Bug Do) and Honam areas (Jeonra Nam Do and Jeonra Bug Do). In the Yeong Nam area (Gyeongsang Bug Do and Gyeongsang Nam Do), the p (gene A) is more frequent,

but the r (gene o) becomes commoner toward northern Korea. For reference q and p (A-gene) show almost equal level in the south Asian region, but p is more frequent than q in the Australian-Polynesian group [37].

3. Language

The origin of the Korean language is linguistically poorly defined; however, it is attributed to the Altaic family due to its agglutinative, polysyllabic features. Instead of the declension and conjugation of Aryan languages, morphologic changes are effected by the addition of postposition and suffix. The concepts of gender and language, rich in vowels and consonants (10 vowels and 14 consonants). The wealth of adjectives and adverbs transmit the various shades of emotion, sensation, and state, but words expressing philosophical concepts and logical terms are less developed.

The Korean alphabet is constructed with scientific and geometric devices and is easily learned, thus greatly promoting the literacy of the general populace. Koreans are noted for their self-motivated, self-effacing, extroverted, and sociable character. Their ways of thinking are quite diverse. They are noted for their emotional inclination for art and music. The traditional religion of Koreans has assumed such forms as bear cult, sun cult, and ancestor worship, all of which are found in central Asia, Siberia, and Manchuria.

4. Chronology

Chronologically, the Korean history backs to Old Stone Age, (300 000 – 10 000 BC) followed by New Stone Age (5000 – 2000 BC), Bronze Age (2000 – 700 BC), Iron Age and Three States (700 BC – 918 AD), Koryo Dynasty (918 – 1392 AD), Chosun (1392 – 1909 AD), Japanese Control Period (1909 – 1945) and Republic of Korea since 1945.

Table 1. AB0 Phenotypes and Gene Frequencies of Korean Compared with other Races {3}

Races	Number tested	Percent of Phenotypes						Gene Frequencies		
		0	A$_1$	A$_2$	B	A$_1$B	A$_2$B	p	q	r
Korean										
(Lee, 1960)	1 208	34.1	30.03	0.27	25.8	9.7	0	0.226	0.197	0.585
(Won, 1960)	322	31.99	31.06	0	28.88	8.07	0	0.2208	0.2069	0.5723
(Sarkisian, 1956)	1 000	27.00	32.00		29.00	12.00		0.2513	0.2315	0.5172
(Furuhata, 1933)	9 434	27.71	31.50		30.72	10.07		0.2365	0.2314	0.5321
Chinese										
(Sussman, 1956)		40.4	28.6	0	25.9	5.5	0	0.195	0.179	0.635
Japanese										
(Furuhata, 1967)	3 523 474	30.58	37.70		22.25	9.47		0.273	0.174	0.553
English										
(Race, 1954)		43.6	34.9	9.7	8.5	2.6	0.9	0.2569	0.0600	0.6831

In the early period of twentieth century, Korea was still noted for being a hermit nation and the western powers and the Japanese competed in the opening of ports for trade. In this international struggle Korea was placed under Japanese control (1905 – 1945). During this period Koreans fought against Japanese control rising up with a nonviolent righteous army, and resisting by tenant farmer and labor strikes. Then great efforts were made for the establishment of national education in terms of a private school system. In 1919 there arose an universal independence movement in Korea against the Japanese Imperial Rule which had considerable impact on the Chinese May 4th movement and Gandhi's resistence movement.

With the Japanese surrender in 1945 in World War II, Korea was liberated from Japanese imperialism; however, she was split at the 38th parallel by the agreement of Soviet Russia and the United States of America. In 1948 the Republic of Korea was established in the southern part of Korea and communist government in the northern part of Korea.

In 1950 the Korean war broke out and Korea became the international fighting ground for the United Nations Army against the Communist North and mainland China. A peace agreement was drawn up in 1953. In 1960 there arose a student uprising against the Sygman Rhee regime.

Since 1961 Korea has adopted a planned economic scheme and made considerable progress in her long-term program of modernization.

II. Geographic Survey

1. General Features

Korea is a peninsula thrust out from the northeast Asia mainland in a southerly direction (see Map 1) and facing three seas: Donghae (the Sea of Japan) to the east, Nam Hae (South Sea) to the south, and Hwang Hae (Yellow Sea) to the west [5, 6]. On the north, two rivers, the Yalu (Abrog) and Duman, form a sharp boundary with Manchuria and Siberia. The peninsula with its associated islands lies within the following extremes of longitude and latitude: the eastern extremity is on the 131° 52′ 22″ east extreme of Dogdo island, Ulreung Gun, Gyeongsang Bug Do; the western extremity is on the 124° 11′ 0″, west extreme of Maando, Yongcheon-Gun, Pyeongan Bug Do. The southern extremity is on the 33° 06′ 40″, the south extreme of Mara-do, Jeju Gun, Jeju Do, and the northern extremity is on the 43° 0′ 39″, northern extreme of Yupojin, Yupo Myeon, Hamgyeong Bug Do.

The Korean peninsula narrows toward the south and projects into the South Sea, which flows into the Pacific Ocean. Its length in a north-south direction is 840 km, but the width is only 360 km, and the total land area comprises about 221 000 km², about 84 800 square miles,

the peninsula accounting for 215 000 km², and the islands, 6000 km².

The agricultural heart of the nation lies in the southern part, which produces chiefly rice and barley.

In North Korea, heavy industry and the mining of coal and iron predominate. It has approximately the same area as Great Britain or New York State. Since World War II the Republic of Korea's effective administrative area (south) by agreement covers 98 757 km², about 38 022 square miles or 45% of the peninsula, an area equal to that of Iceland or Portugal.

Korea is the hard-luck nation of the Chinese cultural realm. It has been the theater of many wars, and its most recent history includes annexation by Japan in 1909 and post-World War II occupation by Soviet and American troops north and south, respectively, of the 38th parallel. Consequently, this dormant, peaceful country was divided into a communist sector in the north and a democratic sector in the south. The land is covered mostly with mountains and hills, presenting one of the hilliest landscapes in the world. Low hills, mostly in the south and the west, give way gradually to increasingly higher mountains toward the east and the north. On the whole, the eastern and southern slope is very steep with no significant rivers or plains because of the high mountains that are very close to the east coast. Plains and basins are rather small and closely associated with rivers; thus, they are valley-plains in nature. At the headwaters of the two rivers, the Yalu and Duman, lies Mt. Baegdu, the highest mountain in Korea, with an elevation of 2744 m. The mountain, an extinct volcano, has a large crater lake. South Korea is dominated by the Taebaeg mountain range paralleling the east coast. This maturely dissected block with summits at heights from 1500 to 1700 m in elevation slopes gently to the west but abruptly descends to the east.

2. Mountain Ranges

Eighty percent of the peninsula is mountainous; however, the mountains are at low altitudes in general, except for the Gaema plateau in the north and the Taebaeg range in the east. The ranges are divided into three groups accordingly to direction: Taebaeg, Macheon-ryeong, and Nangrim range from the northwest to the west; Chang-Baeg, Gangnam, Myohyang, Hamgycong, and Myeolag range from the west to the south; and Masigryeong, Dongryeong, Sobaeg, and Noryeong ranges from the northeast to the southwest. The Gaema plateau, 1500 m above sea level, surrounded by the Chang-Baeg, Nangrim, and Hamgyeong range, is called the roof of Korea. High mountains above 2000 m are distributed about the surrounding area. Of these Baegdu mountain is adored as the main mountain of the land. The Nangrim mountain range and the Taebaeg mountain range together divide the east from the west like a spine. The Taebaeg mountain range with a total length of 500 km forms a steep slope on the east coast side of the peninsula, but is more gradual on the west side. The average height of the mountains in the range reaches 1000 m, long serving as a barrier to transportation between the east and the west. Famous mountains such as Geumgang (Diamond mountain), Seolag, O-Dae, and Taebaeg are located in this range. The Sobaeg mountain range divides the southern part of the peninsula into three regions: Yeung-Nam (Gyeongsang Bug Do and Gyeongsang Nam Do), and Hoseo (Chungcheong Nam Do), and Honam (Jeonra Bug Do, Jeonra Nam Do, and a part of Chungcheong Nam Do).

3. Volcanoes and Hot Springs

Past ages must have witnessed living volcanoes; crater lakes are still found at the summit of some mountains, the most famous being Mt. Baegdu in the north and Mt. Hanra in the south. Many hot springs with some extinct volcanoes give some credence to this assumption. Of these the Jueul in Hamgyeong Bug Do, Baegcheon in Hwanghae Do, Onyang and Juseong in Chungcheong Nam Do, Dongrae and Haeundae in Busan city, and Baegam in Yeong-Il bay area on the Donghae coast are nationally well-known resorts specializing in thermotherapy. The water temperature is 55° – 56 °C, slightly alkaline, but saltish and colorless. It has been frequented by people suffering from digestive diseases, skin diseases, bronchitis, and gynecological troubles. The temperature of the springs at Haeundae is 45° – 50 °C; the water is alkaline and saltish (Map 4) [38 a, b].

Even though the springs have been popularized in recent years by the inhabitants, scientific bases for these claims are fragmentary. Recently Yang [38 a, b] measured the metal and nonmetal levels in these springs by atomic absorption spectrometer and also ran chemical analysis on each with titrisol. The results indicated that most of the springs were weakly alkaline (7.7 – 8.4), except for Baegam, Cheogsan, and Osaeg with pH 8.5 – 9.5 range. Differences in chemical compounds were found; for example, H_2S was found only in Do-go and Baegam, which are located in the north. The hot springs in the southern half contain larger amounts of sodium and chloride than those in the middle part of the country. Boron was found in only trace amounts in the latter area in contrast to the southern part, and the fluoride level is higher in the springs in the northern area. Lithium and strontium levels are also detected at 0.04 mg/liter – 0.38 mg/liter and 0.15 mg/liter – 21.0 mg/liter, respectively, in all the the springs, and a correlation between strontium and calcium was noted [38 b]. The mineral water springs are also not to be excluded. They have been shown to be carbonate in nature. Springs of Sambang in Hamgyeong Nam Do and of Chojeong in Chungcheong Bug Do have been recognized as panaceas by people suffering from many sorts of physical troubles. The latter is famous because King Sejong (1443 AD) almost completed the draft of the Korean alphabet at Chojeong spring while being treated there for diabetes mellitus and eye trouble.

4. Rivers and Plains

Since the Nangrim and Taebaeg mountain ranges, which form the spine of the peninsula, are located on the extreme eastern side, all the main rivers in Korea, except the Duman river, flow in a westerly direction. The Yalu (Abrog) and the Duman originate at the peak of Baegdu mountain and form a natural border between Manchuria and Korea. The big rivers from the north are the Abrog, Duman, Cheongcheon, Daedong Han, Geum, Yeongsan, Seomjin, and Nagdong. Rivers north of the Han are frozen during the winter season, making water transportation impossible. Most of these meandering rivers are surrounded by plains. On the eastern side, only limited plain areas are found along the small rivers: the Suseong plain along the Suseong gang (river), Hamheung plain along the downstream portion of the Seongcheon gang, and the Yeongheung plain along the downstream part of the Yeongheung gang. In contrast, there are many large plains on the western side. Beginning in the North, they are Cheongcheon, Anju, Bagcheon, Suncheon, the basin of the Pyeongyang, Jaeryeung, Yeonbaeg, the Chuncheon basin, the Cheongju basin, Gyeonggi, Anseong, Dangjin, Honam, Naju, Namwon basin, and the Gurye basin. Along the Nagdong river, the Yeongju, Andong, Sangju, Gimcheon, Geum-Ho, and Gimhae basins are plains that contribute to the agriculture of the South. There are other basins along the Namgang, one of the tributaries of the Nagdong river, namely the Hamchang, Jinju, Jinyeong, Geochang, and Habcheon basins.

5. Seashores

Except for the northern border, the Korean Peninsula is surrounded on three sides by seas: the Donghae (The Sea of Japan) to the east, the Namhae to the south, and the Hwanghae (Yellow Sea) to the west. The total length of the coastline, including the coastline of the islands, is recorded at 17 269 km. The length of each coast is: 1727 km on the east side, 2246 km on the west side and 4719 km on the south side. The east coast is rather simple and close to the Taebaeg mountain range, for which reason it has never really been developed. The bays at Donghan, Yeongheung, Yeong-Il, and Ulsan are all well-known bays along the east coast. A few islands are scattered along this coast also. For geographic reasons ports have not been developed here to any great extent in comparison to the other coasts. Only at Weonsan Bay there is a sandy hill along the bay coast. Cold and warm currents meet in this region, thus establishing it as a good fishing area as well as a good tourist spot. The west coast has a good precipitation line with high, favorable winds and many offshore islands. The bays of Seohan, Gwangyang, Haeju, Ganghwa and Mooân, as well as the peninsula of Ongjin, Taean, Usuyeong, and islands such as Ganghwa, Yeonpyeong, Deogjeog, Anmyeon, and Eochong all make the area ideal for good fishing. The well-developed continental shelf favors the growth of good marine products. The tidal difference exceeds

9 m and is usually utilized to make a salt garden cultivatable in many other ways. The Namhae coast, known as the Dadohae (many islands sea) has stronger winds and more precipitation than the west coast. Over 3000 islands add more than 300 m to the circumference, and about 2200 of these islands are scattered off the southern shore.

Among the largest of these are Jeju Do, located at a distance of 140 km from Mokpo coast, Jindo, Wando, Geomundo, Namhaedo, and Geojedo island. Bays are found at Jindo, Gangjin, Boseong, Suncheon, Namhae, and Jinhae cities. Haenam, Yeosu and Goseong are three peninsulas. In this area the tidal difference is only about 2.5 m. The maintenance of clean natural conditions in the region favors the production of rice, laver, seashells, and seaweed. A good wind ensures good harbors for fishing, sea-transportation, and naval stations, located mainly at Busan, Masan, Jinhae, Yeosu, and Mogpo.

6. Islands

Statistically 3421 islands are found surrounding the Korean peninsula. Of these islands 831 are inhabited, 705 in South Korea and 126 in North Korea. The uninhabited islands are widely scattered, 2195 in the South and 395 in the North. The total size of the inhabited islands south of the truce line (38th parallel) is 3653 km², they have 1 100 945 inhabitants according to the 1973 census. Their area comprises 3.4% of the total land area, and the population ratio is about the same. By province, the 402 inhabited islands with the largest concentration of people are found in Jeonra Nam Do. Jeju Do, the largest island of Korea, measuring 1792.06 km², forms an independent administrative province. Next comes Geojedo island, which is 383.44 km², Ganghwa island, 319.82 km², and Jindo island, 319.65 km². The smallest inhabited island is Maeseom (Maedo island), 10 m² in size and inhabited by one family, located in Idong Myeon, Namhae-Gun, Gyeongsang Nam Do. The southernmost island is Marado in Jeju province, 71 km from the Jeju provincial office, 300 m² with 30 families and 130 inhabitants. The easternmost island is Dogdo, an army garrison and a lighthouse are the only inhabitants of this rocky place. The westernmost island is Maan in Yongcheon, Pyeongan Bug Do, and the northernmost is Pungseogdong in Namyang-myeon, Eumseong Gun, Hamgyeong Bug Do. The uninhabited islands with a total size of 11 250 km² range from the smallest island being a rocky 50 m² and to the largest being 900 m², called Sa-do, in Bucheongun, Gyeong-gi Do.

7. Geology

South Korea consists of two related Precambrian schist and gneiss units, the Yeongnam and Gyeonggi massifs, separated by the northeast trend of the Ogcheon Geosynclinal Zone of Precambrian to early Mesozoic metamorphic and sedimentary formations. In the south-eastern part of the areas, the Gyeongsang Basin consists

mostly of Jurassic-Cretaceous sedimentary and volcanic rocks, and tertiary sedimentary basins with basalts are found on the east and southwest coasts and offshore islands (Table 2) [22, 24, 278, 279, 280, 281].

Numerous granitic bodies, elongated parallel to the prevailing northeasterly structural trend, intrude into the Precambrian massifs and the Ogcheon Zone, associated with the Jurassic Daebo Orogeny, and Cretaceous (Bulgugsa) granites scattered throughout the Gyeongsang Basin. Two periods of metamorphism and igneous intrusion are represented in the Precambrian zone by granite gneisses and other tectonic and/or igneous activity recognized in the Paleozoic, the Triassic, and the Tertiary.

Structures are associated with each major tectonic episode, and most follow the northeasterly (sinian) trend except for those in the northern part of the area where trends are north-northeasterly. Metallogenic epochs occurred during the Precambrian, Paleozoic, Jurassic to early Cretaceous, and late Cretaceous to early Tertiary and appear to have coincided with orogenic and igneous activity. Minerals of the latter two epochs include gold, silver, lead, zinc, copper, molybdenum, tungsten, and magnetite, with also some tin, fluorite, and nickel in the Jurassic-Cretaceous period. Mineralization occurred mainly in zones parallel to the prevailing tectonic trends, except in the Gyeongsang Basin. Plate tectonic explanations of the geologic and tectonic development of South Korea have been made, but these are speculative as to their relationships. Rocks of acidic igneous origin constitute roughly 65% (igneous rocks, 35%, metamorphic rocks, 30%) of the total area of the country, with the remaining 35% of sedimentary origin (sedimentary rocks, 25%, metamorphic rocks, 10%) distributed inbetween the igneous stocks [39, 40, 41].

Mining began 2000 years ago during the Sam-Han Dynasties. Gold, silver, copper, and iron mines and the use of these metals have been historically verified. However, modern geologic surveys and mining techniques were only introduced in the early 1920 s. Since systematic long-term exploration plans have been carried out only from the 1960 s, mining exploration of the country has not reached its desired level.

Of the 287 minerals known to occur naturally in the country, 30 minerals are mined for economic purposes, and 52 minerals are defined as legal minerals by the Mining Law. Ore minerals related to the acidic igneous rocks, include gold, silver, copper, lead, zinc, tungsten, fluorite, kaoline, pyrophyllite, and silica sand. Those of sedimentary origin include iron, limestone, anthracite, quartzite, talc, and graphite. Generally, the ore deposits are small in comparison to the large mines throughout the rest of the world.

Characteristics of the domestic mineral deposits are (1) the presence of a few epithermal deposits containing mercury and antimony; (2) the absence of platinum and diamond in basic igneous rocks; (3) the absence of native sulfur of volcanic origin; and (4) no deposits of petroleum and bituminous coal, found in thick sedimentary

Table 2. Generalized Geologic Sequence in South Korea {280}

Age		C. H. Cheong (1956, 1970) System	Series	O. J. Kim (1973) System	Series		
Cenozoic		Quaternary		Quaternary	basalts		
		Tertiary	Yonil	Tertiary	Yonil		
			Janggi		Janggi		
Mesozoic	Cretaceous	Kyongsang	Bulkuksa (granite & porphyry)	Kyongsang	granites (Bulkuksa granites) volcanic rocks		
			Silla ·········· intrusion		Silla		
			Naktong granites (in N. Korea)		Naktong ——— granites (Taebo granites)		
	Jurassic	Daedong	Ryukyong / Sonhyon	Daedong	undifferentiated in S. Korea granite?		
	Triassic	Pyongan	Nokam	Pyongan	Nokam		
Palaeozoic	Permian		Kobangsan		Kobangsan		
	Carboniferous		Sadong		Sadong		
			Hongjom		Hongjom granite?		
	Devonian		Chonsongri?		(hiatus)		
	Silurian						
	Ordovician	Chosun	Great Limestone	Chosun	Great Limestone		
	Cambrian		Yangduk		Yangduk granite?		
Precambrian	Proterozoic	Sangwon	Kuhyon	Okchon	Kunjasan granite?		Late-Pre-cambrian
			Sadangwu (in N. Korea)		Hwanggangri		
			Jikhyon		Changri		
					Munjuri		
					Hyangsanri		
					Kemyongsan		
		granite gneiss			Taebaeksan granite gneiss		
	Archeozoic			Yulri	Kosonri	Chunsong	Middle-Pre-cambrian
					Kakhwasa granite gneiss	Jangraksan (Chunchon)	
		Yonchon		Ryongnam	Wonnam	Yangpyong	Early-Pre-cambrian
					Kisong	Sihung	
					Pyonghae	Puchon (Yonchon)	

sequences. However, unlike the land area, distribution of thick sedimentary beds has been detected by geophysical survey in 120 000 km² of the 350 000 km² of the offshore area. This fact encourages the continued detailed survey for natural gas and/or petroleum.

The production of copper and iron ore does not meet domestic demand and must be imported from other countries, but lead, zinc, tungsten, silica, talc, kaoline, pyrophyllite, graphite, and fluorite have been exported. Ores such as limestone, gold, and silver are exceeding domestic demand and can be exported in the future. Especially, flux materials for iron refineries (limestone, fluorite, silica), refractor materials (clays, pyrophyllite, dolomite, silica), mould materials (silica, sand, bentonite, oliving), construction materials (sand, gravel, stones, expanded shale), rocks (limestone, marble), glass materials (silica, sand, quartzite), cement materials (limestone, gypsum, clays), ceramic materials (porcellanite, kaoline, fire clay, potassium, quartz trachyte, pyrophyllite, feldspar), fertilizer materials (limestone, dolomite, serpentine), fertilizer material (bentonite), electrode material (graphite), absorbents (acidic clay, diatomite), filling materials (kaoline pyrophyllite, talc, sericite), insulating materials (asbestos, diatomite), polishing materials (silica, sand, garnet) are to be developed in the future technology and mining.

Offshore deposits including sand, gravel resources, and offshore energy resources, such as natural gas and petroleum which do not occur in the land area, should be explored more effectively since the country is surrounded by the sea on three sides.

A soil survey in Korea, begun in 1963, was organized in 1964 by a joint project with UN-FAO and is still underway today. The classification is based on morphologic, physical, and chemical features of the samples. Presently 138 kinds of soil have been identified, but surface soils in South Korea have been estimated to comprise about 300 kinds.

The soil texture is not uniform, but rather diverse depending on area. More sandy and sandy-loam soils are found in the mountainous and hilly regions, whereas fine clay or clay over fine, silty soils are distributed in the low lands [278, 280].

III. Climate

1. General Features

The climate of Korea more nearly resembles that of central and northeast China than that of Japan, and the year is clearly divided into four distinct seasons, spring, summer, autumn, and winter. There are considerable climatic contrasts between winter and summer. Because of the location of the peninsula on the margin of the Siberian

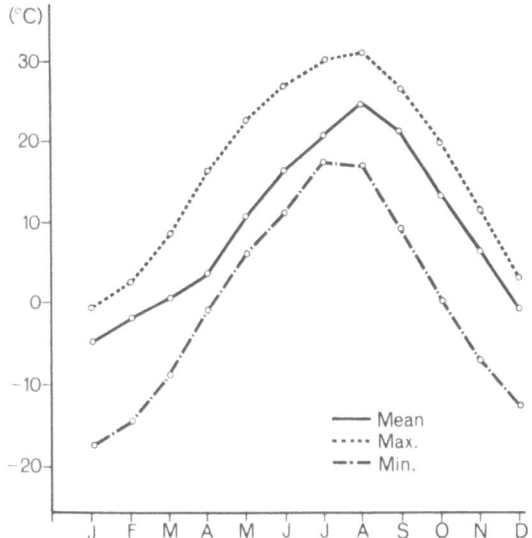

Figure 1. Monthly temperature in Seoul by means (1931 – 1960)
(Source: Climatic Table of Korea Part 1. Published by the Central Meteorological Office, R.O.K., 1968)

winter high pressures, monsoonal drifts of cold air masses move from the north and west during the cold season [1, 2]. Cyclonic storms, especially strong in the spring and fall, bring some snow occasionally, adding nonperiodic variations to the cold and dry winter. Snow may stay long in the north, but it melts quickly in the mild temperatures of the south. The northern interior has bitterly cold winters. Only the extreme southern fringe of the peninsula has had January temperatures above freezing. The frost-free period varies from 130 days in the northern interior (Jung-gangjin), to 179 days in the center (Seoul), and to 226 days in the south (Busan). Summers are hot and humid with a marked concentration of the annual rainfall. Much of this rainfall is convectional in origin and occurs in thunderstorms. Regional temperature contrasts are not so sharp in summer as in winter, although the northern interior and northern littoral are cooler than in the south.

Precipitation varies mainly according to orographic position: the highest amounts (over 150 cm) are in the south in the Sobaeg range; the least (less than 62 cm) are in the sheltered upper Yalu basin of the north. Since most of the precipitation occurs during the growing season, agricultural needs are normally well supplied. Torrential storms, caused by typhoons, may do much damage, but they are rare. Sharp seasonal contrasts are typical, as shown by temperature extremes in Seoul where the maximum is above 30.5 °C (mean) and the minimum below – 17.1 °C (mean) (Fig. 1). The proximity to the continent explains these extremes. Naturally they are greater in the interior than in the south or along the coast.

In general, the eastern and western parts of Korea show climatic differences due chiefly to monsoon, climatic relations between land and sea, and the influence of ocean currents. Similarly, the islands of Japan, located near a vast continent, are considerably affected by the

climate of the continent, although surrounded by water. However, since Korea is connected directly with the Asian continent, the difference between warm and cold is more severe than in Japan. A comparison with the peninsula of Italy, situated at the same latitude as Korea, reveals even more remarkable thermal differences between north and south. Furthermore, Korea is so long in the north-south direction that great climatic differences are found between the northern and southern parts.

On the other hand, as in the case of all maritime countries, the climate of Korea is profoundly affected by the prevailing oceanic circulation. Two principal ocean currents flow in the waters adjacent to Korea. One is the Daima warm current, which flows from the East China sea to the northeast along the east coastline of Korea, and the other is the Liman cold current, which flows southward along the coastline of Korea from the Okhotsk sea. These two currents meet in the neighborhood of Ulreungdo Island. The sea where the two currents intersect is one of the best Korean fishing grounds, which is fast becoming world-reknown.

2. Temperature and Humidity

According to the climatologic standard mean of the 25 stations in Korea during the period 1931 – 1960, five temperature zones cover the country, from the northern part of the land (Pyeongan Bug Do and Hamgyeong Bug Do) to southernmost Jeju Do. The annual mean temperatures are 2.5° – 5.0 °C in the north, 5.0° – 7.5 °C in Hamgyeong Bug Do and Nam Do, 7.5° – 10 °C in Pyeongan Bug Do, Nam Do, and Hwanghae Do areas, 10.0° – 12.5 °C in the middle portion of the land (36° to 38° N), and 12.5° – 15.0 °C in the southernmost zone. The mean maximum temperatures in roughly these same area, progressing from north to south are 10.0 °C, 12.5 °C, 15.0 °C, and 17.5 °C. Annual soil temperatures at 50 cm also divide roughly into five zones from north to south: 7.5 °C and below in the northmost of Hamgyeong Bug Do, 7.5° – 10.0 °C in Hamgyeong Nam Do, 10.0° to 12.5 °C in Pyeongan Do and Hwanghae Do, 12.5° to 15.0 °C in Gyeong-gi Do, Gangweon Do, Chungcheong Do, Jeonra Bug Do, and Gyeongsang Bug Do, 15.0 °C and above in Jeonra Nam Do, Gyeongsang Nam Do, and Jeju Do.

Atmospheric humidity is high during the summer months (June – August), within the range of 71% – 89% and low in winter months (December – February), within the range of 49% – 78%. The annual atmospheric humidity varies by area within the range of 65% – 70%. Jeonra Bug Do and Gangweon Do show 65%, but 70% is found in other part.

a) Air temperature: The annual mean temperature distribution of Korea shows the warmest temperatures (14 °C) in the southern part, and the coolest temperature (3 °C) in the Gaema plateau and its neighboring area.

The temperature difference between South and North Korea is usually 7° – 11 °C (Fig. 2, 3). A comparison of

MEAN AIR TEMPERATURE(°C) ANNUAL

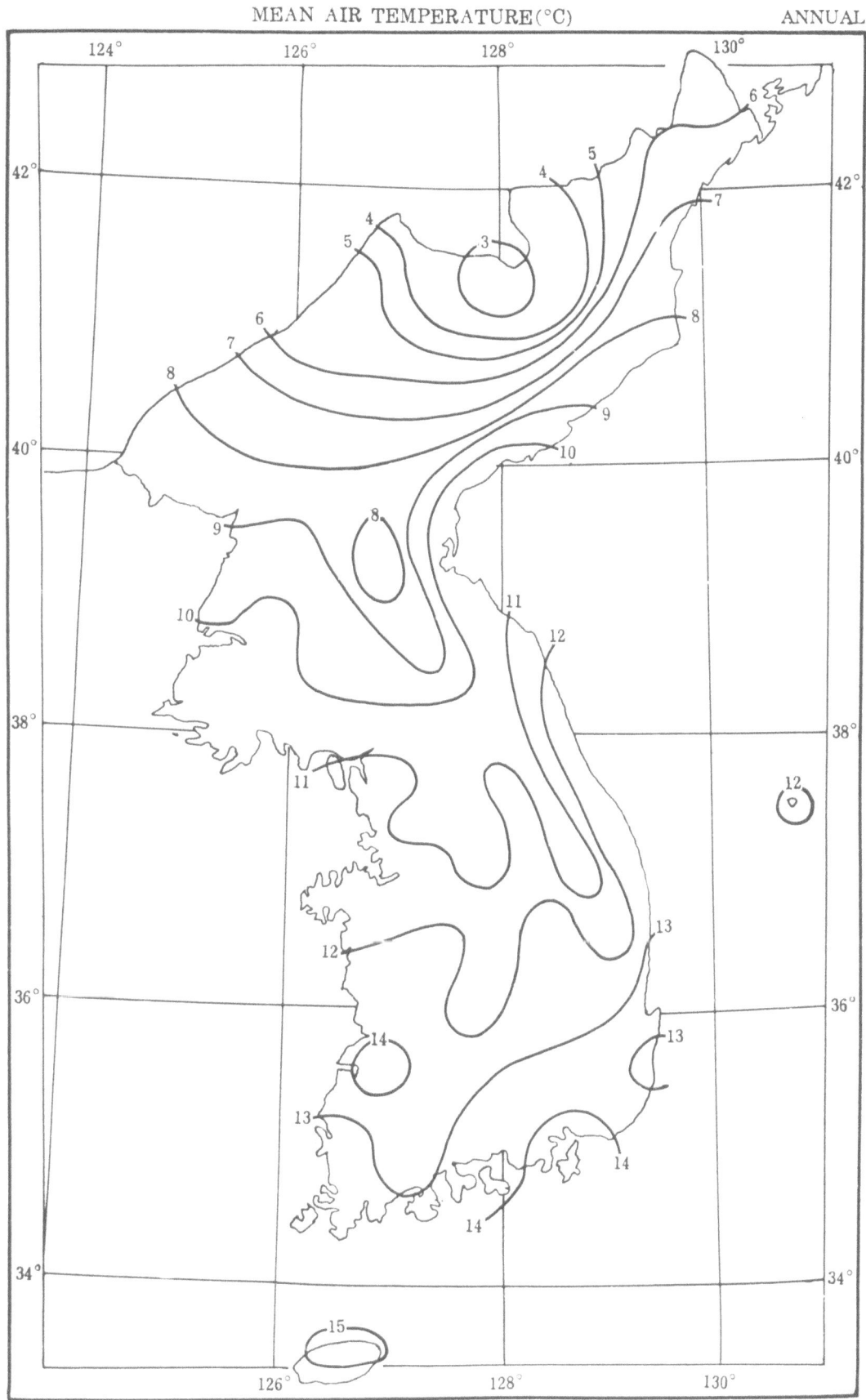

Figure 2. Mean air temperature (°C) Annual
(Source: Climatic Atlas of Korea (1931 – 1960) Central Meteor. Office, Seoul)

MEAN AIR TEMPERATURE (°C)   JANUARY

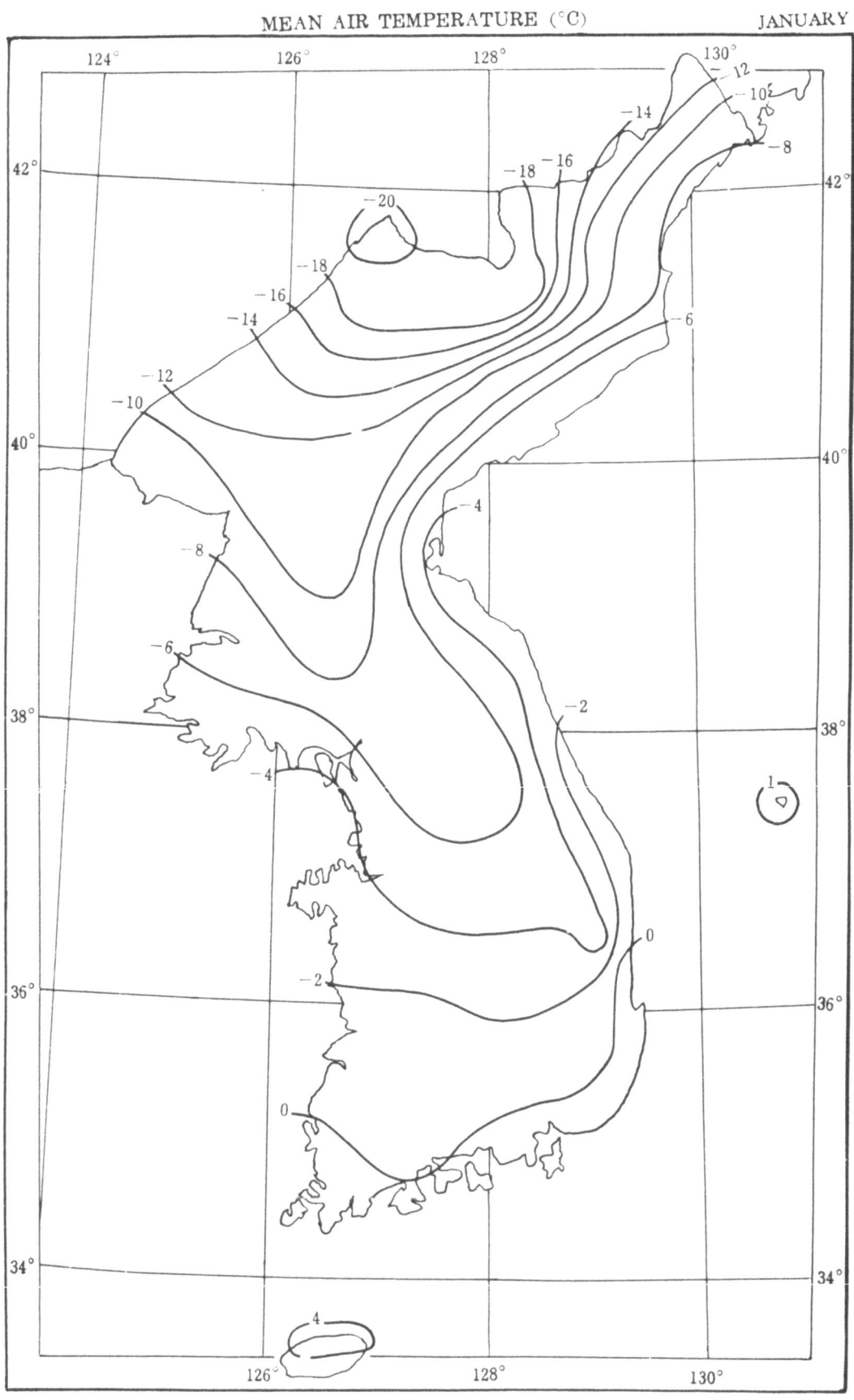

Figure 3. Mean air temperature (°C) January
(Source: Climatic Atlas of Korea (1931 – 1960) Central Meteor. Office, Seoul)

MEAN AIR TEMPERATURE(°C)   AUGUST

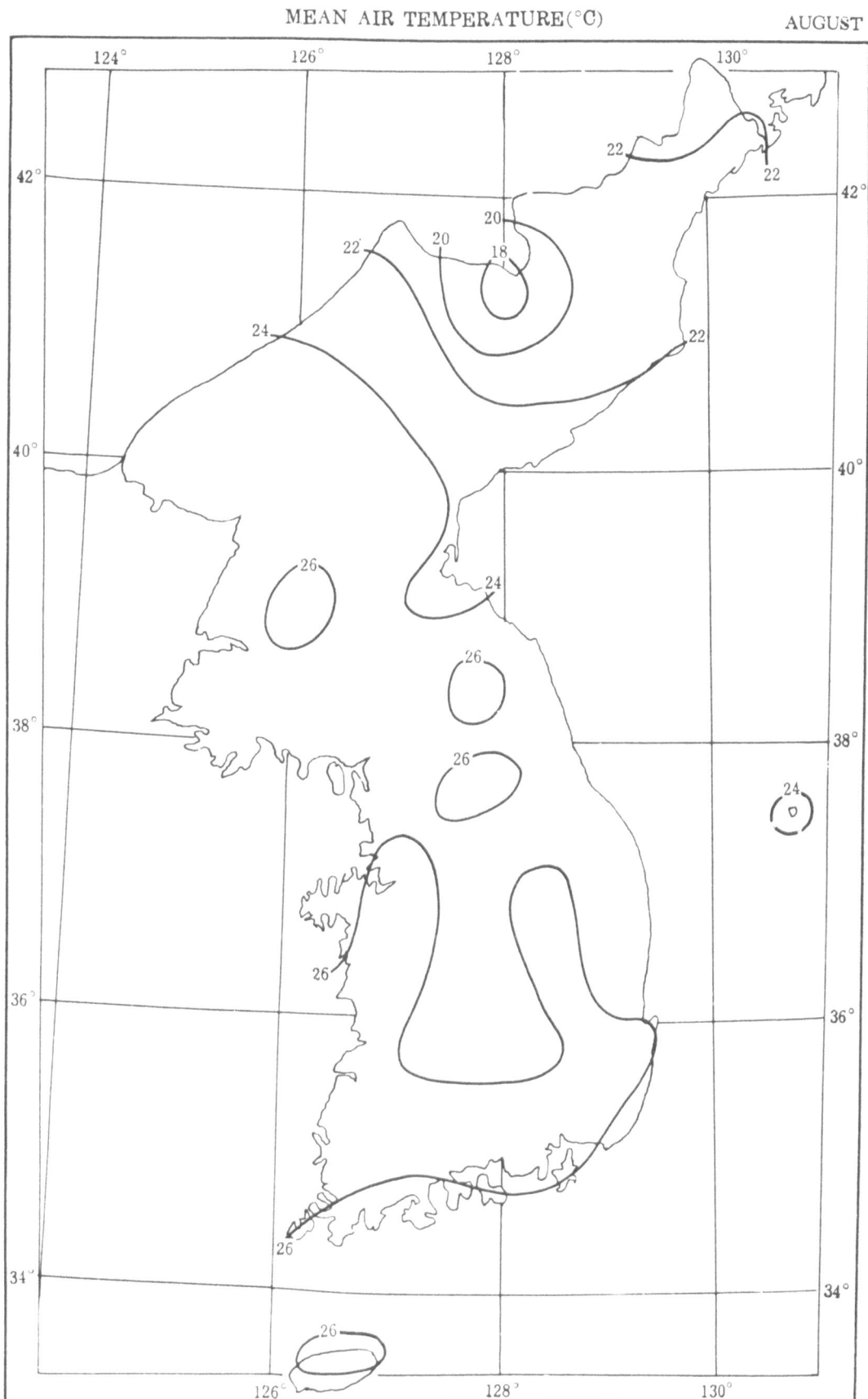

Figure 4. Mean air temperature (°C) August
(Source: Climatic Atlas of Korea (1931 – 1960) Central Meteor. Office, Seoul)

the temperatures of the eastern and western parts shows that the temperature of the east coast is generally higher than that of the west coast, and that in the island is generally lower than in the coastal areas. Such phenomena are caused chiefly through the influence of mountain ranges and the monsoons. As Korea is bounded on the north by Manchuria and Siberia, parts of the largest continent in the world, the climate of Korea has a modified continental character. Winter and summer are rather longer than spring and autumn.

b) Winter: One of the climatic characteristics of winter in Korea is its relatively long duration below 0 °C. The duration and coldness of winter show pronounced differences between the southern and northern parts, and between inland and shore areas. The number of days with minimum temperature below 0 °C is about 210 days in the northern inland sections, 165 in the middle inland and north coast area, 135 in the middle coast areas, and 120 in the south. The coldest month is January, in which the warmest temperature is 5 °C on Jeju island, 1 °C on the south coast, and – 3° to – 5 °C in the Gyeongin district (Seoul area).

The average minimum temperature in January is about 2 °C on Jeju, – 2 °C on the south coast, – 5 °C in the southern inland and east coast sections of middle Korea, – 7.5 °C in the Incheon area, which is 2 °C higher than that of Seoul, – 13 °C to – 14 °C on the west coast of North Korea, – 10° to – 14 °C on the west coast of South Korea, – 10° to – 13 °C in Pyeongyang, Sineuiju, and the east coast sections of North Korea, and – 29° C in the Jung-gangjin district, located inland near the north boundary. The absolute minimum temperature recorded to date is – 43.6 °C at Jung-gangjin on 12 January 1933. This is a very rare phenomenon for this area, which belongs to the temperate region geographically. In general, winter temperature is responsible for the land's harsh climate.

Wind velocity in winter is relatively weak, and there are many clear days. The rise and fall of continental high pressure systems are so periodical that such phenomena in winter show a unique climate that is popularly described as "Three days cold and four warm." The winter climate is generally healthy because it is relatively dry with abundant sunshine and short-term temperatures which fluctuate relatively little.

c) Summer: When the winter is past, the surface temperatures of the land increase gradually so that plants begin to bud in early spring. Plum, peach, and cherry trees bloom on the whole during the same period. Cherry trees usually bloom early in April on the southern coast, in the middle of April in middle Korea, and the latest is early in May in Hamgyeong Nam Do. These dates are 10 – 15 days after the last day with minimum temperature below freezing. If we define summer as days when the maximum temperature is above 25 °C, summer comes first at the end of April in Gyeongsang Bug Do and the southern coast of Gangweon Do, in the north part of Gyeongsang Bug Do and south part of Hamgyeong

Nam Do early in May, and on the northeast coast early in June.

The last day on which the average maximum temperature rises above 25 °C is early in September in the northern part of Korea, in middle Korea, from the end of September to early in October, and in the southern part of Korea, in the middle of October. Thus, summer continues about 5½ months in the central portions, and in the northern parts of Korea, usually 3 – 4 months. The annual absolute average maximum temperature is about 37 °C in the Daegu area. Consequently spring and autumn are much shorter in Korea than winter and summer. Summer months typically have a muggy climate (Fig. 4).

3. Wind

a) Monsoons: The influences of land and water due to the unequal heating of the continents and the oceans make the land warmer than the water in summer and colder in winter. Likewise a low over the land in summer and a high in winter is produced. The air circulation by such natural phenomena produces unique seasonal variations of the prevailing wind directions. Since Korea is a peninsula located between Asia and the Pacific Ocean, wind from the northwest prevails in the winter and from the southwest in summer.

The northwest monsoon prevails about 8 months of the year, from September to April. It is especially predominant in January and is also so dry, stable, and arid that the winter weather of Korea is relatively clear and cold. The southwest Monsoon from the sea in summer commences about May and predominates from the middle of July through August. As southwesterly winds are very moist, unstable, and rainy, the weather generally includes cloudy skies, rainy days, and high temperatures. The annual average wind velocity over the country is 3 – 5 m/s. Regionally, the annual average maximum of 5 m/s in Jeju shows the maximum within the country, and the minimum, 1.5 m/s, is found in inland areas. A survey of the monthly variation of wind velocity shows that the east and south coasts have a maximum in January. For example, Jeju has an average of 6.7 m/s in January, and Unggi in the east, 6.3 m/s. The lowest monthly wind velocity occurs in June in the coastal areas across the country and in October in inland areas. The lowest wind velocity is 0.6 m/s in January at Jung-gangjin.

b) Cyclones: Cyclones affecting the weather of Korea usually originate in two regions. One is in northwestern China, Mongolia, and Siberia and the other, in the Yangtze river basin. These cyclones move eastward and pass through Korea or her neighboring waters and then head for the Japanese islands. Most of the lows originating over northern China, Mongolia, and Manchuria appear in winter, and while these lows stay inland, their intensities are weakened and the occurrence of lows below 980 mb is very rare. However, after passing over the Korean peninsula, the cyclones often develop to 960 mb due to the influence of the warm sea surface. When they arrive over

Hokkaido island and the northwest part of Japan facing the Sea of Japan, they are frequently accompanied by heavy snowstorms.

On the other hand, the lows originating over the upper parts of the Yangtze river basin appear chiefly in the spring and summer, only occasionally in winter. These lows move gradually eastward along the Yangtze river. When they approach the Yellow Sea or the East China Sea, they develop rapidly and move toward Korea or the western areas of Japan. These lows are accompanied by heavy precipitation. Consequently, the rainy season in Japan commences in late June and the rainy season in Korea begins soon after it ends in Japan.

c) Typhoons: Typhoons that strike Korea usually originate in the area covering 7° N through 25° N and 120° E through 180° E, including the Marshal islands, Guam, and the Mariana island areas. The majority of typhoons move slowly northwestward at first, but after they reach the East China Sea near 30° N, they turn to the northeast and finally pass over the Ryukyu islands, Japan and its adjacent seas, or over Korea or China.

The minimum pressure recorded at the Busan weather station in Korea up to the present time is 951.5 mb during the passage of typhoon Sarah on 19 September 1959. The typhoon season of Korea extends through July, August, and September. An average of about three typhoons pass by Korea every year. However, since most of these are weak when approaching Korea, the damage caused by typhoons is not as heavy on the average as in Japan.

Nevertheless, one severe typhoon accompanied by rain causes severe damage on the average of at least once every other year in southern Korea, once every 4 years in middle Korea, and once every 6 years in northern Korea. For example, typhoon Carmen, which passed over Korea in August 1960, was very strong, having a diameter of about 1288 km, covering an area of more than 1 280 000 km², moving at 36 KTS, and carrying maximum winds of 48 KTS. Furthermore, along the Korean coast Carmen created 50-ft waves, which submerged 1500 houses in Busan, sank one ship, and caused a flood that stranded 2000 persons. The death toll was 24 persons, and property damage caused by this storm was estimated at more than two million U.S. dollars.

4. Precipitation

The usual method of representing rainfall is by means of an average annual total (Fig. 5).

The average annual precipitation in Korea is about half that of Japan. Hamgyeong Nam Do and Hamgyeong Bug Do and the Daedong river basin are the areas with minimum precipitation, the annual total of this area being less than 800 mm. In the central part of the Duman river basin, many parts of Hamgyeong Bug Do, and the northwest parts of Hamgyeong Nam Do, precipitation does not exceed 300 mm. The Abrog-Gang (Yalu river) basin and upper parts of the Nagdong river have less than 1000 mm, but the rest of the area mentioned above

receives 1000 mm or more. Further, the Chuncheon, Seomjin, and the middle parts of the Han river basin, and the Honam plains, the coastal area connecting Wonsan and Kangnung city, receive more than 1300 mm. The mouth of the Seomjin river gets the maximum amount in this area with 1500 mm annually. The amount of rainfall is determined chiefly by the paths of cyclones and the effects of local topography. Weather phenomena are greatly influenced by the height and horizontal extent of the mountain ranges as well as the angle between fronts and the main direction of the ranges.

Owing to the predominant Föhn phenomena caused by the Taebaeg mountains running in a north-south direction near the east coast, we find minimum rainfall east of the mountains and inland in Gyeongsang Bug Do. Much rainfalls in the area containing the southern parts of Changbaeg and Nangrim mountain ranges and west of the Taebaeg mountain range. Further, due to the Noryeong ridge and Jiri mountain, which rises to a height of 1815 m, the Seomjin river basin west of Jiri mountain is the maximum rainfall area. As to the seasonal distribution of precipitation, only about 15% of the average annual rainfall occurs in the dry season, the period from October to March, and about 85% occurs from April to September. The rainfall in July and August, the rainy season in Korea, reaches up to 40% – 60% of the average annual rainfall.

The record of heavy rains observed in Korea shows that in six areas the rains have exceeded 400 mm in 24 h. Accordingly, in summer season the heavy rains cause quite significant damage to the economy of the country.

5. Evaporation

The transfer of water from the ground to the atmosphere by evaporation and transpiration from various crops under different atmospheric conditions can be calculated by measuring the water vapor and the transfer properties of the atmosphere near the ground.

However, the evaporation data shown herein was determined from evaporation-pan measurements. The annual average record of sunshine in Korea is roughly 500 h, more than that in Japan. Also the vapor content is 20% less than that in Japan; thus, the atmosphere is comparatively dry except during July and August. The average amount of evaporation is about 1000 – 1500 mm, and its distribution shows that evaporation over the northern part and east coast is smaller than over the western district and the west coast.

The amount of evaporation on Jeju island is around 1500 mm, the greatest for the whole country and the next highest is the Daegu area. In these districts, the annual total evaporation is much greater than the amount of precipitation. Evaporation is extremely small in January, gradually increasing to reach a maximum value in May on the east coast of middle Korea, in June on the northern border district and inland, and the maximum on the northern coast and southwest coast occurs in August. The maximum to date recorded in one day in

AMOUNT OF PRECIPITATION (mm) ANNUAL

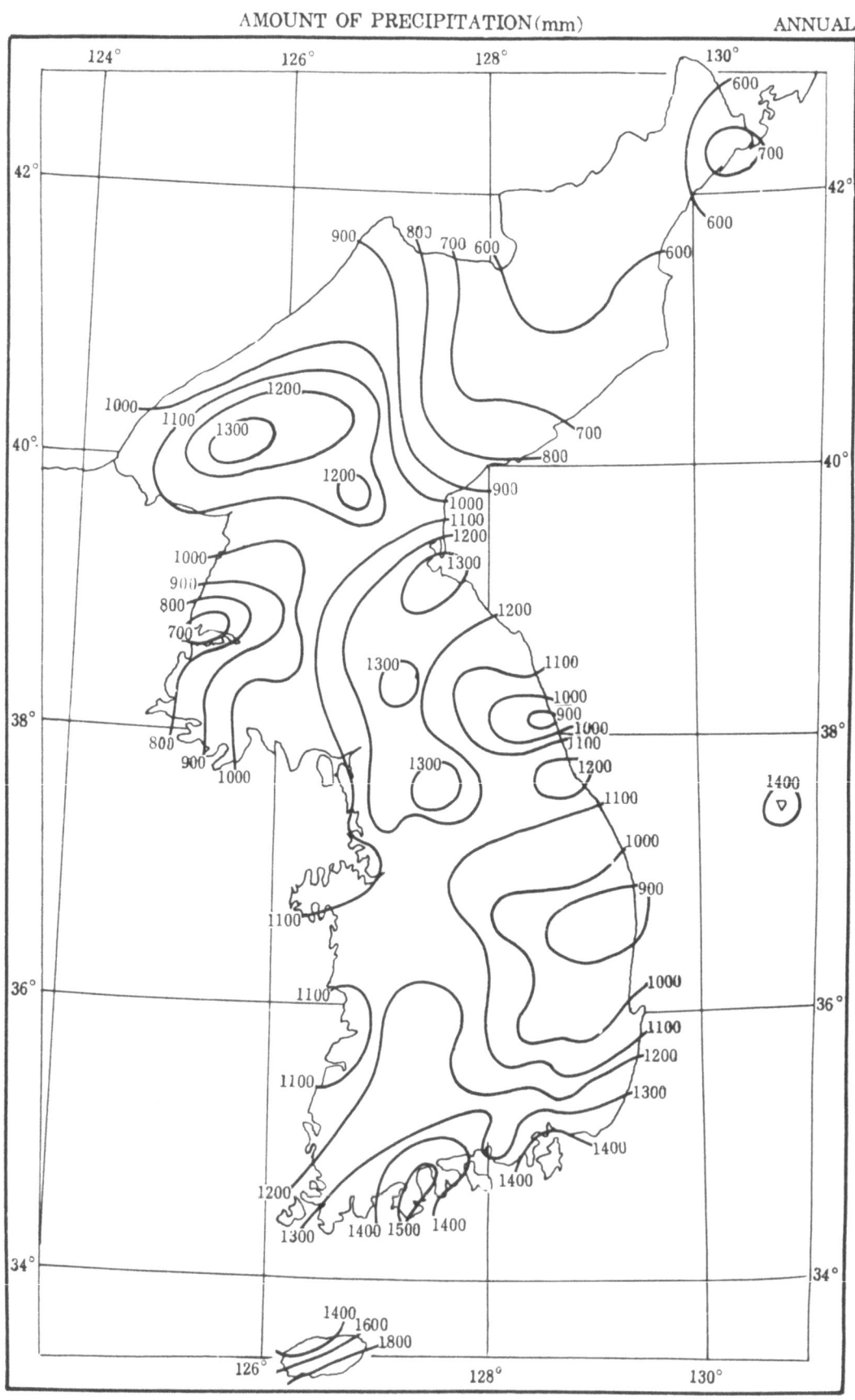

Figure 5. Amount of precipitation (mm) annual
(Source: Climatic Atlas of Korea (1931 – 1960) Central Meteor. Office, Seoul)

Korea is more than 200 mm, observed on Jeju island. In the Munsan district of Gyeong-gi province about 17 mm of evaporation in one day has been recorded.

6. Frost and Snow

The dates of first frost in Korea show large regional differences. The Gaema plateau has the earliest frosts in the whole country, in early September. In other areas of northern and middle Korea the first frost occurs early in October; in the inland parts of southern Korea, about the middle of October, and in the southern coastal areas it occurs early in November. On Jeju island the first frost occurs in the middle of December, later than in any other area. The last frost of the year is just the opposite of the date for the first frost; the last frost on Jeju island is the earliest in Korea and is normally early in March. On the southern coast the last frost of the season usually occurs from late in March to early in April, and in the other areas on the whole, during middle to late April. However, in the Jangjin and Gabsan areas, the centers of the Gaema plateau, the earliest recorded date of the first frost in Korea was 21 August 1916, and in Pungsan the latest recorded frost occurred on 22 June of 1921.

The average date of the first snow in Korea also shows regionally notable differences similar to that of frost. The average first snow in Korea appears generally about 1 month later than the first frost. The Gaema plateau has the earliest snowfall, on the average about early or middle October. In the northwest area the date is early in November and in the southeast coastal area, late in November or early in December. The average date of the last snow on Jeju island and the coast is early in March, over the inland parts of the southern areas, in middle March, over the middle areas and west coast of the northern part of Korea, in late March, and over the Gaema plateau, early in May. Therefore, the Gaema plateau experiences not only the first snowfall of the season but also the last.

IV. Flora and Fauna

1. Natural Vegetation

The peninsula, narrowing toward the south, has a length in a north-south direction of 840 km, but the width is only 360 km. The northern latitudes lies between 33° 44′ and 43° 2′.

Because of the long north-south extension of the land and its complicated topography, there are great variations in temperature and rainfall. The average temperature through the four seasons ranges from 5° to 14 °C, and the rainfall, from 500 to 1500 mm. Such a natural environment yields a diversified floral region with 201 families, 1102 genera, 3347 species, 50 subspecies, 1012 varieties and 168 forms of higher plants, including pteridophyte. Thus more than 4500 plants grow in the country

as compared to some 1500 species in Denmark or some 2000 species in England [13].

Many northern plants have common elements with those growing in Manchuria. Whereas in the north and high mountain areas many alpine plants are seen, the central part and the western lowlands show the prominent vegetation of the temperate zone such as broad-leaved deciduous trees. The southern coast and the offshore islands of Jeju and Ulreung are warm regions where temperate plants grow abundantly. Many evergreen plants growing in the southern parts are identical or similar to those seen in the southwestern part of Japan.

Although an enormous number of plant species grow in Korea which have common elements with those growing in neighboring countries, the aforementioned unique natural environment has caused the emergence of many endemic species. The study of flora in Korea was initiated by the Russian botanist, Palibin, in 1899 and was all but completed in 1945 by the Japanese botanist, Nakai, with the cooperation of many Korean and Japanese scholars. The study was taken over by a group of Korean botanists led by professor Cheung Tae-Hyon after World War II.

Warm-temperate vegetation: A synopsis of Korean vegetation shows climatic characteristics, because of the high average annual temperatures (14 °C) prevailing over the southern part and the offshore islands — Jeju, Soheugsan, and Ulreung — where numerous plant species grow. On the shorelines of Jeju island more than 70 species of broad-leaved evergreens in diversified growth, e.g., *Camillia japonica, Cinnamomum camphora, Ficus pumila,* and *Citrus nobilis,* are distributed.

Ulreung island, located at 37° 30′ N. Lat., is also the growing area of many plants of the warm zone such as *Daphniphyllum glaucescens, Camellia japonica,* and *Neolitsea serica* (Blume). Because of the high contour of the temperate zone, *Camellia japonica* and *Neolitsea serica* (Blume) are distributed as far north as Daecheong island of Hwanghae province due to the warm sea currents.

Temperate vegetation: The Korean peninsula belongs to the temperate zone, except for the high terrains of Mt. Hanra on Jeju and the Taebaeg mountains, which traverse the land in the east, and those islands where plants of the warmer zone grow. The temperature averages between 5° – 14 °C over the land, which lies between 35° and 43° N. Lat. In this temperature zone *Pinus densiflora* and other deciduous and broad-leaved trees grow abundantly. Typical deciduous broad-leaved trees growing here are *Quercus sp., Carpinus tschonoskü, Rhododendron sp.,* etc. As for the endemic species of plants, there are *Abeliophyllum distichum, Hylomecon hypomeconoides* (Nakai), and *Aconitum Chiisanense.*

Cold-temperate vegetation: Cold-temperate plants grow in northern Korea and on high mountains such as Mt. Seolag, over 1000 m high, Mt. Jiri, over 1300 m, and Mt. Hanra, over 1500 m with temperatures averaging 5 °C. Typical of these plants are such needle-leaved trees as *Abies nephrolepis, Pinus pumila,* and such broad-leaved deciduous trees as *Quercus mongolica* and some six other species. As for endemic plants, *Echinosophora koreanensis* predominates in Hamgyeong Bug Do and also near Yang-gu, Gangweon Do. Myeongchon Gun, Hamgyeong Bug Do is the habitat of *Sasa coreana* and forms the northern boundary of the bamboo growing region. Endemic herbal plants are *Terauchia anemarrhenae folia* and *Hanabusaya asiatica,* which grow in the northern part of the country. *Rheum coreanum* is limited to the Changbaeg mountains and the Bujeon plateau, Hamgyeong Nam Do. Several species of plants adorn the country with their beautiful colors: *Forsythia coreana, Rhododendron mucronulatum, Prunus yedoensis, Abeliophyllum distichum, Pulsatilla cernua* (Thunb.) *spreng, var. Koreana (Nakai), Viola albida, Chamaephylloides* W. Becker, *Viola mandshurica* W. Becker, *V. seoulensis, Hanabusaya asiatica, Miscanthus sinensis,* and *Mis. sacchariflorus, Chrysanthemum spp., Acer pseudo-sieboldianum Komarov var. Koreanum Nakai.*

Forestry: About 70% or 16 million hectares of the land is covered by forests. In the past, forestation was not practiced conscientiously for several reasons. A major influence was the traditional custom of using wood for cooking and warming the room; this greatly hindered forestation projects. In addition, two forestry disasters occurred. First, toward the end of World War II, all available trees were cut down by the occupying Japanese forces for ship building. Then during the Korean war of 1950 – 1953, refugees from the north cut down young trees for fire wood in the winter. Even afterwards, trees in unsecured areas were felled to prevent communist guerilla infiltration. However, in 1973 the government promulgated a forestry-protection order imposing heavy penalties for felling trees in any designated area without authorization. Moreover, the 5th of April of each year was designated the official planting day. Now the land is becoming ever greener by the year.

Despite these vicissitudes, several areas are still islanded with deep forests. The area north of the 41° latitude along the Yalu river and Duman river is, for example, known as a primitive jungle zone in Korea, 70% of which is composed of needle-leaf trees and 30% of broad-leaf trees. The trees are transported to Sineuiju and Hoeryeong on rafts and utilized for pulp and lumber. In the middle part of the land, at the 35° – 41° N latitude, pine trees comprise almost 70% of all trees. In accordance with the development of agriculture and population, increased artificial forestation has been practiced especially in the hilly areas. Pine trees, chat-trees, fir, oak, and birch are characteristic of this zone. In the southern area, south of the 35° N latitude, pine trees, camphor trees, and bamboo are the main tree species. Jeju Do is especially famous for its abundant flora. As concerns timber production, however, North Korea prevails over the south with a ratio of 83 : 17. The annual amount of timber needed in South Korea is estimated of about 1 million cubic meters, but the native production of timber covers only one-quarter of the total.

Medicinal plants: Crude natural drugs, commonly used in Korea, are classified by their origin: 476 are plant derivatives, 184, animal drugs, and 57, mineral drugs.

The plant drugs include such items as herbs, trees, shrubs, vines, poisonous plants, melons, fruits, and grains. Those herbs obtained locally and commonly used as vermifuges are *Xanthoxylum piperitum, Meliae cortex, Torreya nucifera,* pomegranate trees, and ficus trees. These are distributed mostly in the warmer parts of the land and the islands. Also, *Polygonii aviculareae, Ficus rhizoma,* and *Lisimachia clethroides* are commonly used for removal of round worms and tapeworms. *Panax ginseng* has the status of a panacea in the country. Ingredients in animal drugs include fish, shellfish, animal organs, snakes, etc. And the mineral drugs are prepared from ores, metals, inorganic salts, etc.

2. Zoogeographic Distribution

Korea belongs to the paleoarctic zoogeographic realm. Korea's geographic history, topography, and climate divide the country into highland and lowland districts. Included in the former are the Myohyang mountains, the Gaema plateau of the Bujeon-ryeong mountains, and the more rugged terrain of the Taebaeg mountains, high in altitude with a similar climate to the Amur-Ussuri river region. Most of this district lies above 1000 m in the Baekdu mountain region on the Korean-Manchurian border [23]. The native animal life is closely related to that of the boreal zones of Manchuria, mainland China, Siberia, Sakhalin, and Hokkaido. Representative species are deer, roe deer, amur goral, Manchurian weasel, brown bear, tiger, lynx, northern pika, shrew, water shrew, muskrat, Manchurian ring-necked pheasant, black grouse, hawk owl, pine prosbaek, Jankowski's bunting, and the three-toed woodpecker. The remainder of the country, the lowland peninsular area including the islands of Jeju and Ulreung, has a milder climate. The native fauna, closely related to that of southern Manchuria, central China, and Japan, includes black bear, river deer, mandarian vole, Tristrem's woodpecker, fairy pitta, and the ring-necked pheasant.

There are 6 orders, 17 families, 48 genera, and 78 species of indigenous mammals in Korea. These include 28 species of *Chiroptera,* 18 *Rodentia,* 16 *Carnivora,* 11 *Insectivora,* 2 *Lagomorpha,* and 7 *Artiodactyla.* On record 29 endemic subspecies inhabit the land, but this has yet to be reexamined. Other wildlife species include 25 species of reptile, 14, of amphibian, and 129, of freshwater fish in South Korea. Seventeen species of terrestrial mammals have been found on Jeju island. Wild boar, deer, and wildcat are now extinct, whereas the island is still inhabited by roe deer, weasel, hamster, field mouse, and house rat. There are also 198 species of birds, 8 amphibian species, and 8 reptilian species on the island. Ulreung-Do island is devoid of any endemic mammals. The mammals known to inhabit the island are also found on the mainland and comprise six species, two species of bats, one of shrews, and three of house rats, commensals of man. No amphibians or reptiles are native to this island; frogs and snakes were artificially introduced.

Birds recorded in the Republic of Korea include 18 orders, 67 families, and 366 species. Of the 366 species of birds, 52 species are vagrant visitors, and Kuroda's sheldrake is probably extinct. The remaining birds represent 313 species of birds, 48 of which are permanent residents and 265, migrants [23].

Of the migratory birds, 111 species visit the country in the winter, 64 in the summer, and 90, in the spring and autumn with 112 species that breed in Korea, including 48 indigenous species and 64 species of summer visitors. Eighteen other species of birds have been recorded in North Korea. Of these five are boreal residents of the high terrain of Mt. Baegdu: the black grouse, hawk owl, lesser spotted woodpecker, three-toed woodpecker, willow tit; the remaining 13 are vagrants. Among the birds, only domestic fowls have proven to be carriers or transmitters of human parasites (Forestry Research Institute, 1969).

Of the approximately 2400 species of snakes in the world, about 300 are known to be poisonous to man. According to reports by Lah [42] 3 of the 14 species of Korean snakes are venomous. These species are the *Agkistrodon saxatilis, A. blomhoffii brevicandus,* and *A. calaginosus.* To date, their venoms are known to contain ten different enzymatic proteins and three nonenzymatic proteins.

Lah [42] suggest that poisoning from Korean venomous snake bite may be partially caused by the enzymatic component of the venom acting in concert with the hematoxin, neurotoxin, and cytolysin. Of the 192 species of freshwater fish recorded in Korea 29 species are suspected of being intermediate hosts of *Clonorchis sinensis.* Of these, 27 species belong to either *Cyprinidae, Bagridae,* or *Clupeidae. Plecoglossus altivelis,* known as the second intermediate host of *Metagonimus yokogawai,* are distributed in the clean streams of the mountains or valleys, thus establishing an endemic focus for the intestinal fluke. Since the fish inhabit Milyang, Seoguipo, and along the banks of the Seomjin-gang (river), these are the high endemic areas of the parasite [43]. Marine fish, salmon, and trout, which are found in the east coast area, are known to be the intermediate hosts of fish tapeworm.

As for crustaceans, 8 families and 25 species of Anomuran tribe and 18 families and 168 species of Brachyurans have been recorded in Korea [12]. The *Cambaroides similis, Eriocheir japonicus,* and *E. sinensis* are known to be important intermediate hosts of human paragonimiasis. The former inhabit the clean water in mountainous streams, and the latter is distributed in brooks and ditches in paddy fields. *Palaemon nipponensis* has also been suspected of being the second intermediate host of paragonimiasis in Goheung peninsula, Jeonra Nam Do [44].

V. The Population

1. Demographic Situation

In recent decades the population of South Korea has experienced major demographic changes in growth, mortality, fertility, and distribution. The country is undergoing an unusually rapid demographic transition, which has been accompanied by important social and economic changes [5, 6, 45, 46, 47, 48, 50, 51].

The 1970 census listed 31 430 000 persons, a density of 319 persons per square kilometer. The population projection for 1976 was 35 256 632 persons, a density of 357 per square kilometer, whereas the actual population was 34 706 620 (Table 3). A sharp decline in mortality since the end of the Korean War has contributed to a population growth rate of about 2.1% per year. During the decade 1960 – 1970 alone, the crude death rate fell from an estimated 11 to 8 deaths per thousand population. The mortality decline is likely to continue for some years, due to a further increase in life expectancy, which in 1970 was about 65 years for males and 69 years for females.

Fertility has also declined, especially in recent years. During the decade 1960 – 1970 the birth rate decreased from 43 to 29 births per thousand, and the total fertility rate, from 6.2 to 3.9 children per woman. Although fertility has continued to be higher in rural than in

Table 3. Republic of Korea: Area and Population

Special cities and provinces (do)	Area		Population		
	sq. km	sq. mi	1966 census	1970 census	1975 census
City of Seoul	613	237	3 803 000	5 525 000	6 889 502
Busan City	373	144	1 430 000	1 876 000	2 452 173
Gyeong-gi Do	10 958	4 231	3 108 000	3 353 000	4 039 132
Gangweon Do	16 712	6 453	1 833 000	1 865 000	1 861 560
Chungcheong Bug Do	7 437	2 871	1 550 000	1 480 000	1 522 203
Chungcheong Nam Do	8 699	3 359	2 905 000	2 585 000	2 948 553
Jeonra Bug Do	8 051	3 108	2 523 000	2 431 000	2 456 403
Jeonra Nam Do	12 060	4 656	4 050 000	4 004 000	3 984 123
Gyeongsang Bug Do	19 798	7 644	4 477 000	4 555 000	4 858 551
Gyeongsang Nam Do	11 948	4 613	3 176 000	3 118 000	3 280 052
Jeju Do	1 835	706	337 000	365 000	411 732
Whole country	98 477	38 022	29 192 000	31 430 000	34 703 984

Source: Official Government figures 1975

urban areas over the last decade, the two sectors have shown about the same rate of reduction. Most of the decline was due to higher age at marriage and to lower marital fertility. The national family planning program, combined with the widespread and now legalized practice of induced abortion, appears to have played an important role in reducing marital fertility, particularly among older women. However, it is still too early to predict whether the recent fertility decline will continue.

The decade 1960 – 1970 showed unprecedented increases in the urban, especially the metropolitan populations, while the rural population remained almost constant. Thus, most of the recent national growth has been absorbed by the large cities. Furthermore, the pace of urbanization increased during the decade.

Assuming continuation of the recent trends in mortality and fertility, plausible population projections estimate the South Korean population between 54 and 57 million in the year 2000. Because the rural population has already achieved very high density relative to that of rural areas in other countries, most of the future growth will have to be absorbed by metropolitan and urban areas. This means that in the year 2000, some 34 – 39 million persons (or 70% of the total population) would live in the cities, compared with 13.6 million (or 43%) in 1970. Urban metropolitan growth of this magnitude has serious social, economic, political, and environmental implications. Further analysis of current and future trends

will be needed to improve our understanding both of the Asian demographic transition and of the causes underlying the rapid fertility decline that has taken place in South Korea.

2. Population Growth

At the time of the first census in Korea (1925) there were 19 020 000 people, e.g., a density of 88 persons per square kilometer. During the following 20-year period approximately 6 500 000 people were added, thus reaching a density of about 120. Of the total population 3% was foreign, the majority of these, Japanese. The average calculated annual rate of natural increase during the intervals between census for Koreans varied from 2 to 20 per thousand, and it may be said that Korea's population increased approximately 15 per thousand annually prior to 1945 (Tables 4 and 5).

Population growth in South Korea from 1945 presents a different picture than that described above. Even in the Japanese days, South Korea was far more densely populated than North Korea; the estimated density below the 38th parallel in 1944 was 173 per square kilometer as opposed to 113 for the entire peninsula that year. Between 1944 and 1949 the population increased from 16.244 to 20.167 million, corresponding to the extremely high annual increase rate of 4.4%. The rate well exceeded 40 per thousand. Although the rate of increase

Table 4. *Population Growth in Korea, 1900 – 1970*

Year	Number (in 1000)	Density per km²	Annual Rate (in 1000) of			Net reproduction rate
			Natural increase	Migration	Total increase	
1900 (Oct. 1)	17.082 [a]		2	–	2	–
1910 (Oct. 1)	17.427 [a]		4	– 1.4	2.6	–
1915 (Oct. 1)	17.656 [a]		7	– 2.3	4.7	1.17
1920 (Oct. 1)	18.072 [a]		12	– 1.8	10.2	1.25
1925 (Oct. 1)	19.020	86.1	18.7	– 4.3	14.4	1.40
1930 (Oct. 1)	20.438	92.5	20.2	– 3.6	10.6	1.68
1935 (Oct. 1)	22.208	100.5	20.6	– 8.9	11.7	1.77
1940 (Oct. 1)	23.547	101.1	20.2	– 2.2	18.0	1.86
1944 (May 1)	25.120	113.7				1.93 (1940 – 45)
		All Korea				
		South Korea				
1945 (Sept. 1)	16.136 [a]		18.9	41.9	60.8	
1949 (May 1)	20.167	205.1				
	19.904 [b]		7.9	6.6	14.5	1.97 (1945 – 50)
1955 (Sept. 1)	21.502	218.4	28.6	–	28.6	–
1900 (Dec. 1)	24.989	253.5	20.5	–	26.5	2.27
1966 (Oct. 1)	29.160	290.1	18.8	–	18.8	2.19 (1960 – 65)
1970 (Oct. 1)	31.435	319.2				1.77 (1965 – 70)

[a] Estimated

[b] for 1955 Boundary

Source: 1) The figures for 1910 – 25 are estimated from census results of 1925 and 1930 and the observed trend of population growth in the late Yi Dynasty.

2) The figures for 1925 – 66 duplicated from Tai Hwan Kwon. *Population Change and Its Components in Korea* 1925 – 66 (Unpublished Ph. D. thesis, Australian National Univ.), 1972.

2) The figures for 1966 – 70 or 1965 – 70 are recent estimates of the author.

From: Kwon T. H. et al. (1975) [46]

Table 5. Estimated or Projected Rates of Population Growth, Fertility and Mortality, 1965 – 90

					(Per thousand)
	1965 – 70	1970 – 75	1975 – 80	1980 – 85	1985 – 90
Annual growth rate	19	17.0	16.6	15.8	14.2
Crude birth rate	32	29.1	27.7	26.3	24.2
Crude death rate	13	12.1	11.1	10.5	10.0
Total fertility rate	4.63	4.23	3.82	3.42	3.01
Gross reproduction rate	2.26	2.06	1.87	1.67	1.47
Net reproduction rate	1.77	1.68	1.58	1.45	1.31
Expectation of life at birth					
Male:	51.0	53.5	56.0	58.0	60.0
Female:	56.5	59.0	61.5	63.5	65.5

From: Kwon, T. H. et al. (1975) [46]

was considerably lower following the 6-year period including the Korean War (1950 – 1953), the rate has again accelerated to about 30 per thousand in recent years.

Thus, during the 15-year period following liberation, South Korea added nearly 15 million people. Since the Republic of Korea was established in 1948, the government has conducted five censuses: in 1949, 1955, 1960, 1966, and 1970. The relatively low increase before 1945 was primarily the result of two factors: a high emigration rate and a high death rate. For instance, in 1940 nearly 1.5 million Koreans lived in Manchuria and 1.2 million in Japan, indicating that more than 10% live outside Korea. Available records show that in the early 1940s approximately 150 000 emigrated per year. The United Nations estimates the emigration rate of those days at 10 per 1000 annually. It is true that the death rate slowly decreased, but until the end of World War II, the death rate of Koreans was estimated at about 20 per 1000 as against the birth rate of 40 or more.

After 1945 emigration ceased. Moreover enormous numbers of repatriates and refugees came from North Korea. From 15 August 1945 through 1948, 1 200 000 repatriates from abroad and 2 190 000 refugees from northern Korea were reported. On the other hand, 887 000 foreigners (mostly Japanese) have been repatriated from Korea in the meantime, leaving 1 400 000 as the natural increase. However, this may be an underestimation, since exaggerations have been suspected in the reports on refugees. In any case, no great change seems to have occurred in the natural rate of increase immediately before the Korean War.

During the Korean War, South Korea lost perhaps 700 000 people as military casualties and missing persons, whereas also an influx of 996 000 refugees arrived from the north. The natural increase might have been far less than 10 per 1000 in 1950 – 1952. Since 1953, the high death rate decreased rapidly, which was reflected in the increased rate of population growth in 1955 – 1960. During this period only negligible international migration took place, hence the increase rate was nothing more than the natural increase rate.

The computed annual increase rate from the two censuses of 1955 and 1960 was 2.88%, one of the highest in the world. The density, which was over 250 per square kilometer, was more than that of Great Britain, a highly industrialized country. If the present rate, based on the figures mentioned, continues, Korea's population will double every 25 years.

How closely this rate represents the actual situation is yet to be seen. However, the natural increase rate has been climbing during recent years; the death rate is rapidly dropping, and the birth rate shows a slow decline.

3. Sex Distribution

Traditionally Korea has had an abundance of males. However, due to sex differentials in mortality and migration, the excessively male population gradually decreased until finally in 1944 there were slightly more females than males. After 1945, with the influx of repatriates and refugees, the situation was reversed and South Korea reached the sex ratio of the late 1930s. During the Korean War, 1950 – 1953, Korea suffered many war casualties. Thus, in spite of receiving a large number of refugees from North Korea, presumably mostly men, almost a balance between the sexes was observed in 1955. However, in certain age groups, especially ages 20 – 29, a large discrepance between the sexes was recorded, for example, as low as 80 men to 100 women.

During 1955 – 1960 there was far more of an increase reported in males than in females, probably mostly due to the higher ratio of male births. However, there may have been less accurate reporting for men than for women of certain ages at the time of the 1955 census. The difference in reporting, if any, may have been closely related to the implication of military service. If so, there would have been more of an increase in the reporting of men than of women in the 1960 census because of the settlement of the military situation, thus, the apparent increase of males.

Cities used to draw more males than females because of opportunities for work, education, etc. However, the situation was reversed in the 1960 census: more females lived in cities than in the rural area. A traditional distributional feature is that more females live in coastal areas and islands, while more males live in mountain areas.

4. Age Distribution

The Korean population has always been characterized by a young age structure. Children under 15 years of age account for about 40% of the total population, and people 60 and over account for about 6% (Fig. 6 and 7). Obviously, the dependency ratio is very high; nearly half of the population being dependents. A low per capita income of the adult population further complicates aspirations of raising the standard of living.

The problem of an aging population is probably not serious in Korea, since there has been no relative increase in the proportion, although the actual number is increasing. However, it should be noted that the population of children under 15 is gradually increasing; it was 39.6% in 1925, 40.9% in 1935, and 43.2% in 1944. After liberation the increase has halted at a level slightly over 48%. The increasing percentage probably comes from a rapid decrease in children's mortality. The young adult (15 – 29) group has fluctuated greatly, ranging from 23.5% in 1944 to 25% in 1949, and to 26.8% in 1955.

This age group is the most vulnerable to social and military conditions. The decrease in 1940 – 1944 was probably due to manpower mobilization for industrial establishments outside Korea and for military service overseas. However, it should be noted that there was no decrease in 1955 despite heavy casualties in the male population during the Korean War. The influx of young adults from North Korea before and during the war must have contributed to maintaining the level; however, there was a great decrease in the male population in their 20s.

The middle aged, 30 – 44, comprise about 17% of the population. A gradual decrease during the Japanese regime reached a minimum of 16.1% in 1944. However, after liberation this figure rose to 16.7%. The age group 45 – 59 lies between 10.3% and 11.0%. The record after 1945 reveals a slight decrease as compared with prewar figures.

As may be expected, cities attract more young people. According to the 1955 census, in urban areas (Shi — city) children under 15 and adults 40 and over accounted for 39.8% and 17.7%, respectively, but in rural areas (Gun), 41.7% and 21.9%, respectively. The situation was reversed for the 15 – 39 age group: in the urban area it was 42.5% and 36.4% in the rural. This tendency was also found in the 1944 census. In urban males the period of work productivity lasts until the age of 50, but in females, only to 30. No doubt the opportunity is greater for work and schooling in urban than in rural areas. However, job availability for women in the cities is more or less confined to housekeeping, in Korea generally restricted to young girls. Probably this, together with the education facilities, is reflected in the concentration of young women 15 – 29 in urban areas. However, the 1960 census showed that more women lived in urban than in rural areas, a contrast to previous censuses. When the age distribution is tabulated, perhaps some of the changing patterns of Korea urbanization may be explored.

When the age distributions between North and South Korea were compared, the 1944 data revealed fewer of the group aged 15 – 39 in the south than in the north among males (33.0% in the south vs 34.5% for all Korea) and no difference in the female population. The difference in males was probably due to migration of young people from the south to the north with its heavy industrial plants.

Examination of the provincial distribution of population by three age groups in 1940 and 1944 reveals fewer children in Gyeong-gi Do (including Seoul) and the

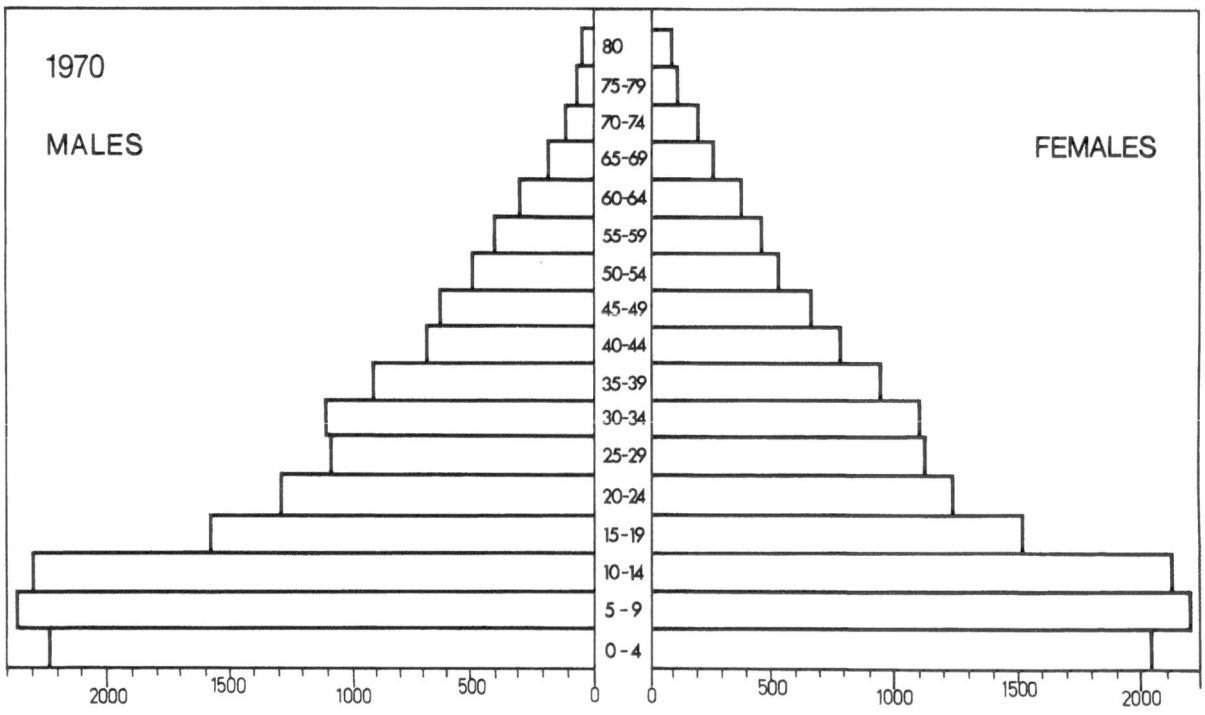

Figure 6. Age distribution 1970 (Source: Korea Statistical Yearbook 1977, Table 13)

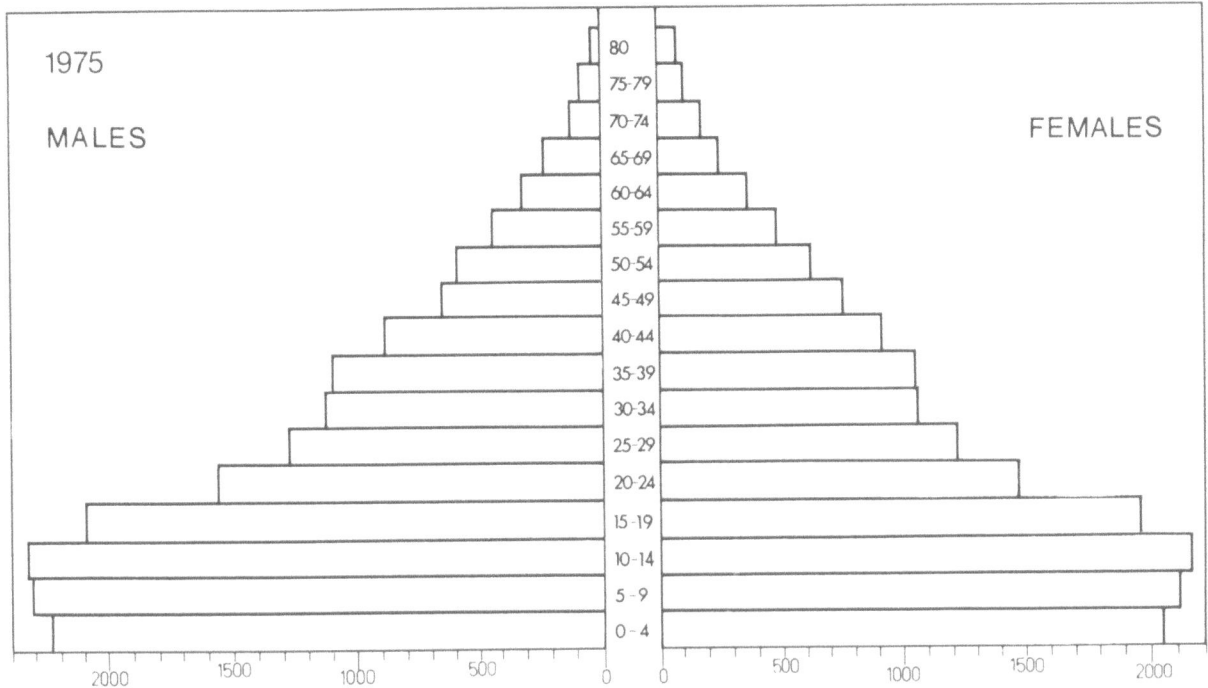

Figure 7. Age distribution 1975 (Source: Korea Statistical Yearbook 1977, Table 13)

northeastern industrial provinces and relatively few children in a stretch of land from the west coast of the middle section of Korea (including Seoul) across to Gangweon Do and Gyeongsag Nam Do.

According to the 1955 census the tendency remained the same, although the relative children's population decreased markedly on Jeju Island. However, the concentration of the productive age group in Gyeonggi Do, Gangweon Do, and Jeju Do may to some extent be due to the military installations situated there. The large proportion of aged on Jeju Do should be noted. In 1944, old people, 60 and above, accounted for more than 11% of the population. Probably emigration from the province was the cause rather than mortality.

5. Economically Active Population

In 1976 out of a total population of 35.86 million 22.55 million were 14 years or over; of these 13.06 million were economically active and 9.49 million not economically active. Due to the prolongation of education and greater participation of females in education (in 1965 0.769 million male and 0.428 million female students of 14 years or over attended school, in 1976 1.814 million male and 1.191 million female students) the proportion of the not economically active population has been steadily increasing in recent years (cf. Fig. 8).

As a result of the rapid industrialization, the pattern of employment has changed profoundly since the early sixties. Whereas in 1963 58.0% of the economically active population were still employed in the primary sector (agriculture and fishing), their proportion dropped to 42.9% in 1976. Simultaneously the proportion of the population employed in the secondary sector (mining and manufacturing) increased from 8.0% in 1963 to

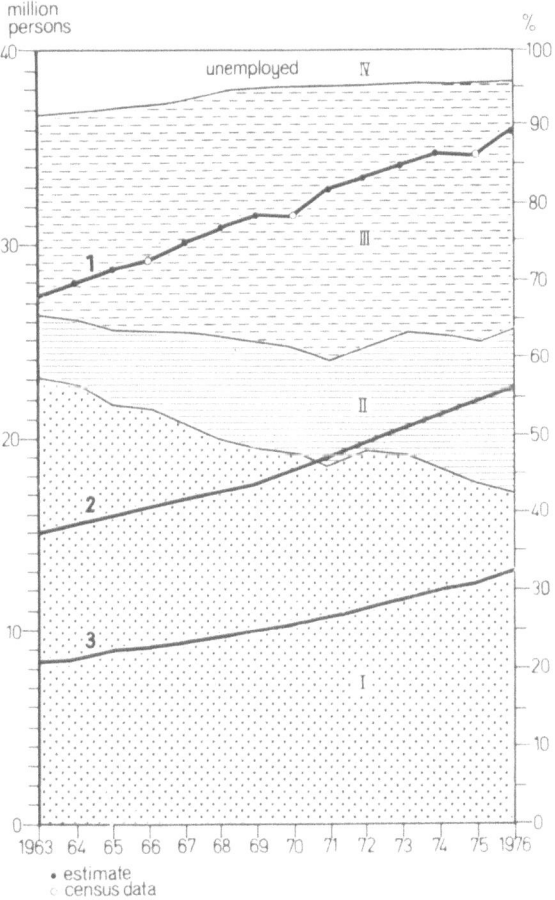

Figure 8. Development of total population, population of 14 years and over and economically active population, employment by industry (Design: Dr. Eckart Dege)

Left scale: population growth. 1 total population; 2 population 14 years old and over; 3 economically active population.

Right scale: economically active population by occupation. I farming and fishing; II mining and manufacturing; III services; IV unemployed
(Source: Korea Statistical Yearbooks)

21.0% in 1976 and that of the tertiary sector (services) from 25.9% to 32.2%, while the proportion of unemployed dropped from 8.1% to 3.9%.

6. Fertility

Pre-World War II birth registration statistics, like the registration statistics on deaths, were incomplete, but they nevertheless indicated a moderately high crude birth rate prior to 1944. More reliable estimates based on the census data suggest a crude birth rate of 40 – 50 between 1925 and 1944 [40, 52]. As already noted, the registration of vital statistics has greatly deteriorated since 1945, and therefore no reliable birth rates for the postwar period can be calculated from registration statistics.

The crude birth rate in South Korea declined from 43 in 1960 to 33 in 1966 and then to 29 in 1970. This represents a decline of 23% and 12%, respectively. The necessary data are available to isolate the three components of change in the crude birth rate during the 1960s change in age-sex structure, change in marital status, and change in age-specific marital fertility rate.

Changes in the age-sex structure and in age at marriage taken together account for approximately 40% of the decline in the birth rate during the decade 1960 – 1970 [52]. The largest proportion, about 60%, of the total decline is therefore due to a reduced marital fertility rate, out-of-wedlock births being virtually nonexistent.

The total fertility rate declined from 6.1 to 3.9 during this period. Although only small reductions in age-specific fertility rate are visible in ages 25 – 34, drastic reductions are observed in the age groups 15 – 19 and 20 – 24 and in the older age groups of 35 and over.

The fertility decline in the younger ages was largely due to a sharp drop in the proportion married; but the lower fertility among women 35 and older was to a large extent due to their adoption of contraception and abortion procedures.

Although rural women continue to have higher fertility than urban women, the extraordinary fact is that the rural fertility has declined at virtually the same rate as urban fertility and remains higher than urban fertility only because it has declined from a much higher level. In 1960 the total fertility rate of rural women was 6.8 compared with a rate of 5.0 among urban women. The age-specific fertility rate in rural areas was also higher for each age group than for corresponding age groups in urban areas. By 1970 the total urban fertility rate had dropped by 36% to 3.1%, and the total rural fertility rate showed almost an identical decline of 15% to 4.4%.

The total fertility rate of married women shows a sharply inverse relationship to education. The incidence of induced abortion has been increasing at a very rapid rate in recent years. Although induced abortion was illegal until very recently, the government has not attempted to enforce the law prohibiting the practice since the early 1960s. In February 1973, all legal restrictions on induced abortion were abolished. The proportion of currently married women practicing contraception has

increased steadily from only 0% in 1964 to 25% in 1971. In 1966 significantly more urban than rural women were using contraceptive methods. By 1971, however, rural women had reduced the urban-rural difference in contraceptive practice. Since the government's family planning effort has concentrated on rural areas, much of the urban-rural difference in fertility may be accounted for by the higher proportion of women practicing induced abortion in urban areas.

7. Mortality

Death rates during the period 1925 – 1945 are based on more complete registration data than have been collected since the end of the Japanese colonial rule in 1945. The crude death rate declined from about 28 in the mid-1920s to about 20 in the early 1940s. On the basis of registration data, life tables have been calculated for the period 1925 – 1930 which indicate a life expectancy of 32.4 years for males and 35.1 for females. Life tables calculated for the period 1938 – 1943 suggest a considerable improvement in longevity; life expectancy increased by 10.1 years to 42.5 for males and by 9.9 years to 45.0 for females.

After 1945, the registration of vital statistics in the south deteriorated to the point that it was no longer a reliable indicator of birth and death rates. Despite some recent improvement in registration, vital statistics continue to suffer from a serious lack of registration even today, and therefore estimates of mortality for recent years must come from census of survey data. Consequently only a tentative mortality trend for the last few decades can be established. From the censuses of 1955, 1960, 1966, and 1970 estimates can be obtained of the level of mortality and direction of change [48]. Using the age structure derived from the 1955 and 1960 censuses, estimated survival probabilities for an abridged life table have been made. The calculations suggest a life expectancy of 51.5 years for males and 53.7 years for females. This compares closely to another set of life tables based on the same census information prepared by Lee [53], who used the methods recently developed by Brass [54].

To identify the mortality pattern of a rural community, a total of 125 deaths occurring in two townships (Naega and Seonweon Myeon) in Ganghwa County during 1 April 1975 – 31 March 1976 were studied by Kim et al. [55]. All deaths were reported by family health workers, and cause of death as identified by public health physicians. The families of the deceased were interviewed by two trained nurses using a pretested questionnaire. A sociologist visited the ceremonial places and made case studies. The family health workers, public health nurses, and public physicians were all members of the Yonsei University Community Health Teaching Project (Table 6), [283]. The study focused on various death rates, causes of death, cultural factors related to death, the economic burden imposed on the family by death, and other sociologic aspects. The age-adjusted crude death rate was 10.6

per 1000. The neonatal death rate was 23.4 per 1000 live births, and the infant mortality rate was 27.2 per 1000 live births. No maternity deaths were observed. (For the most common cause of death see Table 6.)

These statistics imply that the causes of death in rural Korea have changed rapidly in recent years from infectious to noninfectious diseases.

Table 6. 10 Leading Causes of Death and Disease Specific Death Rates (Dr. I. S. Kim, 1976) {55}

Disease category	No. of death	Rates (per 100 000)
1. Cerebrovascular diseases	28	221.8
2. Malignant neoplasms	20	158.4
3. Ill defined, unknown senility	15	118.8
4. Suicide	12	95.1
5. Pulmonary tuberculosis	7	55.5
6. Liver cirrhosis	7	55.5
7. Neonatal death	6	47.5
8. Pneumonia	6	47.5
9. Accident	4	31.7
10. Nutritional deficiency	3	23.8
11. Others	16	
Total	124	

The present level of mortality in South Korea is roughly equivalent to that of Japan in 1954 – 1955. By 1970 the Japanese crude death rate had declined to 6.8, and life expectance had increased to 69 years for males and 75 years for females. Barring unanticipated calamities, the South Korean mortality level will most likely decline along about the same pattern as the Japanese mortality level during 1955 – 1970. This would mean a gradual decline in the South Korean death rate to about 5.4 in 1985, resulting in life expectancy comparable to that of Japan in 1970.

8. Settlement

The recent transformation of industrial structure has accelarated the urbanization of Korea, bringing about a rapid growth of urban population and an absolute decrease in rural population. However, Korea still had a large rural population, amounting to 44.1% of the total in 1973. Most of the rural population engages in agriculture, and its distribution is very closely associated with that of cultivated land, e.g., it is concentrated in the lowlands along the west and south coast and along the major river valleys (Examples see Maps 5 a – 5 c).

Rural houses are located mostly along the southern side of low hills where highly agglomerated villages form. Such a location can give protection against the cold winter monsoon from the northwest. Maximum use of lowlands for rice cultivation is another reason for selecting this location. Villages usually consist of from ten to one hundred or more individual houses, which are built in close proximity to each other within the villages.

The explanation for this agglomeration has something to do with protection from outsiders and, above all, the system of large families. Quite a number of villages are still found throughout the country which consist of many families with the same family name, such as Kim, Lee, or Park. In such cases, the family name is reflected in the village name.

There are a small number of scattered or isolated houses in mountain areas. Shifting cultivation, which demands a larger field for subsistence, is the main agricultural practice in the mountains.

9. Internal Migration and Urbanization

Like most countries of Asia, South Korea has incomplete information on population movements. What follows, therefore, is an analysis based on available data that is limited in variety as well as in quantity. (Tables 7 and 8, Map. 3 a and 3 b).

Over the years Koreans have exhibited a growing tendency to change their place of residence. According to the 1930 census data on place of birth (Government-General of Chosen, 1935), most Koreans were living in the province of their birth. Fewer than 10% of the inhabitants of most provinces had ever crossed a provincial line. In the south, the largest gain through lifetime migration occurred in the province of Gangweon Do, located in Gyeongsang Bug Do province.

Considerable redistribution of the population of the Korean peninsula took place toward the end of World War II, with the direction of redistribution being predominantly from rural to urban areas and from south to north. Although the urban population was growing fairly rapidly, the urban proportion of the total population was still rather small. Despite the percentage decrease in rural inhabitants, the actual number increased slightly, i.e., the agricultural density of the South Korean population still ranks among the highest in the world [5, 6, 45, 46, 47, 48, 49, 50, 51]. Urban population growth was primarily the result of rural-urban migration, natural increases contributing only marginally to this growth.

a) Internal migration: By 1960 the volume of movement away from native provinces had increased substantially. Gyeongsang Nam Do, which contained Busan, and Gyeongsang Bug Do with the city of Daegu, showed slight gains through net migration, but the nine other predominantly rural provinces of the south lost a considerable proportion of their populations. The Seoul metropolitan area acquired 39% of its 1960 population through migration from the provinces. By 1966 the volume of migration to the metropolitan areas of Seoul and Busan was even more pronounced: more than half (56%) of Seoul's inhabitants had moved there from the provinces. The largest net loss of population occurred in Gyeongsang Bug Do, 41%; and in Chungcheong Bug Do, 30%. A breakdown by urban and rural areas indicates that the cities in most provinces had attracted large volumes of immigrants from other provinces, but the urban areas of two provinces, Chungcheong Bug Do and

Table 7. Intercensal Population Growth Rate and Density by Metropolitan Area and Province: Republic of Korea, 1955 – 70

Metropolitan area and province	Population growth rate (%)			Population density	
	1955 – 60	1960 – 66	1966 – 70	1966	1970
Seoul	9.2	7.6	9.9	6187.2	9014.1
Busan	2.2	3.5	7.1	3820.7	5029.1
Gyeong gi Do	3.1	2.0	1.0	283.2	306.1
Gangweon Do	2.2	1.9	0.4	109.6	111.6
Chungcheong Bug Do	2.8	2.1	1.1	208.3	199.0
Chungcheong Nam Do	2.6	2.3	0.4	333.7	328.6
Jeonra Bug Do	2.4	0.9	0.9	313.2	302.2
Jeonra Nam Do	2.3	2.2	0.3	335.7	332.1
Gyeongsang Bug Do	2.7	2.5	0.5	225.9	230.2
Gyeongsang Nam Do	2.1	0.9	0.4	265.7	261.0
Jeju Do	– 0.6	3.0	2.0	184.2	199.6
Whole country	2.9	2.7	1.9	296.1	319.3

Source: Institute of Population Problems (1972; 1977)

Table 8. Percent Changes of Rural, Urban and Metropolitan Population, Republic of Korea, 1955 – 1970

Area	Percent change			
	1955 – 60	1960 – 66	1966 – 70	1955 – 70
Rural	+ 9.5	+ 8.0	– 7.0	+ 9.9
Urban	+35.2	+32.5	+ 7.3	+ 92.3
Metropolitan	+37.5	+44.6	+62.9	+223.9

Sources: Republic of Korea, Ministry of Home Affairs, Bureau of Statistics (1959); Republic of Korea, Economic Planning Board, Bureau of Statistics (1963, 1969, 1972 b)

Jeju Do, had lost virtually half of their native-born residents.

A special demographic survey conducted in 1966 by the Bureau of Statistics [51] shed light on the migration streams of the 1961 – 1965 period. The largest volume of interprovincial migration was from rural to metropolitan areas: approximately 521 000 persons, or 35% of the total volume of migration. Yet only about 67 000 persons, or 5%, moved from rural to urban areas, compared with 201 000 migrants (14%) who made lateral moves from one rural area to another. Some 253 000 or 18% of the total moved from urban to metropolitan areas, and another 6% moved from urban to other urban areas. As expected, only a small proportion (4%) of the migrants moved from urban to rural areas, but a somewhat larger proportion (7%) moved from metropolitan to rural areas. About the same percentage moved from one metropolitan area to another as moved from a metropolitan area to an urban area, i.e., 6% and 5%, respectively.

The greatest proportion of all migrants (55%) was from rural areas. Migrants originating from urban areas were the next largest group (28%) followed by migrants from metropolitan areas (18%). Destinations were metropolitan areas (60% of migrants), rural areas (25%), and urban areas (15%).

b) Urbanization: The population in the *shi* or *si* (city) is rapidly increasing; in 1925 it was only 4.4% of the entire population, but by 1940 it rose to 11.6%. The first census of the Republic of Korea (1949) showed that 17.2% of the people (or over 3 400 000) lived in Seoul (Special) city and 14 sis. According to the census of 1955, there were 5 300 000 urbanites or 24.5% of the total population. In 1960 the percentage increased to 28.8% with an urban population of 7 000 000, and in 1970 to 43.2%.

However, the term "urban" population must be treated with caution. Any local administrative unit of more than 50 000 population can become a si regardless of the general features of the area. Thus, in some cities nearly half of all households are farmers. Similarly contradictory is the jurisdiction of Jeju Si, which includes Mt. Hanra, the highest mountain in South Korea. Further, a si can absorb adjacent administrative units as it grows. For instance, Jeonju Si added 50% of its population during 1955 to 1957, but the increase resulted largely from the absorption of part of Wanju Gun, which showed a 10% decrease.

Nonetheless, the tendency towards an urban population is obvious. Before 1935 no city had a population of more than 500 000. At the last stage of World War II, Seoul city finally attained the level of one million. Unfortunately, the population data for individual administrative units in 1944 is not available for all Korea. However, before liberation, units under 10 000 gradually decreased, while middle-sized units between 10 000 and 50 000 were increasing.

Throughout 1930 only five cities had populations of more than 50 000. In 1925 Seoul had a population of 343 000, and Busan, 107 000. In 1935 there were 12 cities registered of more than 50 000, and 4 cities, more than 100 000, the largest being Seoul with 444 000.

An extraordinary increase was observed for Seoul city during 1935 – 1940, when it added nearly half a million (a large part of the increase was due to the annexation of neighboring fringe areas).

In 1949 there were eight cities in South Korea alone with more than 100 000 population, accounting for almost 15% of the total population. Of these, only one city (Seoul) had a population of nearly 1.5 million; all others were less than 500 000. There were 11 additional small cities and towns (*eub*) which ranged from 50 000 to 1 000 000, but there were no moderately large cities of 500 000 – 1 000 000.

The Korean War (1950 – 1953) changed urban distribution to some extent. Refugees concentrated in the temporary capital city, Busan, resulting in a more than 100% increase. Two large cities with more than a million population were thus created in 1955, and 12.2% of the entire population lived in these two cities, Seoul and Busan. The number of medium-sized cities (100 000 to 500 000) remained the same as in 1949, but the percentage of inhabitants was somewhat reduced, from 7.7% to 7.4% in 1949. The small cities and towns increased in number to 18, and the percentage of inhabitants also increased from 3.1% to 5.5% in 1949. Two such places were directly related to the establishment of training centers for military draftees. For instance, Gujagog Myeon, Nonsan Gun, Chungcheong Nam Do where the main army training center is located, raised its status from 11 000 in 1949 to 57 000 in 1955.

The postwar concentration of population in urban areas further accelerated. The two cities Seoul and Busan of over a million composed more than 14% of the population. Six cities ranged in size from 100 000 to 500 000 with 5.7% of the entire population. Another 5.7% lived in 20 small cities and towns. Only 20% lived in local units of less than 10 000.

10. Major Cities

a) Seoul, the capital of Korea, is the largest city in the country with a population of 8 114 000 as of October 1979.

Originally it was a river port, to which seagoing vessels navigated prior to the days of big ships. Presently, Incheon has assumed the role of the port for Seoul. Incheon is located on the west coast of the Korean peninsula approximately 25 miles to the west of Seoul.

The first major development of Seoul took place in 1394, when the founder-king of the Yi Dynasty chose it as the new capital. Its central location in the peninsula is considered one of the best assets for the administration of the nation. Previously the Koryo Dynasty also recognized the importance of the location and maintained a separate palace there with the name of Namgyong (southern capital).

Seoul has a rich heritage, being the capital of the Yi Dynasty for 518 years from 1392 to 1910. Originally it was a walled city with eight gates, most of which were torn down in the course of modern expansion. Namdaemun (south gate) and Dongdaemun (east gate) are still preserved as national treasures. Palaces, pagodas, bell towers, and other historical sites in the town attract increasing numbers of tourists throughout the year.

Like many other cities, Seoul has many traffic problems, although public transportation is well developed.

To solve the problems, a number of plans are being put forward, such as perimeter roads connecting the radiating lines from the city and subways. Part of the subways went into operation in August 1974, and Korea became the second nation in Asia (Japan was the first) to have an underground public transportation system.

Seoul today is not only the chief administrative center of a highly centralized country, but also the largest cultural, commercial, and industrial center. As a result, it has become highly overcrowded, and is one of the most densely populated cities in the world with nearly one-quarter of the total national population.

The mass drift of rural population into the capital inevitably creates such social problems as unemployment, housing shortages, inadequate water supply. Over the past 10 years, the government has made great efforts to decentralize its functions, especially by establishing industrial centers throughout the country. Some success has been achieved.

b) Busan is the second largest city and the largest sea port of Korea with a population of 1.6 million. It is located on the southeastern end of the peninsula.

Busan was originally a fishing hamlet that began to develop as a port with the opening to foreign trade in 1443.

Some 450 km from Seoul, Busan is the main gateway to Korea from Japan and western countries by sea and the southern terminal of Korea's main railway networks. The harbor is well protected by many large and small islands, and the port presents a forest of colorful masts of both foreign and domestic ships at all times.

The streets were developed along the foot of Mount Yongdu (dragon head), which commands a view of the picturesque southern seascape. Songdo (pine tree island) about 2 km south of the civic center is renowned for its summer beach. Heaundae, east of the civic center, is one of the most famous resorts in the country, providing a long beach and hot springs with good facilities. On the eastern outskirts is located the United Nations' Cemetery. There lie the valiant soldiers from 16 UN member nations, who gave their lives during the Korean War (1950 – 1953) for the defense of freedom and democracy.

Considerable manufacturing industry has grown up recently in the Busan area. Textile, food processing, ceramics and flour mills are some examples.

11. Projected Trends in Population Growth (projected population growth and composition, 1970 – 1990)

The population of the Republic of Korea is expected to pass the 40 million mark by 1985 with an increase of 890 000 people yearly in the 15 years following 1970 [48, 49].

The density is likely to rise to 380 persons per square kilometer in 1980 and 440 in 1990. The annual rate of population growth for 1985 – 1990 is projected to decline by 5 points to 14 per thousand from the level for 1965 to 1970.

If the present differentials in mortality are maintained, the sex ratio of the total population is expected to reach a balance around 1980. After that point, the number of females in the population will exceed that of the males. Owing to the declining fertility rate since 1960 and the assumption of further continuous declines during 1970 – 1990, a series of declines in the proportion of population aged 15 and below can be expected. In contrast, the proportions for the age group 25 – 29 and above show a continuous increase. This can be explained by the improving health and medical conditions expected in the coming years, as was the case in the past. These expected trends clearly suggest that the age structure of the Korean population will undergo a substantive and consistent transformation in the next 20 years at least. The degree and pattern of change can be amply exemplified by the comparison of age pyramids for 1970 and 1990.

The proportions of the population of preschool, age 0 – 5 and primary school, age 6 – 11, will decline substantially during 1970 – 1990 to the extent that the present demographic trends continue. On the other hand, a slightly upward trend in the proportion of the population of secondary school age 12 – 17, is expected during 1970 – 1980 due to the post-Korean War baby-boom generation. Subsequent declines will be observed during 1980 – 1990 with the replacement of that age group by the post-1960 birth cohorts.

Although the crude birth rate is projected to drop, the proportion of women of reproductive age, 15 – 49, in the total population is expected to increase without interruption during the entire 20-year period. In other words, if age specific fertility rates are constant, this would result in a rise in the crude birth rate. Similar trends are expected with women in the crucial reproductive ages, 20 – 34.

The dependency burden of the population of working age (15 – 59) will be lessened to a significant degree during this period. The expected drop in the total dependency ratio is mostly accounted for by continuous declines in the youth-dependency ratio. The extent of change in the aged will be very minor in the next 20 years.

It is an easy matter to project future levels of growth for the South Korean population, given its present age composition and with the assumption that current rates of fertility and mortality will continue. As described before, the present total fertility rate is estimated to be 4.0 and life expectancy to be about 65 years. If these rates continue, the population would increase to 46 million by 1985 and to 65 million by the year 2000. It is highly improbable, however, that fertility and mortality rates will remain constant over this period. The process of industrialization, urbanization, and modernization are likely to continue and to have a sustained negative effect on both mortality and fertility. The national family planning program, combined with recently legalized abortion, should also contribute to a further decline in the birth rate.

VI. The People and Their Way of Life

1. Family Structure

Except for the cities or industrialized towns, most of the rural villages still hold to the custom of the large family system. Many of the village units are composed of only one, or possibly two, family names. This village custom originated from Confucian thought, according to which the eldest son is authorized to decide and control all small and large matters in the family. In this system, the right of the female to participate in any way is generally ignored. Confucian thought emphasized that the female should be obedient in three areas of her life: obedient to her parents in childhood, obedient to her husband in adulthood, and obedient to her son in her later years. In this way the housewife is forced to be dependent on her family and is not allowed to become a controlling influence in the home. These three obedient behaviors are recognized as true female virtues. Likewise the eldest son is valued highest among the children and receives the inheritance of the father.

However, as the country became industrialized and feudal customs were replaced by democracy, the large family system begun to crumble especially in the urban areas. The large family systems, previously run only by the eldest in a family, are now changing to smaller systems composed of couples, the nuclear family. Similarly, the power system is consequently converting from a longitudinal to a transverse mode. As the female groups have marched out to enter politics, offices, and businesses in recent years, their rights to equality have also been expanded not only at home but at the state level as well. Moreover, the newly established law on inheritance approaches that found in the west. Accordingly the eldest son can receive only half of the inheritance, and the other half is to be shared with his mother and the rest of his siblings [5, 6, 13, 14, 17, 21].

2. Social Structure

The village is the social unit and is mainly composed of the large family systems, especially in the rural areas. The ancient saying sheds light on this traditional organization in Korea: "the neighbor is more than a cousin", meaning that neighbors are sometimes more intimate and have a deeper relationship with the family than remote relatives who are not involved in the life of the family. This nonunique organization arises from the existing living circumstances. Each village has developed according to the natural conditions that provide a means of living, e.g., agriculture, fishery, forestry, and industry.

In contrast to the rural villages, towns or cities manifest other characteristics. One major feature is that the people in towns have formed a new society organized along lines of interests. These interests create different relationships or ties according to meeting opportunities, e.g., in their profession, alumni, social activities, all of which are influenced by standards of living. A member of this society is unstable due to his frequent mobility, i.e., the member can leave one realm of urban society to enter another, even though the urban society itself does not change. The cities in Korea are categorized according to their main functions: Seoul and each provincial capital city are political centers; Yeongdeungpo, Ulsan, and Masan, industrial; Seoul, Daegu, and Busan, commercial; Daejeon, Iri, Cheonan, transportation centers; Busan, Incheon, Gunsan, trading harbors; Nonsan, Jinhae, Weonju, military cities. Recently the concept has been verified that urbanization is accelerated by rapid industrialization. But considerable problems have followed in its wake, i.e., public health problems and pollution, housing, transportation, and delinquency — all have escalated in the past few years [6].

3. Dwellings

In central Korea (Province of Gyeong-gi Do) the traditional house consists of an L-shaped inner building. This is often supplemented by an outer house (gate house), which together with a high wall or an additional wing encloses a central court yard (cf. Fig. 9). In southern parts of Korea the traditional house shape is an elongated rectangle with two or three rooms in a row and a covered veranda in front of them (cf. Fig. 10). The houses of the upper classes were constructed with heavy squared wooden posts, resting on corner stones and supporting large wooden beams, over which round rafters form the ceilings of the rooms [6].

When well constructed, the result is quite attractive. The houses are generally surrounded by stone walls with two gateways, an outer and an inner. Almost all the houses are only one-story high, because under the old Korean tradition, buildings higher than those in the King's palaces were forbidden. This restriction no longer exists, of course, and two-story and even brick houses are favored, especially in urban districts. The houses have a large center room, serving as a parlor and office, at either end of which are small rooms for the use of male members of the family.

Women live in inner compartments in accordance with the custom of keeping the sexes apart. These inner compartments consist of three portions: the women's living or bedrooms, the hall (veranda), and the kitchen. The hall (*maru*), which is open to the yard, is used as a living, dining, and sewing room in the summer, and sometimes wedding or funeral services can take place there. The relatively large number of rooms is explained by the Korean large family system; frequently three or four generations still live under one roof. The houses of the lower classes were built with the same basic floor design, but smaller and were covered with thatched rooves made of rice straw.

In recent years the straw-thatched roof is being converted to the tiled roof, a result of the practices of the "New village movement" (*Sae-Maul* movement). The distinctive feature of a Korean house is its age-old heating system, *ondol* (hot floor). Ducts under the cement floors circulate heated air from kitchen fires. A room heated in this way is most comfortable in the old Korean winter.

Shoes are never worn indoors, and with modern exceptions most houses have no chairs or beds. People sit on the floor, eat from low tables, and sleep on soft, thick quilts or mattresses, which are stored during the daytime. A dwelling unit consists of an entrance, an exclusive room for living, and separate cooking facilities. The total number of dwelling units in Korea in 1968 was estimated at 4 146 526.

The provisions for drinking water are classified into private well, public well, public piped-water system, private piped-water system, etc. The number of dwelling units provided with a private well are 434 146; public well, 3 123 828; piped-water system (public), 430 288; piped-water system (private), 192 419: and other, 283 845.

Housing projects yielded in 1970 a total of 4 340 000 housing units, an increase by 500 000 since 1966. In 1970 demand for 1 200 000 residences, 22.2% of the total demand, was not met. The problem of housing shortage has been worse in urban areas than in rural and coastal areas, and has become increasingly serious due to population growth, the dissolution of the traditional large family system, and the inflow or rural population to urban centers. The rising prices of land and construction materials have even further impeded the supply of new houses. Nevertheless, the number of households in Korea was expected to increase from 5 574 000 in 1970 to 6 349 000 in 1976. To solve the housing problem about 800 000 new housing units will be constructed during the planned period. This will increase the nationwide total of housing units to 5 008 000 by 1976. In the urban areas, where the shortage of housing has been the most acute, new housing lots will be developed for resale to homeless people at lower prices. To increase the use of land, the construction of inexpensive, standardized apartment houses will be encouraged in the cities, whereas in rural areas existing residences will be improved.

To lower construction costs, construction materials will be produced according to unified specifications, and standardized blueprints will be distributed. This will make it possible to construct a large number of housing units at one time.

Table 9. Demand for and Supply of Houses

	1970	1976	Net increase (%)
Total population	31 317 000	34 345 000	9.7
Total households	5 574 000	6 349 000	3.9
House available	4 338 000	5 008 000	15.4
House to be built	120 000	183 000	52.3
Super-annuated houses	40 000	45 000	12.5
Housing shortage	1 236 000	1 341 000	8.5
Rate of shortage (%)	22.2	21.2	–

Source: Government of Republic of Korea [45]

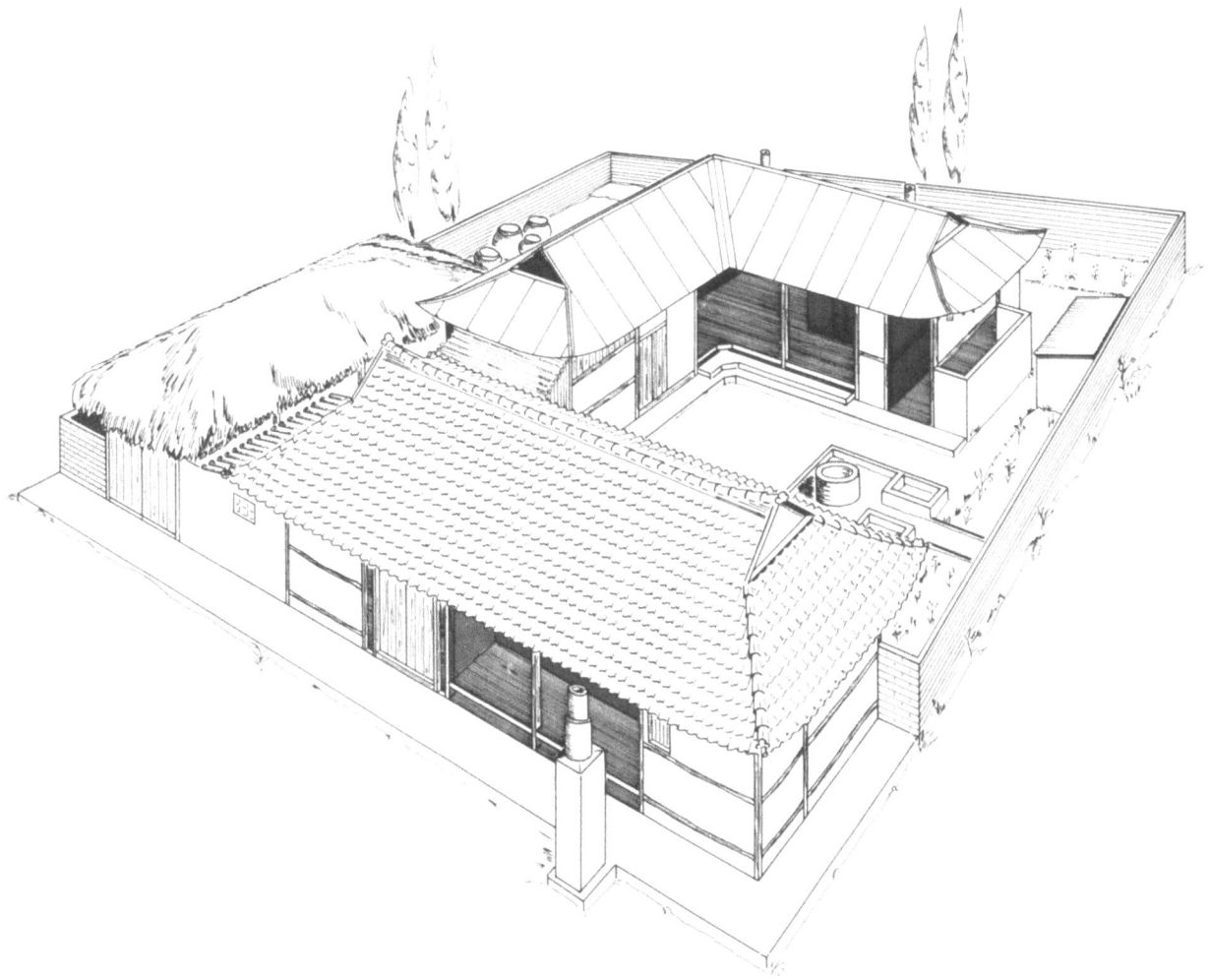

Figure 9. Korean farm house type: Central Korean type (Gyeong-gi Do)

The total funds for this program during the plan period amounts to 382.7 billion won (Table 9).

4. Clothes

The majority of the population in urban cities wear western-styled clothes. However, many of the housewives, members of elder generations, and farmers in rural areas still wear Korean-style clothing. The weather in Korea is continental due to Korea's location on the easternmost part of the Asian continent. Consequently, Korean dress is intended to adapt to both the cold and the hot seasons [5].

The fashion is quite different from the western style. The upper (jacket) and trousers during the cold season are stuffed with cotton between the inner and outer layers of the garment. The female wears a padded skirt in addition to thick cotton long underwear. Cotton socks (*boson*) cover the feet and ankles almost up to knee and are tied to the bottom of the trousers to keep out the cold air. Wrists are covered with cylinderlike gloves (*tosi*) to prevent loss of body warmth. In general, Korean dress is not the open type of the Japanese or western-style dress. Summer wear is devised to prevent and reduce build up of hot air. Clothes are made of hemp fiber, which is stiffly starched to enhance ventilation. The color of the clothes varies with the season. During the hot season, white is preferred to reflect direct sunlight, and black in winter to absorb sunlight. The variety in colors seen usually does not deviate much from either one of these extremes.

5. Livelihood

More than 80% of the total national product used to derive from agriculture until the opening of the ports in the latter part of the nineteenth century. However, throughout Korea's long history, skillful and dexterous craftsmen have engaged in melting and refining metals, in shipbuilding, brewing, weaving, and ceramics. They have continuously spread their technology to neighboring countries. As soon as Korea opened its ports to the western world, the trend to industrialization was accelerated through railways, modern communication networks, and urban electrification projects [5, 6, 13, 14, 17, 21].

In the early twentieth century when Korea lost its independence to Japan, the industrial structure began to undergo changes. Due to the exploitative colonial policy, its industrial structure was gradually fossilized to fit into a subordinate role and complementary component for

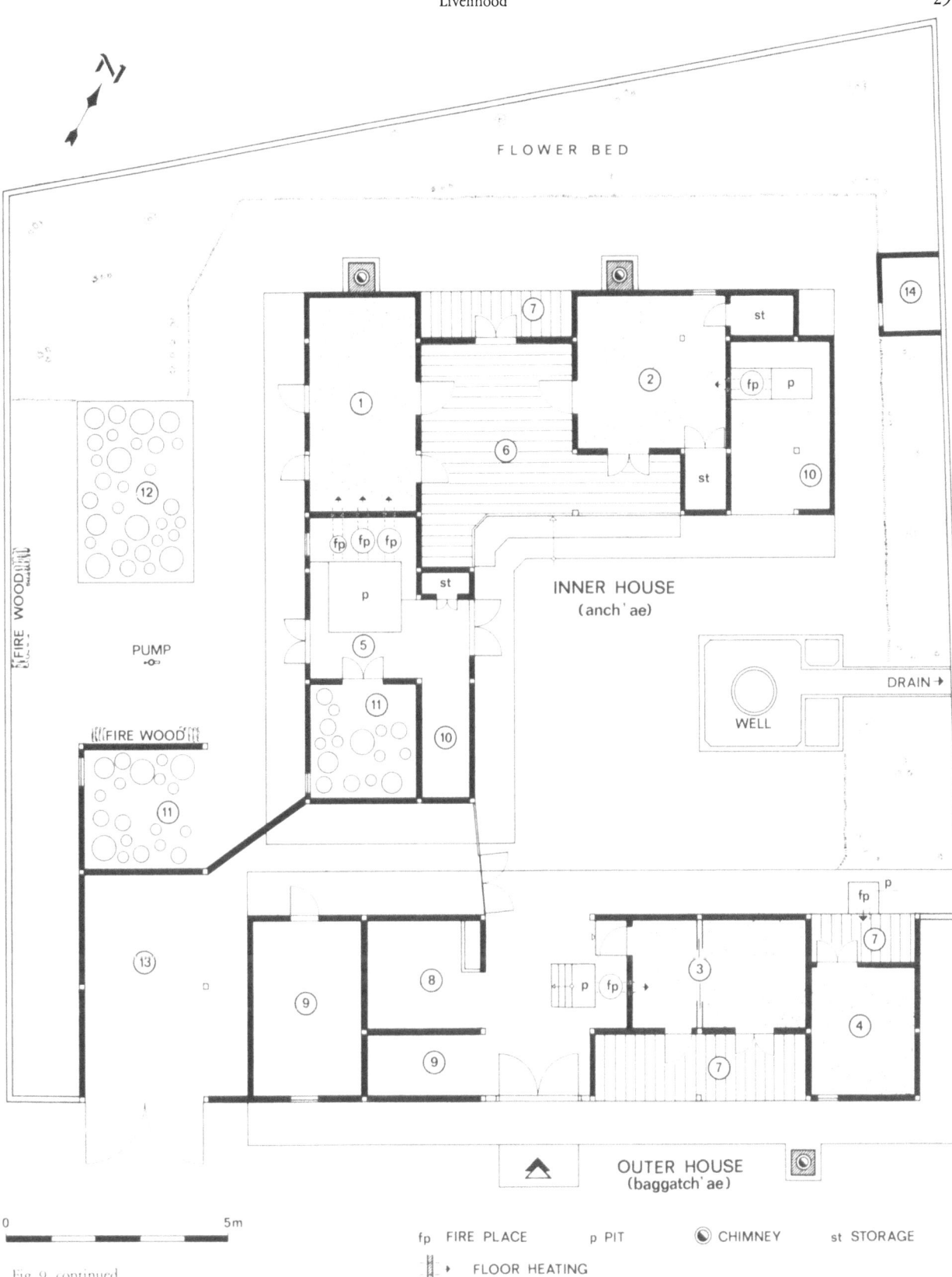

FLOWER BED

INNER HOUSE
(anch'ae)

PUMP

WELL

DRAIN →

OUTER HOUSE
(baggatch'ae)

0   5m

fp FIRE PLACE   p PIT   ◐ CHIMNEY   st STORAGE

▮→ FLOOR HEATING

Fig. 9 continued

1 family room, dining room in winter (an-bang)
2 separate room, often used by eldest son and wife (kŏnnŏn-bang)
3 room of head of household, room for (male) guests (sarang-bang)
4 room of farm hand (mŏsum-bang)
5 kitchen (puŏk)
6 large veranda, dining room in summer (taech'ong maru)
7 veranda (t'oen maru)
8 cow stall (oean-gan)
9 corn crib

10 straw for stove
11 store-room for pickled vegetables (kimchi-kwang)
12 platform for pickled vegetables (changdok-dae)
13 cart shed
14 chicken shed (tak-jang)
15 pig shed (twaeji uri)
16 shed
17 gate
(Design: Dr. Eckart Dege)

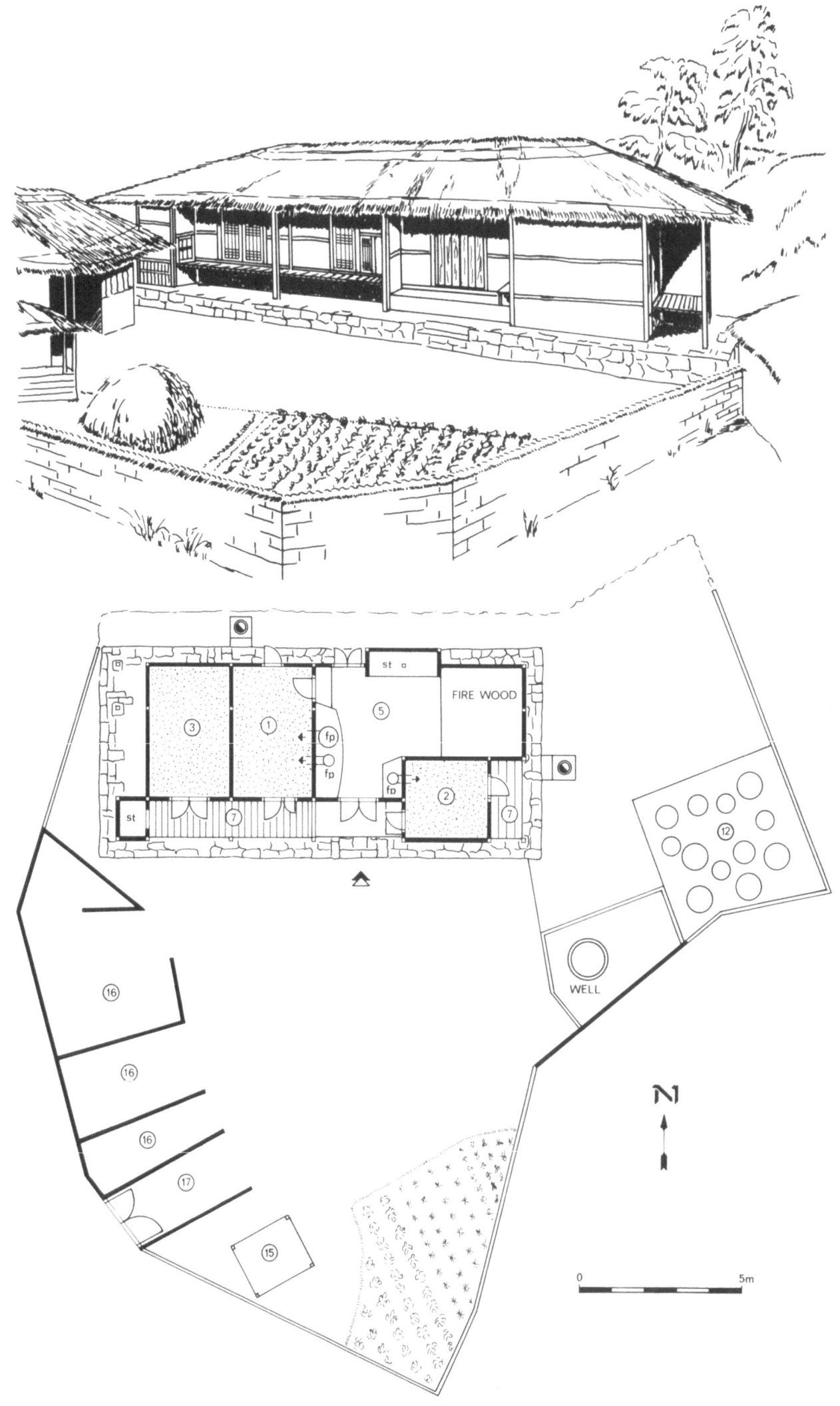

st □

FIRE WOOD

③

①

fp

fp

⑤

②

fp

st

⑦

⑦

⑫

WELL

N

⑯

⑯

⑯

⑰

⑮

0 5m

Figure 10. Korean farm house type: Southern Korean type (Jeonra Nam Do). Legend see Figure 9

the Japanese economy. Whereas the rice production expansion program enforced by the colonial government increased the productivity of rice paddies, the rice was exported to Japan in the 1930s at enforced low prices. In the midst of good and plentiful harvest, peasants were fed mostly on Manchurian soybean cake, formerly used as cattle fodder.

The manufacturing industry proved no exception to this exploitative tendency. Even though the number of plants and the level of industrialization increased every year, the country played a supplementary role to Japanese industry. The Japanese defeat in 1945 brought Korea complete independence, but the newly reborn country was crippled by its division into two parts, the southern half of which was endowed with agricultural potential, the other half being more suitable for industry. The Korean War in 1950 brought further misery.

However, after overcoming various kinds of political and economic difficulties following the cease-fire in 1953, Korea has made enormous progress. Especially the Third Republic, born in 1963, has witnessed great achievements in the economic field as well as in social welfare through successfully implementing three consecutive Five-Year Economic Development Plans. A modern highway system has been constructed, railroad and internal air transportation facilities have been expanded and improved, and with the help of foreign loans a number of new industries, like shipbuilding, manufacturing of glass, rubber, electronic equipment, plywood, and oil refining, have been added. These economic developments have been accompanied by various changes in the livelihood of the people as well.

6. Farming

The cultivated land area comprises only 23.5% of the total land, even though the population in South Korea engaged in agriculture includes about 62% of the total population. Each respective agricultural household has only 0.7 ha land. Due to the narrowness of the fields, many of the farmers still use cattle and farm workers as labor sources instead of machines. This small land tract is not enough to meet the living expenses of the family. Because the main staple of Koreans has been rice, the principal agricultural production is focused on rice. Wheat is the second most important product. For this reason, rice and wheat provide about 80% of all the agricultural products (Examples see Maps 5 a – 5 c).

Farming patterns in the agricultural areas differ somewhat between the north and the south. In northern Korea, the ratio of field to paddy is 8 : 2. Due to less rain during the summer season in the north, field agriculture has been more extensively developed. In Hamgyeong Nam Do and — Bug Do, potato, pea, corn, and barley are the main products, but millet, pea, potato, red bean, bean corn, indian millet, and apples are the main products in Pyeongan-Do. In the middle part of Korea, the ratio of field to paddy is almost the same, its agricultural products being intermediate between north and south. In southern Korea, however, the ratio of field to paddy is 6 : 4. Rice production has priority in the south, but wheat production follows closely. Cotton, tobacco, and beans are also important products of the area [5, 6].

The main overall products of the country are rice, wheat, beans, and millet. The annual mounts of these grains produced are about 30 million sums (sum = 47.6 U.S. gallons) of rice, 8 million sums wheat, and 4 million sums barley. Beans include the bean, red bean, small green pea, and peanut. The plants also contribute to fertilization by absorbing the nitrogen from air and replacing it in the soil. Bean provides the raw material for soyabean sauce, beanmash, and bean curd. An important food stuff in addition to rice and wheat is the potato. There are two kinds of potatoes, the irish potato and sweet potato. The former is produced more in North Korea, and the latter, in South. Other agricultural products are apples, pears, grapes, persimons, oranges, peaches, and chestnuts.

Nevertheless, food shortage still remains one of Korea's problems. To avert it the government has established a plan to expand the land by cultivating hilly areas, thus practicing land reclamation. Reformation of land, establishment of an irrigation system, and improvement of agricultural techniques have also accompanied these new government programs. In 1976, 36 million sums rice were produced, double the amount produced during the Japanese occupation before 1945. In those years the annual amount of rice production from the land, including North Korea, was only 18 million sums. With 20 million sums meeting the annual needs of the nation about 10 million sums can still be reserved for emergency use. In this way, the nation is eager to engage in "Green Revolution".

Panax ginseng is one of the most famous products of Korea. Ginseng is grown in Gaeseong, Ganghwa Island, Geumsan, Buyeo, and Punggi. Tobacco was imported into this land about 360 years ago during the era of King Kwanghae. Cotton, hemp, ramie, and silk provided the main materials for the drapery industry in the past, but now all except the last are used less and less due to massive production of chemical fibers.

In contrast to crop farming, stock farming was not a prosperous project. Several reasons may account for this: the ignorance of the people and the influence of the Buddist doctrine prohibiting eating of meat. For these reasons stock farming remained at the level of a subsidiary business in the rural area. However, it became a booming industry two decades ago. Cattle are now the most popular domestic animals. In 1960 their number was estimated at about one million in South Korea. They are used for ploughing as well as providing a meat source, but the concept of cattle is changing from a labor power to a good source of meat and milk. The horse is raised mostly in Hamgyeong provinces, Jeonra Nam Do and Jeju Do in the south. Their total number was estimated at about 17 850 head, about 50% of which are bred in Jeju Do. Pig and fowl are raised as supplementary income sources by most rural families and are also popular protein sources in Korea. The excrements and droppings provide good material for compost, but also serve as distribution factors for pathogens. Although goats, sheep, rabbits, and ducks are raised in Korea, their number is not very high.

7. Water Control

a) Annual rainfall: The nation's average annual rainfall is 1160 mm, and the total volume of water resources in South Korea is estimated at 114 billion metric tons. Although relatively rich in water resources, the country has suffered almost every year from drought and flood, because approximately two-thirds of the yearly rainfall is concentrated in a three-month period, from mid-June to early September. This unbalanced seasonal distribution of rainfall confines the available volume of water resources to 8.8 billion metric tons, only 7.7% of the total. The utilization of water resources is also unbalanced in different areas of the country, even though many dams have been constructed. The construction of river banks up to now covers 5202 km, only 25.3% of 20 600 km, the total length of river embankment extension projected [5].

b) Comprehensive development: To provide a stable all-weather farming foundation and improve water resources, comprehensive development has been promoted case by case. Surplus and shortage of water resources should be offset by transferring water from one area of the country to another. To this end, comprehensive development of the nation's major rivers — the Han, the Geum, the Nagdong, and the Yeongsan — is planned. These projects will include erosion control, construction of multipurpose dams, river embankments, installation of waterworks and sewerage facilities, industrial waterworks, and irrigation facilities. In as much as the projects will deal with water resource problem areas from upstream to estuary, they will need to be systematic and comprehensive.

The Soyang river multipurpose dam was constructed as part of the development of the drainage area of the Han river basin. Potable waterworks and industrial waterworks with daily capacities of 65 900 and 150 000 m³ are still to be constructed.

The Andong multipurpose dam was completed in October 1976 in the Nakdong river basin and will supply 186 million m³ of irrigation water and 740 million m³ of industrial water annually.

The Jangseong dam in the Yeongsan river basin was completed in October 1976, and the Daecheo and Damyang dams followed. River embankment totaling 183.5 km will prevent floods. With the additional installation of a potable waterworks plant having daily capacity 42 000 m³ the implementation of erosion control projects, and the construction of estuary weiers, the drainage area of this river will be comprehensively developed.

The Daecheong dam is under construction in the Kum River basin, and a waterworks plant with a daily capacity of 62 000 m³ will be installed.

In addition, projects along the other major rivers will be expedited. Erosion control and afforestation projects will be conducted on 517 000 ha during the Fourth Five-Year Economic Development Plan period (1977 – 1981), including 274 000 ha in the drainage areas of the four major rivers. Of the 12 dams to be developed, eight multipurpose dams will be constructed during the period, and 1366 km river bank will be repaired, including 989 km in the drainage areas of the four major rivers. The daily supply of both potable and industrial water will be increased by 1 850 000 metric tons, including an increase of 1 773 000 metric tons in the drainage areas of the four major rivers. During this period, irrigation facilities in the drainage areas of the four major rivers will be installed on 102 500 additional hectares.

Erosion control projects will be conducted on 40 000 ha, and 477 000 ha of forests will be created in upstream lands to preserve water resource, prevent erosion, and control flooding. Following the completion of three multipurpose dams, namely the Soyang river, Andong, and Jangseong dams, five other multipurpose dams, the Habcheon, Daecheong, Damyang, Daecheo, and Inha, are underway thereby reducing flood damage.

c) Water supply: In 1964, the ratio of potable water supply to its demand was 18.6% and the daily per capita water supply was 104 liters, falling short of the demand. During the Second Plan period (1967 – 1971), potable waterworks were installed with funds from foreign loans in large cities such as Seoul, Incheon, Daegu, Daejeon, Gwangju, and Cheongju. As a result, the ratio of potable water supply to its demand increased to 35.5%, and the country daily per capita water supply increased to 175 liters in 1970.

At the moment, emphasis is being placed on providing water supplies for large and medium cities and towns through the construction of multipurpose dams and other methods.

Through the establishment of water facilities, the effective use of potable water and industrial water will be furthered. The ratio of potable water supply was thus increased to 51.4% and the daily per capita water supply, to 200 liters in 1976.

8. Underground Resources and Mining

Korea is a museum of minerals with about 210 species being reported. Of these gold, iron, tungsten, molybdenum, fluorine, black lead, magnesite, phosphorus, and kaolinite abound. Zinc, cobalt, asbestos, silica, etc., follow in amount, but the rest of the minerals are not sufficient to meet the national needs [5, 6].

Of all the minerals, gold and iron are the most abundant, with over half the amount produced originating in the northern part of Korea.

In South Korea the total amount of gold is estimated to be about 750 thousand tons. The estimated total amount of iron has been set as about 1.2 billion tons. Coal, one of the main resources of energy, occurs as smokeless coal, lignite, and pitch.

The mining industry has grown at an annual average rate of 7.3% during the 1964–1970 period. The ratio of mining to the GNP, however, has gradually decreased from 1.8% in 1964 to 1.2% in 1970. Coal production continued to increase in the first half of the 1960s and reached 12.4 million tons in 1970.

Throughout the 1960s the metal mining sector of production remained inactive, while the nonmetalic mining sector (excluding coal mining) operated at a brisk rate. The annual production of copper, for example, has increased from 12 000 tons in 1964 to 27 000 tons in 1970 and that of lead and zinc ores from 12 000 to 80 000 tons.

9. Marine Products

Being a peninsula, Korea offers favorable conditions for a prosperous marine industry. The warm and cold currents along the coastal line support an abundance of various fauna and flora. Islands scattered along the southernmost end of the Korean peninsula also favor the propagation of various fishes. The good condition of the continental beds and an abundance of plankton add to these advantages. In spite of this, in the past the industry was not very prosperous for several reasons: the low social prestige attributed to fishery by people in general and the lack of appropriate equipment and techniques. The situation now seems to be changing somewhat due to improvement of the shipbuilding industry over the past decade. It has brought a better understanding of the importance of the fishing industry, and this improvement has extended even to deep-sea fishing as well [5, 6].

The fishing grounds on the Sea of Japan, beginning from the Busan port and extending to the mouth of the Duman Gang, have a total length of 1850 km. Cold currents from the north seas and warm currents from the south seas meet in this area. For this reason, various species of fishes from both north and south are found in abundance in this region. Alaskan pollock, herring, codfish, mackerel, yellowtail, anchovy, sardine, and cuttlefish are among the most delicious and favorite species found not only in the Korean market but also abroad. Sardine, pollack, herring, and codfish are the main species found in cold currents, and cuttlefish inhabit the warm currents. During the early cold season, the fish are caught in abundance in the waters around Ulreung Island, and are exported to Southeast Asia. Namhae (South Sea) fishing grounds extend from Busan to Mogpo, a length of 6000 km. About 2000 islands are scattered along the winding coastal line. Herring and codfish are caught in the eastern areas where the cold currents prevail.

The major part of the coastal area bordering the South Sea is influenced by the warm current. For this reason, anchovy, mackerel, flatfish, and hairtail are caught around Jeju island. Laver and jelly plant, the main products along the coast, are exported. Agar is produced from the jelly plant. The yearly production of agar amounts to more than 500 tons, making Korea the second largest producer in the world.

Hwanghae (Yellow Sea) fishing grounds lie between the mouth of the Yalu river (Abrok-gang) and Mogpo. The length of its coastal line is about 8500 km. In addition, there are about 1500 islands scattered in this region. Yellow covina, sea bream, shrimp, mackerel, squeteague, and flatfish are the fish most often caught here. Even though the amount of fish caught does not differ much from that in the eastern and southern coastal areas, still seashells, shrimp, oysters, and laver are harvested in abundance along the west coast. The large difference between flood and ebb tide provide favorable conditions for the propagation of these sea creatures. Yellow covina is the most famous species. Being a warm current species, it is best caught from early spring to autumn. The best season for the shrimp industry is July and August. During these seasons, Eochongdo and Yeonpyeong-do function as temporary markets. Also sea mammals are fished for they can be caught off-shore in Korea at various locations. Jangsaengpo in the South Sea and Daeheugsan islands are the main habitats for this species. The total annual catch is estimated at 200–300.

Salt (NaCl) is another product of the western coast. A broad estuary resulting from the large difference between flood and ebb tide provides good conditions for deriving salt from brine. Ganghwa, Incheon, Gwangyang Bay, and the mouth of Cheongcheon-gang are famous areas having estuaries suitable for salt production.

10. Industry

In spite of sufficient manpower, industry in the feudal ages was very primitive. From a chronological viewpoint, modern-style industrialization was started in Korea in 1930, the year a water power plant was established in Bujeongang. However, during the Japanese occupation from 1909 to 1945, a few of the power plants and other factories were exclusively used for the benefit of Japanese administration in Korea. Most Koreans were not taught the various skills and techniques needed for running a factory. This was unfortunate because it forced Koreans to restart their industrialization in accordance with the reconstruction plans made after the Korean War. In general, the modes of the industries in the north and the south were different. In the north, heavy industries such as metal-refining and chemical plants were developed because of the abundant water and underground resources available. In the south, light industries, such as spinning, paper mills, and foodstuff production, were more prosperous. But within the past decade, several heavy industrial areas have been established in the south, e.g., Ulsan, Yeosu, Biin, Gumi, Masan and Iri. Large-scale oil refineries, producing over 1000 barrels of refined oil daily, have helped to accelerate heavy industries such as steel, chemicals, and even shipbuilding. The industries of Korea are largely divided into home industry and factory industry. The former is represented by cotton woven goods, hemp woven goods, silk woven goods, breweries, bamboo goods, wooden handicrafts, embroidery, and straw-bag manufacturing. As regards the latter, factory-level industry includes iron manufacturing and refining, light metal industry, mechanical industry, soap industry, polyethylene products, artificial fibers, rubber industry, paper mills, leather production, cement, ceramics, tobacco, glass, and chemical fertilizer production, all of which are items that sell well on the Korean market as well as abroad [5, 6].

11. Food Production

a) The demand for food: During the 1960s the demand for food crops rose at an annual average of 2.5%. The demand for rice, the staple food, also climbed at an annual average of 3% against an annual average increase of 2.3% in its domestic production. Consequently, the supply shortages had to be met with imported food grains.

The demand for food is expected to rise continuously because of population growth and increases in the national income. In 1970 the per capita consumption of rice was 140 kg. It is assumed that this level of per capita rice consumption will not rise in the future, since it is expected that the pattern of food consumption will change due to increasing income and the government's policy of an optimum rice price [5, 6]. Consequently, the national demand for rice is expected to rise at a rate of 1.5% a year in proportion to population growth.

The demand for barley will fall as the level of income increases. However, the government will adopt a policy to broaden the relative price gap between rice and barley and to encourage the use of barley for animal feed. Therefore, the demand for barley is expected to rise at an annual average of 3.3%.

The demand for wheat is expected to increase at an annual average rate of 4%, largely owing to the increase in the demand for food processes from wheat. The demand for corn is expected to rise at an annual average rate of 20% because of a sharp increase in the demand for corn as animal feed. Rice production is projected to increase from 3 939 000 tons in 1970 to 4 860 000 tons in 1976, an increase of 23.4%, and barley production from 1 974 000 tons to 2 406 000 tons, an increase of 21.9%. To achieve the rice increase, the annual production must grow at 3.5%, and the yield per 300 pyeong (pyeong = 3.2 m²) must reach 400 kg. An intensive policy for increased rice production is being carried out.

To attain self-sufficiency in rice production, lowering consumption is as important as increasing production. Therefore, a policy to encourage people to eat food other than rice has been pursued. Wheat production is expected to rise from 357 000 tons in 1970 to 488 000 tons in 1976, a 39% increase, and corn production, from 68 000 tons to 149 000 tons, a 119% increase during this period. However, demand will still surpass supply, making unavoidable the import of these two grains.

Growing new varieties of rice, including the "Tongil" (IH-667) is encouraged, and programs to guide farmers in this endeavor are being carried out intensively. Cooperative cultivation will continue to be encouraged be-

cause it has increased production in the past. Also, the application of fertilizer per hectare of farmland is to be increased from 160 kg in 1970 to 220 kg in 1976. The supply of calcareous and silicate fertilizers will be increased to complement other fertilizers.

The efficiency of insect damage control will be raised by simplifying the variety of insecticides used an improving the quality of such chemicals. To increase food production, irrigation facilities will be expanded and improved, and suitable cultivated land will be rearranged for mechanized farming. The use of farmland for purposes other than farming is being restricted. Even though yet far from the final goal, the above efforts succeeded in producing 36 million sums rice in 1976, thus exceeding the annual amount of the nation.

b) Agricultural water resources: The development of agricultural water resources has been one of the agricultural projects most intensively promoted during the last decade. Of the total rice paddy area of 1.3 million ha in 1970, the area of irrigated paddies has been expanded to more than 1 million ha.

During this period, agricultural water resource development projects have been continued to complete the irrigation of 1 176 000 ha. In addition, water resource development has been reinforced with the comprehensive development of four major river basin areas, thus creating cultivated farmland free of droughts and floods.

c) Expansion of rice paddy rearrangement: Rice paddy rearrangement is needed to expand farm mechanization, to increase farm labor efficiency through the completion of farm roads and waterways, to increase the utility of farmland, and to improve soil quality and fertility. Rice paddy rearrangement projects are carried out intensively with a view to increasing the area of rearranged paddies to 450 000 ha or 75% of the total 600 000 ha capable of being rearranged.

d) Promotion of farm mechanization: The supply of farm labor began to decrease gradually in the latter half of the 1960s. Therefore it became increasingly urgent to mechanize farming to replace the decreasing supply of farm labor to increase labor and land productivity, and to promote soil fertility through deep plowing.

e) Development of four major river basin areas: The four major river basin areas (the Geum, Yeongsan, Nagdong, and Han Rivers) occupy 62 800 km² of 64% of the total area of land, and 1 243 000 ha or 54% of all farmland under cultivation in the country. These areas must by all means be developed for future agricultural development in this country.

12. Livestock Industry

Between 1964 and 1970 beef consumption increased from 32 000 tons to 37 000 tons, milk consumption from 7000 tons to 40 000 tons, and egg consumption from 943 million to 2600 million. During the Third Plan period (1971 – 1976) beef consumption increased to about 74 000 tons, milk consumption to 112 000 tons, and egg consumption to 4000 million. Therefore, prospects for the development of the livestock industry are promising. Korean cattle have heretofore been raised chiefly as draft animals. Since fewer draft animals will be needed as mechanized farm work increases, a policy will be necessary to ensure that farmers continue to raise beef cattle [5, 6, 14].

During this period an effort was made to improve the breed of Korean cattle and to secure a sufficient number of bulls for breeding. In addition, artificial insemination has been practiced extensively. As a result, the country was expected to have 1.3 million head of beef cattle in 1976 to ensure a constant supply of beef. To develop the dairy industry, 9600 head of dairy cattle were imported between 1962 and 1970. Also, many facilities were installed for processing various dairy products, including milk. As a result, dairy farms have become a thriving enterprise, and milk no longer needs to be imported. The number of dairy cattle was increased from 24 000 head to meet the domestic demand for milk, estimated to be about 112 000 tons in 1976. In an effort to meet the increasing demand for meat and eggs, farmers are encouraged to raise pigs and chickens as a means of additional income.

Raising livestock has in the past been only a sideline among small farmers. Therefore, the policy for developing the livestock industry will emphasize the development of farms engaged exclusively in raising livestock. But this policy will be executed flexibly so that there will be neither a surplus nor a shortage of livestock products.

13. Education and Illiteracy

Koreans have always had a high respect for education. According to Confucian thought the respectability of a person was judged in the past by his occupation. Taken in hierarchical order, these were scholar, farmer, engineer, and merchant. Even though this concept has largely disappeared, people still put more stress on being a scholar or a bureaucrat than on being an engineer. Moreover, the opportunity for education has traditionally been given to males more often than to females due to the erroneous thought of Confucius that "men are respectable but women insignificant". In 1944 one-half of all Korean males aged 15 and above were reported to have had some formal education, whereas this was true for only 20% of the females. In the same year there were 26 000 male and 3600 female college graduates. In 1955, 87% of males over 15 were literate, whereas 66% of the females were [17]. Likewise there were over 100 000 male and 16 000 female college graduates. The inequality based on sex still remains. However, traditional Confucian thought regarding segregation of males and females has now almost disappeared, and all schools, even universities, are open to both sexes.

The total number of schools in Korea of various levels in 1974 were 10 182 with an enrollment of 8 953 083 (Table 10).

Table 10. Summary of school types (except Kindergarten) in 1974 (6)

Schools	Number	Number of students enrolled
Primary school	6 315	5 618 768
Middle school	1 935	1 929 975
General high school	613	530 177
Vocational high school	476	451 032
Junior technical college	86	50 328
Junior college	11	3 798
Junior teacher's training college	16	11 176
College and university	72	192 308
Graduate school	80	12 289
Undergraduate institute	13	1 966
Nurse's training school	4	1 614
Civic school	30	3 029
Civic high school	257	52 532
Trade school	62	11 867
Higher trade school	48	12 991
Middle school institute	37	8 182
High school institute	84	55 184
Special school	43	5 867
Total	10 182	8 953 083

Literacy in Korea may be defined as the ability to read and write Korean. Illiteracy on the other hand, refers to the inability to read or write one's own language; that is, those who cannot express their own simple ideas in written form are illiterate. With this in mind, illiterates in 1960 were estimated at 4 450 230 (male, 1 239 106; female, 3 211 124), or about 17% of the total population. The government has allocated about 18% of the budget to solve this problem. Through various endeavors the point has been reached where it may be reasonable to say there are no illiterates except for some members of the older generation.

14. Trade and Transportation

The first attempt at trade was made with the signing of the Ganghwa treaty with Japan in 1875. Subsequently, trade has been developed through treaties with a number of foreign countries, the United States, England, Germany, and China etc. However, the trade business suffered a setback during the Japanese colonial regime from 1909 – 1945. Goods for export were mainly rice, beans, fish, timber, ginseng, leather, iron, and some other raw materials. Almost all mechanical and chemical products were imported. However, in accordance with the recovery from damage suffered during the Korean War and the industrial development which followed, materials for export are changing to chemical products and construction materials. Synthetic textile goods, cement, chemical fertilizer, and some machinery are now exported. Nevertheless, the ratio between exports and imports still remains unbalanced, with imports exceeding exports by almost threefold.

Roads: As in other parts of the world, roads are valued as arteries of development of the nation's culture and industry. The term road as used in this chapter signifies a width of over 6 m. The total length of roads in South Korea, about 27 170 km, provides the means for 20% of the total transportation. National roads originate in Seoul and reach as far as Busan and Mogpo to the south and Sinweuiju and Najin to the north [5, 14].

Expressways: Besides roads, expressways have been established since 1968. Nowadays the expressway forms a network over the land of southern Korea with a total length of over 1400 km. With the improved convenience of road travel, the number of automobiles has increased. The total number of licensed cars was 7326 in 1945, 31 139 (excluding military traffic) in 1960, and 311 000 in 1976. Before the *Saemaul* (new village) movement started in 1968, many of the rural villages were not accessible to automobiles. But nowadays a car can drive to any remote village. Consequently transportation means have been changed from the A-framed wooden carrier and coach to truck, bus, and automobile, thereby reducing time and manpower.

Railroads: The first railway with a length of 32 km was constructed in 1889 between Noryangjin (Seoul) and Incheon, 65 years after the first railroad construction in England. In 1905 the Gyeongbu line (Seoul-Busan) was established; it was followed by the Gyeongju line (Seoul-Sineuiju) in 1906, and then by the Honam (Seoul-Mogpo), Gyeongweon (Seoul-Weonsan), and Hamgyeong (Weonsan-Najin) lines. In 1945 total length of railroads on the Korean peninsula was 5369 km (2557 km in South Korea). Statistics in 1970 report 3000 km in South Korea. About 11 470 locomotives (including 100 diesel engines) are in operation these days. A total of approximately 100 million passengers are transported annually, and maximum capacity is 1148 tons.

Shipping: Transportation by water has been mainly limited to agricultural products and people. Rivers used for these purposes are the Yalu (Abrok), Duman, Daedong, Han, Imjin, Geum, Nagdong, Seomjin, and Yeongsan. However, except for the Daedong river, only a 3000-ton ship can sail up to Bosanpo, 66.5 km from Chinnampo at the mouth of the river. Raft transportation on the Yalu river has been one of the romantic attractions in this country. Coastal or sea transportation is a relatively simple matter since Korea is a peninsula. Good ports are especially situated at the mouths of rivers: Sineuiju, Yongampo, Jinnampo, Haeju, Incheon, Gunsan, and Mogpo are the main ports on the west coast; Yeosu, Masan, Jinhae, and Busan are located on the southern coast; and Pohang, Mugho, Weonsan, Seongjin, Cheongjin, Najin, and Unggi, are on the eastern coast. Before 1950 ships were usually only 140 000 tons, but now in South Korea, their tonnage capacity may be as much as one million tons.

The demand for coastal freight shipping is expected to increase by 118% from 4232 million ton-kilometers in 1970 to 9228 million ton-kilometers in 1976. Coastal shipping's share of the total domestic freight transportation thus increased from 31.6% in 1970 to 34.6% in 1976.

Airlines: Transportation by air travel has become popular. Korean Air Lines has formed a network throughout the land which has shortened all available communication and transportation between cities. Nowadays it flies all over the world, to New York as well as to the European continent.

The domestic demand for airline passenger transportation is expected to increase by 467.3%, from 257 million passenger-kilometers in 1970 to 1458 million passenger-kilometers in 1976. The airline's share of the total domestic passenger transportation is thus to be increased from 0.9% in 1970 to 2.2% in 1976. Meanwhile, the demand for international airline passenger transportation is projected to increase sharply by as much as 375%, from 393 000 passengers in 1970 to 1 867 000 passengers in 1976.

B. Health Facilities, the Health Professions, and Public Health Services

I. A History of Korean Medicine

The history of Korean medicine is based upon a traditional inheritance emanating from the dim past before recorded history.

The legendary story begins with Sil-longse, the second of the five ancient rulers of China, who founded a dynasty lasting from 2823 – 2648 BC. This personage has been honored with the title of "Father of Medicine" and is reputed to have written an original manuscript on medicine called Bon-cho, which literally means , "first manuscript" (*Bon,* 1st; *Cho,* manuscript).

This tradition is commemorated in the form of a sign on all retail drug shops that prepare and sell native medicines. The final Syllable *Se* (an honorific) is dropped, and the word for inheritance, *You-aup* is added; the sign reads *Sil-long-you-aup* (inherited from Sil-long-se or inherited medicine), which in point of significance corresponds to the English word pharmacy. However, it is never employed to indicate the science of pharmacy, because Korean medicine makes no such distinction. As time elapsed, the traditional inheritance was influenced and modified by Confucian, Buddhist, and other native cultures. Eventually the title was reformed and represented by the Chinese characters, which read oriental medicine [56].

Oriental medicine is based on a philosophical belief in the dualistic cosmic theory of the Yang and the Um.

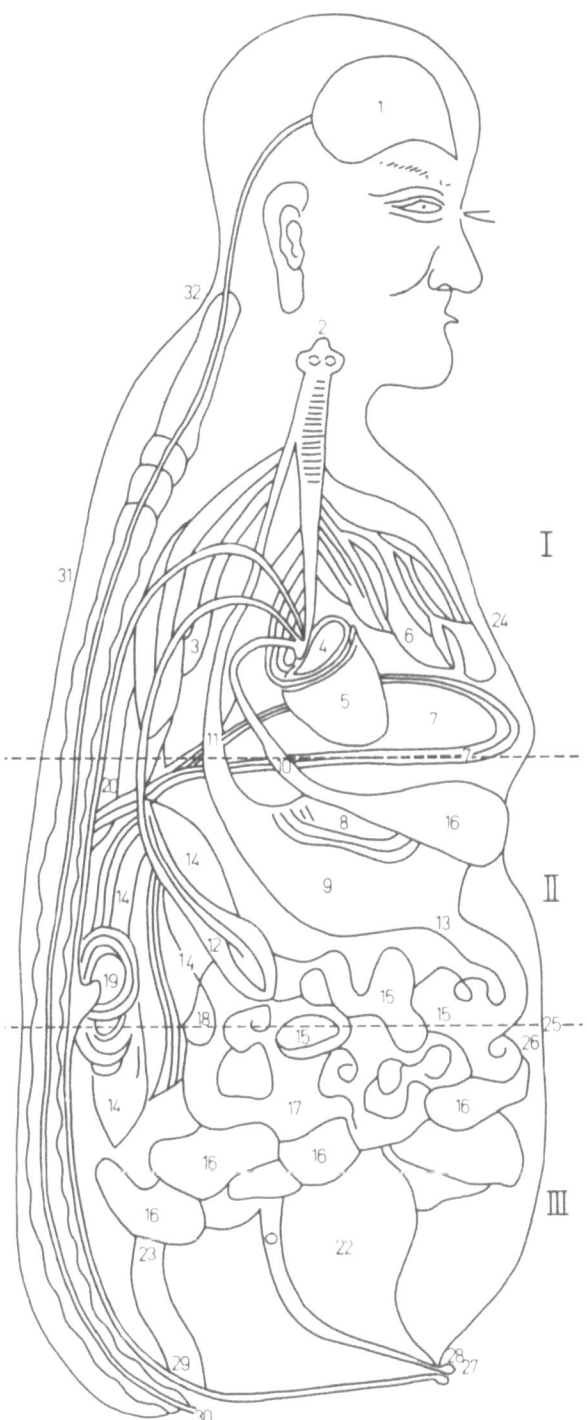

Figure 11. Anatomical chart in Ancient Korean Medicine
(Source: N. H. Borman [56]), (Explanation see page 122)

The Yang, the male principle, being active and full of light, is represented by the heavens; the Um, the female principle, being passive and dark is represented by the earth. The human body is controlled by the Yang and the Um, and the harmony of these two principles is essential for the maintenance of individual health. According to the concepts of oriental medicine, the patho-physiology of all diseases is based on the disturbance of these two principles. Thus, diagnosis and treatment have been directed to determining and restoring the disturbed balance of diseased states (Figure 11).

Diagnosis in oriental medicine is largely based upon examining the pulse, and treatment involves herb medicines and acupuncture.

The scientific basis of oriental medicine, however, has never been clearly established, perhaps because of the ancient Korean social trend that emphasized learning Confucian philosophy and depreciated the study of the natural sciences. Nevertheless, oriental medicine prospered for many centuries and continued to maintain its popularity even after western medicine was introduced into Korea. Its popularity has gradually diminished in urban areas, but it is still favored in rural areas. The Korean National Assembly has been persuaded to legalize a college of oriental medicine and pharmacy in spite of strong opposition by the Korean Medical Association as well as by the government. This school presently admits 100 students every year. After completing 4 years of study at this college, one is granted a licence to practice oriental medicine.

There were 2852 herb doctors throughout South Korea in 1978, their ratio to the population being about 1 to 15 000. Oriental herb medicine has little place in a modern health care program, and it must be replaced by modern medical care. However, we must admit that oriental medicine responds to the demands of a relatively large part of the population, which has no access to modern medical care. Therefore, oriental medicine, unscientific as it is, will probably linger on in Korean society for some time and will continue to pose a problem for modern health care programs. This is a medical-cultural conflict situation similar to that seen in other countries of the Third World. Recently the World Health Organization (WHO) has developed a new strategy for integrating the rest of traditional medicine or folk medicine into the health care planning and health care systems of the Third World countries. Oriental medicine will, however, gradually subside as the socioeconomic situation improves, and modern medical care programs reach the entire population.

II. Distribution of Health Personnel

Health personnel include physicians, dentists, veterinarians, pharmacists, midwives and nurses, X-ray and laboratory technicians, etc.

1. Physicians

The total number of medical doctors in 1974 was 15 722 (male, 13 682; female, 2040). The number includes 15 188 physicians and 534 limited physicians, and the ratio to the population averages roughly 1 : 1999. In 1978, the number increased to 20 079. However, about 60% of the medical doctors are concentrated in Seoul and Busan (Tables 11 and 12, Annex II). By city or province, Seoul city shows the lowest doctor-to-patient ratio, 1 : 1140; but Chungcheong Bug Do shows almost nine-fold the difference of Seoul, 1 : 13 550. This is strong

Table 11. Distribution of Medical Facilities

	Total	Hospital and clinic				Dental hospital clinic	Herb clinic	Dispensary	Sanitaria	Midwifery	Health center
		Sub-total	General hospital	Hospital	Clinic						
1966	9 866	5 233	18	203	5 012	1 129	2 316	83	16	900	139
1969	10. 117	5 392	12	217	5 163	1 219	2 434	117	11	752	192
1972	11 055	6 073	17	260	5 796	1 466	2 419	5 93	12	799	193
1974	11 296	6 191	36	127	6 028	1 369	2 367	201	12	749	197
Seoul	4 543	2 319	22	29	2 268	847	1 071	61	1	233	11
Busan	1 192	681	3	12	666	155	235	25	1	88	7
Gyeong gi Do	1 107	607	3	17	587	99	246	22	2	104	27
Gangweon Do	385	207	–	11	196	29	66	14	–	50	19
Chungcheong Bug Do	301	155	–	4	151	28	78	8	–	20	12
Chungcheong Nam Do	725	388	–	7	381	85	188	8	1	38	17
Jeonra Bug Do	459	271	1	10	260	40	76	21	1	32	16
Jeonra Nam Do	641	416	3	10	403	60	60	10	2	67	26
Gyeongsang Bug Do	1 162	694	4	16	674	150	211	18	2	54	33
Gyeongsang Nam Do	678	392	–	8	384	64	127	14	2	54	25
Jeju Do	93	61	–	3	58	12	7	–	–	9	4
1978	12 801	6 317	61	212	6 044	1 790	2 306	249	11	572	202

Source: Ministry of Health and Social Affairs (1975; 1979)

Table 12. Medical Personnel

	Total	Physician	Herb doctor	Dentist	Pharmacist	Midwife	Nurse
1972	37 880	10 405	2 442	1 909	13 221	1 603	8 300
1973	40 752	10 922	2 404	1 997	13 868	1 847	9 714
Seoul	18 223	5 122	1 172	1 058	6 933	513	3 425
Busan	3 853	1 069	221	147	1 260	212	944
Gyeong gi Do	2 939	782	241	103	1 009	197	607
Gangweon Do	1 288	285	68	42	361	106	426
Chungcheong Bug Do	1 113	140	59	23	566	77	248
Chungcheong Nam Do	1 403	395	169	71	253	109	406
Jeonra Bug Do	1 703	355	71	39	515	144	579
Jeonra Nam Do	2 558	682	60	52	957	177	630
Gyeongsang Bug Do	3 611	980	220	152	1 251	150	858
Gyeongsang Nam Do	1 302	355	113	52	403	124	255
Jeju Do	256	57	9	11	96	21	62
Over sea countries	1 146	284	1	23	150	7	681
Military	1 357	416		224	114	10	593
1978	86 531	20 079	2 852	3 102	22 371	4 455	33 672

Source: Ministry of Health and Social Affairs (1975; 1979).

evidence that the distribution of medical doctors in Korea still presents a real imbalance between urban and remote rural areas.

During the Japanese regime, several licence examination systems existed for nonmedical graduates, the examination systems of the central and provincial governments. Their purpose was to supplement the shortage of medical doctors in the rural or remote mountain and island areas. Applicants who passed the examination of the central government were permitted to practice anywhere, but applicants who passed the provincial level examination were permitted to open clinics, as so-called limited physicians, in the area designated by the governor of the respective province. With the establishment of the Republic of Korea in 1948, this system was abolished. The local physicians and medical doctors who came from North Korea as refugees were given an opportunity to take the national examination. Those who passed were authorized to receive a licence as a regular physician.

Medical doctors in Korea can be categorized as medical graduates, nonmedical graduates who have passed state examinations, and nonmedical graduates practicing as limited physicians. Medical graduates (University or College) predominate with 15 188, including those who have passed the state examination; 534 nonmedical graduates work as limited physicians. There are 11 foreign medical doctors in Korea, most from the United States and Canada and working as missionary doctors in medical schools and mission hospitals.

Among the licensed doctors, 880 are retired from professional duty for reasons of old age, sickness, etc. The actual number who are still practicing in the country is 11 310 (male, 9900; female, 1410), including 5262 specialists. The number of specialists in 1977 was about 30% of the total number of medical doctors. These included general surgery, 1267; obstetrics and gynecology, 901; internal medicine, 869; pediatrics, 604; ENT, 387; orthopedic surgery, 417; preventive medicine, 274; ophthalmology, 270; urology, 240; neuropsychiatry, 212; dermatology, 185; neurosurgery, 178; radiology, 164; tuberculosis, 126; anesthesiology, 153; thoracic surgery, 92; surgical pathology, 84; and clinical pathology, 71. The number of medical graduates from the 17 medical colleges in Korea is increasing year by year. The number in 1970 was 642, but 1250 in 1977. A considerable number of medical graduates are also working, studying, or training abroad. Statistics in 1975 reported that 3135 are distributed in 32 foreign countries: United States, 2834; Canada, 72; Japan, 65; Uganda, 34; Jamaica, 22; Malaysia, 18; West Germany, 17, and so on.

The geographic distribution of physicians in 1973 within the Republic of Korea is shown in Tables 11 and 12. The geographic distribution of physicians in rural and urban areas reported to the ministry of Health and Social Affairs in December, 1973, indicates that over 60% of the physicians and dentists are concentrated in the urban areas. Of the total number of physicians, 10 922, who engage in professional practice, 6191 (57%) are concentrated in the cities of Seoul and Busan alone. This uneven distribution of medical personnel is one of the serious problems facing Korea at this stage.

2. Other Medical Personnel

a) Dentists: Statistics in 1978 showed 2852 dentists. Dental college graduates with an academic background occupy the majority. The ratio of dentist to population is roughly 1 : 15 000.

By province or special city, in 1974 about 41% (843) were concentrated in Seoul. The least concentrated province was Jeonra Bug Do with only 39 dentists serving 2 431 892 people. Ratio of dentist to population shows a big difference between urban and rural areas. In Seoul the ratio is 1 : 6738, but only 1 : 75 997 in Jeonra Bug Do province. Other provinces range in between these two extremes.

b) Herb doctors: The total number of herb doctors registered in 1974 was 2738 (male, 2690; female, 48). Of these 2504 were formally licensed, and 234 had limited licences, allowing them to practice only in designated villages. In 1978, the number increased to 2852. Statistics

in 1973 reported that 1172 (male, 1144 and female, 28) herb doctors were registered in Seoul.

c) Midwives: The total number of midwives in 1974 was 3445, and the ratio of licensed midwives to population was 1 : 9125. Most of them were practicing in the urban areas. In 1978, the number increased to 4455.

d) Nurses: The total number of licensed nurses in 1974 was 19 842. Their ratio to population was 1 : 1504. Of the 9714 (male, 4; female, 9710) registered in 1973 almost half (3425) were registered in Seoul. In 1978, the number increased to 33 672.

e) Pharmacists: The total number of licensed pharmacists as of September 1974 was 18 729, including 43 state-examined pharmacists. About half of the total 8906 are female pharmacists. The pharmacist-to-population ratio is 1 : 1680.

f) Paramedical personnel: In 1974 there were 4497 (male, 3751; female, 746) paramedical professions: laboratory technicians, 1774; sanitarians, 623; X-ray technicians, 989; physiotherapists, 261; dental technicians, 692; and sanitary laboratory technicians, 159. Persons who fulfill the 2-year course requirements are authorized to try the national licensing examination.

III. Medical Education

1. Historical Outline

Western or modern medicine was introduced into Korea at the end of the nineteenth century along with western culture. In 1885, Dr. Horace N. Allen, an American physician, came to Korea as a Christian missionary. He was summoned to the palace to treat a prince with serious stab wounds inflicted by an assassin; the herb doctors of the Royal Court had failed to cure him. Dr. Allen treated his wounds and saved his life. This so pleased the Emperor that he gave Dr. Allen a hospital where he could practice western medicine. He also appointed him physician to the royal household. Dr. Allen was later appointed the United States Minister Plenipotentiary to Korea. Dr. O. R. Avison, A Canadian Presbyterian medical missionary, succeeded Dr. Allen as the director of the hospital [57, 58].

Dr. Avison, being interested in medical education, began teaching western medicine to young Koreans. In 1899 Dr. Avison attended the World Missionary Conference held in New York and gave a speech on the necessity of medical education in Korea. Mr. L. H. Severance, a wealthy philanthropist of Cleveland, Ohio, impressed by Dr. Avison's speech, made a generous gift to build a medical school and a hospital. Thus, the first modern medical school was started in 1893 in Korea and was named Severance. It was renamed Yonsei University College of Medicine in 1958. In 1907, following the Japanese occupation of Korea, the Japanese government founded the second modern medical school in Seoul. This was called Keijo Imperial University Medical College and later renamed the Seoul National University Medical School after the end of the Second World War.

In 1945, the number of modern medical schools in Korea had increased to eight, two of which were in North Korea. These six were supported by the government. The number of students enrolled totaled 2189; staff members numbered 855. About half of the students and staff were Japanese. The government-supported medical schools had many more Japanese students than Koreans. During the Japanese occupation, the medical profession was favored by Korean people because it offered political as well as financial security. Consequently, medical science was the most advanced field among the natural sciences in Korea at the end of the Second World War.

During this period, the system of medical education essentially followed the German system as modified by the Japanese. The medical curriculum did not greatly differ from what we presently have, but teaching methods differed considerably. Compulsory didactic lectures with few laboratory and clinical hours were the rule during undergraduate years. The postgraduate training differed entirely from the present system of internship and residency training program in the United States. It was primarily a form of prolonged apprenticeship, through which specialty training was accomplished. During the apprenticeship of 5 years or more one was required to do research, mainly animal experiments, and write a thesis. Upon approval of the theses the degree of Doctor of Medical Science was granted.

2. Education Courses

a) Medical college: Following the end of the Second World War, along with social changes, all the Japanese-styled educational systems were altered in the direction of the American system, including that of medical education. Unfortunately, the Korean War broke out before the change was well underway, and everything remained at a standstill during the war. The real change in Korean medical education took place after the Korean War, as South Korea was recovering. The system of medical education is very similar to that in the United States. After 12 years of elementary and high school education, one becomes eligible for admission to the 2-year premedical college.

Undergraduate medical education has also closely followed the American system. Existing colleges of medicine in 1979 in South Korea number 17. Eight of the medical colleges are in Seoul. A total of about 4500 students are enrolled in medical schools, and every year about 1000 students graduate. The medical curriculum is very similar to that of the American medical school. Upon completion of 6 years of education — 2 years in preparatory courses and 4 years in college — candidates receive the M. D. degrees, and through national qualification examinations they are authorized to be licensed by the government for medical practice (Table 13).

Table 13. Number of students by course (medicine and pharmacy) {19}.

	Department	Enrollment
Premedical course	17	2947
Medical science	17	4390
Predental course	3	516
Dentistry	3	675
Preherb medicine	2	157
Herb medicine	1	222
Pharmacy	14	3058
Pharmaceutics	6	825
Nursing	12	2318
Health and nursing	–	101
Health and pharmaceutics	1	60

b) Other colleges: There are three dental colleges with a 6-year training period; 2 years of preparatory courses and 4 years of college. Their annual output numbers about 300 graduates.

Herb-doctors are trained at Kyung Hee University, Herb-medical college for 4 years. Currently about 260 students are enrolled.

There are 14 colleges of pharmacy with 1500 graduates each year. After 4 years of education, the Bachelor of Pharmacy degree is given.

Twelve colleges of nursing offer a 4-year education course, and 23 nursing schools of high school level, 3 years of education.

Seven veterinary colleges once existed in Korea with an annual output of 200 graduates. However, after 1961 only four colleges are functioning, providing 4 years of education. As for the paramedical personnel, some institutions, i.e., Korea University Medical College, Yonsei University Severance Hospital, and some other private institutions in Seoul, Incheon, Wijong-bu, Gwang-ju, and Busan, have established 2-year courses. The student enrollment is roughly more than one thousand.

3. Postgraduate Training

Although undergraduate medical education has improved substantially in recent years, postgraduate training has lagged behind in Korea, because of the shortage of funds, facilities, and experts in various fields. Three types of postgraduate training are at present available in South Korea. One is specialty training in the form of internships and residencies. After completing 5 years of internship and residency, usually at university hospitals or a few other accredited hospitals, one becomes eligible for the specialty qualification examination given by the Korean National Board. There were 18 specialty bords in 1974, and the total number of specialists in various fields was 5262, approximately one-third of the total number of physicians. The second type is a degree course patterned on the graduate Ph. D. degree programs in the United States. This consists of 4 – 5 years of course work and research. After two foreign language examinations and written and oral preliminary examinations and upon approval of a thesis, one is granted the degree of Doctor of

Medical Science. The third type of postgraduate training is the old-fashioned Japanese-style degree course in the form of an apprenticeship, requiring 3 or more years of experimental research. Upon approval of a thesis one is granted the degree of Doctor of Medical Science. Clearly the maintenance of three types of degree systems causes confusion. The second type, the one similar to the Ph. D. degree program in the United States, has gradually become more favored.

It is now obvious that Korean medical education in the last two decades has been fundamentally oriented toward improving its standards and producing technically better-trained physicians. Korean medical graduates in recent years appear to be more knowledgeable and better prepared in clinical medicine than in the past. There has also been a marked increase in the number of specialists in various fields of medicine; approximately one-third of all Korean physicians are specialists. As a result of the higher standards of medical education, individual medical care for patients has been remarkably improved. This in turn has brought modern medicine closer to the population, and resulted in the psychological barriers against modern medicine among the population progressively waning.

4. Some Persistent Problems

Since adoption of the new system of medical education, the number of medical graduates seeking their career in the basic science fields of medicine has progressively decreased. Under the old system of medical education, basic science fields attracted many good students, mainly because teaching and research in those days carried remarkable prestige. Professors in basic science fields used to enjoy higher prestige than those in clinical fields. With social change taking place, values have also changed. A lofty idealism and sense of mission no longer inspire young graduates, indeed such words appear to belong to the past. Young graduates naturally see no prestige where there is no money. If this trend continues, very few will undertake the job of teaching in the basic sciences; and this will sooner or later, threaten Korean medical education.

Korean medical education has not adjusted itself to the needs of the society. The Korean socioeconomic situation is such that only a small portion of the population can afford to pay for private medical services. Consequently, a large part of the population, particularly in rural areas, has no access to modern medical care.

Despite the demands of society for general practitioners, the majority of new medical graduates choose to specialize, and the shortage of general practitioners is becoming more and more acute. General practice at present is mainly undertaken by older physicians. Unfortunately, the important mission of medical education to respond to the real needs of society has thus been neglected, while higher standards of medical education have been promoted. It appears that Korean medical education has outrun its socioeconomic development and failed

to train personnel to provide care for the entire population. Since we do not expect Korean socioeconomic development to solve the problems in the near future, some way must be found to adapt medical education to existing social needs.

IV. Health Care Services

1. Historical Background

a) Before 1945: In olden times no logical system of public health was established in the country. The Yi Dynasty (1430 – 1909) substantially marked the dawn of a new age in public health by introducing medical systems: the *Naeyi-won,* specifically directed to the royal medical services; the *Jeonui-Guan* in charge of public medical administration, and the *Hemin-so* and *Hwalin-so,* which is dispensary level, predominatly responsible for public medical services.

To keep abreast, medical officials were assigned to various levels of offices ranging from the *Uijeong-Bu* (corresponding to the Council of State), *Yug-jo* (six ministries), down to the military services, including *Sung-gyun-gwan* (Confucianism institute), and the office of *Jonuiso* by which the public, students, and even prisoners, not to speak of officials, could enjoy the benefits of medical services.

To meet the requirements of rural medical treatment, a variety of medicine were supplied to the respective provinces, and a number of medical officials, proportionate to the size of the administrative districts of Bu, Gun, and Myeon, were assigned and entrusted with the missions of general medical administration and relief of destitute patients.

However, because of the undeveloped and limited medical science, based exclusively on Chinese medicine, no one can say for sure whether such arrangements contributed to public health.

By and large, one might easily deduce that the public was not provided with the necessary medical services for the prevention of infectious diseases. Abandoned, as it were, they had recourse only to the so-called folk remedies, such as the extremely primitive Chinese medicine, witchcraft, personal family remedies, and futile superstitions.

By dint of the *Kab-o* Reform (31st year of King Kochong reign, 1894) a new medical service system influenced by Western medical knowledge was established, which led to the abolishment of the past medical administrative system.

The Hygiene Department, which played a key role in the realm of medical administration, consisted of two subordinate divisions: Hygiene and Medical services.

The applicability of hygiene activities at this stage was limited to the prevention of infectious diseases, encouragement of vaccination, and enforcement of quarantine.

b) Medical training: A government medical school was introduced as the training center for a new medical administration and its academic system, based on western medical science and knowledge. In 1888 the Gyeong-sung Medical Institute (*Ui-Hak Gang Seup-So*) was founded by a proclamation of the public health system with Mr. Chi Suk Yung as the president. With the assistance of two instructors and two clerks, he guided western medical science into this territory.

In February 1909, under a proclamation of new academic regulations, this school supervised a Medical Department (4-year course), Pharmacy Department (3 years), Midwifery Department (2 years), and Nursing Department (2 years). This was the predecessor of the present Medical College of the Seoul National University.

c) Medical treatment: As a token of appreciation to Dr. H. N. Allen, a missionary doctor of the American Presbyterian Church who was in charge of royal medical services to King Kojong, the *Kwanghe Won* (charity hospital), the first modern hospital in Korea was established at Jae-Dong block of Seoul in 1884. Then in 1885 this hospital was moved to Kurige street (the entrance of today's Ulchi-Ro st.) and redesignated *Jejung Won* (salvation of the people) Hospital, under the management of the American Presbyterian Mission.

In 1893, Professor O. R. Avison of the University of Toronto (Canada) Medical College was assigned to this hospital. Subsequent to his acceptance in 1893 Dr. Avision erected a new hospital with funds presented by Mr. L. H. Severance of New York and named this hospital Severance Union Hospital, where he started the training of physicians.

In 1909 this hospital was called the Medical School of Severance Hospital and was the predecessor of the present Yonsei University College of Medicine. The two above-named institutions became the initiators of modern medicine in Korea, and their graduates have continued to be dedicated leading pioneers for the effective development of medical education, medical services, and public health, and its administration in the country.

Following the increase in the number of physicians through the medical training school, the number of private hospitals markedly increased, and the general public could thus benefit from modern medical treatment.

d) After 1945: Following the withdrawal of the Japanese regime in 1945, the U.S. Army took over the land up to the 38th parallel and established a U.S. army government. Among the divisions of the governmental organization, the Ministry of Health and Welfare was formed with several bureaus: preventive medicine, medical affairs, dental affairs, and nursing affairs. But since the Republic of Korea was established in 1948, the organization has been modified over and over again in accordance with the needs of the times.

In the early stages of ROK government, efforts were made to provide an administrative framework more than to prevent and treat infectious diseases prevalent at the time. In the administrative field, medical personnel were classified as medical doctors, dentists, herb doctors, nurses, and midwives. Rules and requirements to get respective licences were fixed. The specialist board system for medical graduates in the United States was also introduced during this period. In the 1960s administrative processes for health and medical affairs were developed and stabilized. Health centers were established in each county (Gun) of the provinces and each district of the special cities of Seoul and Busan. The health centers took part in the practice of the health care delivery system and contributed much in the field of preventive medicine and health education. Control measures were set up at this time for tuberculosis, leprosy, venereal diseases, and parasitic diseases. Sections or divisions for the above diseases were established in the Ministry of Health and Social Affairs. Maternal health, environmental sanitation, and food sanitation have been additional areas with which the Ministry of Health and Social Affairs has concerned itself in the 1970s. To carry out an effective health care delivery system, all the medical personnel are obliged to practice a regular reporting system. The government has supported various voluntary movements in the field of health and social affairs, allocating a part of the national budget to such work.

Disposition of medical doctors to doctorless villages (*Myeun*) has proved the persisting handicap for the past decade. Even in the established health centers at county level, there are still a number of places without any medical doctor. To solve these problems, the government adopted a temporary plan in 1976. Medical graduates who failed to pass the national qualification examination are given another chance, and those who pass this second examination are listed as conditional doctors. They are required to serve for 2 years in a designated doctorless village or health center to be authorized to receive the formal licence as a physician. In this way, the government is focusing efforts to solve the rural health problems through effective utilization of the systems and manpower.

2. National Administrative Organization

With the promulgation of Law No. 734 on 24 October 1961, the Ministry of Health and Social Affairs was set up with six bureaus: Medical Affairs, Health, Pharmaceutic Affairs, Labor, Social Affairs, and Women's Bureaus. Under the Medical Affairs Bureau are the medical affairs, local health, and nursing sections; under the Health Bureau, preventive medicine, sanitation, and chronic diseases control sections; under the Pharmaceutic Affairs Bureau, pharmaceutic affairs, medical supplies, and narcotics sections. The subsidiary organizations of the Ministry are the National Medical Center, National Institute for Public Health Training; Leprosaria (Sorogdo, Bupyong, Chilgog, and Yeongju Hospitals); Tuberculosis sanitariums (Masan and Gongju hospitals), National Mental Hospital, quarantine stations (nine in the country), National Institute of Health, National Chemistry Laboratories, National Herb Laboratories, Leprosy Preventorium, and women's and children's institutions.

Local health administration in Korea falls under the jurisdiction of the Ministry of Home Affairs, and therefore under the direct responsibility of the Mayor of Seoul (special) city, the Provincial Governors, and the heads of the local autonomous governments. The execution of local health administration is therefore under the direct responsibility of local authorities.

In the Provincial government, the Bureau of Health and Social Affairs is reponsible for health administration in which there are Medical and Pharmaceutic Affairs Sections dealing with medical and pharmaceutic administration and the Preventive Medicine Section dealing with preventive medical services as well as health center operations. In Guns (counties) the Social Affairs Subsection deals with health administration and, upon the establishment of a health center network in compliance with the law, health administration and services in cities and Guns will be established in the health centers.

In Seoul city, the Bureau of Health and Social Affairs was created with three sections: Medical and Pharmaceutic Affairs, Preventive Medicine, and Sanitation, each with the responsibilities of their respective fields. In the Gu (district) of Seoul City, health administration is under the Sanitation Subsection of the Social Affairs Section, and health services are provided by the health centers in the Gu.

In Myeons (Myeon, administration unit in a province, see Annex I), the Social Affairs Subsection, being the lowest administrative unit, is responsible for health services. As the local health center is unable to provide satisfactory health services at present, the establishment of substations of the health centers at the myeon level is urgently required.

Though short-term in-service training is being given to local health workers, the indifferent transfer of personnel in lower levels of the government often hampers the local health services. Local health services are being carried out in accordance with instructions and policy of the national government without first considering and emphasizing specific local problems and needs.

3. Medical Facilities

The total number of medical facilities in Korea in 1974 was 11 228, of which 4543 were concentrated in Seoul and 1192 in Busan. Medical facilities are classified as university hospitals, national or municipal hospitals, private general hospitals, private clinics, psychiatric or tuberculosis hospitals, dental clinics, herb-doctor clinics, midwife-nurseries, and district health centers (see Map 4).

In contrast with the surplus number in the large cities, there are still a large number of doctorless myeons (township) in the provinces. Statistics of 1974 indicate that 602 myeons (41%) of 1459 are doctorless myeons. These contain about 5 625 095 people. Villages without dentists are even more numerous. A total of 1321

myeons (90%) with 14 608 231 inhabitants are without dental services. Even the herb doctors show shortages in the rural areas. The fact that 1115 myeons lack herb doctors only serves to endorse the problem of these. There are 536 myeons with a population of 4 954 489 that have neither doctor, dentist, nor herb doctor. By province, Jeonra Bug Do is the most needy province as regards the disposition of medical personnel: 62% of the total myeons have no licensed medical personnel. Two myeons of 13 in Jeju Do have no licensed medical personnel. Such an unbalanced distribution of medical personnel results in many cases of illegal treatment of the sick by unqualified personnel; 1973 statistics report 1725 such cases, and in 1156 of them treatment and injections were given by persons acting as qualified medical doctors.

a) Hospitals, sanatoria, and clinics: General hospitals, according to the National Medical Law, are defined as hospitals with facilities accomodating more than 100 patients, with specialized departments of internal medicine, pediatrics, surgery, obstetrics, and gynecology, ophthalmology, ENT, urology, and radiology (see Table 11).

b) Number of beds: The number of beds in use per 100 000 people in 1974 was 134.3 in Seoul, 14.3 in Chungcheong Bug Do, and 169.4 in Busan, with an average throughout the country of 63. The total number of beds in 1974 was 19 062, of which 14 899 were relegated for general diseases, 2584 for tuberculosis, and 1579 for mental diseases.

The cumulative number of hospitalized inpatients was 4 021 970, and the duration of hospitalization per patient was on the average 14.1 days. Of the total 285 429 patients discharged, 3.1% of the patients died. The bed occupancy rate was 57.8%.

c) Health centers: According to the Health Center Law, each city and county should have a health center for every 200 000 population. The health center has become the most important health unit for public health administration, since direct public health services to the people are carried out here. The functions of health centers, as defined by the law, are as follows:

Communicable disease control,
Maternal and child health services,
School health services,
Environmental sanitation and industrial health,
Public health statistics,
Health education, and
Other public health services.

A total of 1197 health centers is located throughout the country. The personnel in each health center, except for several larger cities, numbers five: one physician, two public health nurses, one sanitarian, and one clerk. This is inadequate to provide health services for all the people in the district, and reinforcement of health center personnel is seriously required. Training of personnel is performed by the National Institute for Public Health Training with facilities and buildings supported by the government.

d) Public physicians: Public physicians are appointed by the government to work in rural areas. They are partially supported by the government and carry out public health services as well as medical treatment. More than two-thirds of the rural areas are covered by public physicians, but many areas are still without physicians. The detailed numbers of doctorless areas (myeon) in the country are shown in Annex II. The government is planning to allocate physicians to all of these areas in the near future.

4. Medical Care Programs

The medical care program is one of the most important activities to improve the nation's health. Since liberation in 1945, Korea has unfortunately had insufficient political and social stability to study her own nationwide health needs and problems. Practically no systematic country-wide studies were conducted on morbidity and medical care costs except for a few observations by workers among segments of the Korean population (Table 14).

a) The concept of a medical care delivery system is changing from an individual basis to a community basis. In the past, medical care built on a patient-physician relationship, but in recent years the scope of such care has broadened to include paramedical workers and auxiliary community health workers, as in the case of community medicine. Medical science itself shows a change from comprehensiveness to specialization. The specialist system has raised the standard of medical care, but at the same time has caused medical fees to rise. In Korea medical care had previously been proclaimed a

benevolent art. Physicians were proud to devote themselves to the care of the sick, and patients in turn recognized the authority of the physician. In this way the physician was trusted by patients even in those cases with adverse results. However, paralleling industrial development, this humanity-based benevolency is gradually fading away and being replaced by mechanization.

With this in mind, a brief look at the medical care system in Korea will be worthwhile.

The existing system has been largely influenced by the system in the United States as regards education and practice. All the medical students in the 17 medical schools in Korea are required to take a 2-year premedical course and 4 years of undergraduate study; then they must pass the national examination to be licensed as a physician. By law medical care institutions are categorized as hospitals or clinics and dispensaries; the former are authorized by permission, but the latter only by report. The specialized divisions in medical care are internal medicine, psychiatry, general surgery, orthopedic surgery, neurosurgery, thoracic surgery, plastic surgery, anesthesiology, obstetrics and gynecology, pediatrics, ophthalmology, ENT, dermatology, radiology, surgical pathology, clinical pathology, thoracic diseases, health administration, and preventive medicine. To be authorized as a specialist a 1-year internship and 4 years of residency training are required at a qualified training hospital, and the doctor must pass a board examination by the Korean Medical Association. Due to the fact that 80% – 90% of the medical graduates take the training courses for their Boards, the ratio of specialists to general practitioners is

Table 14. Existing and Foreseen Health Facilities and Medical Personnel in Korea by Year {6 and 17}

Item \ Year	1967	1971	1972 – 1976	1977 – 1981	1982 – 1986
Clinics, private	8 694	9 564	10 520	11 572	12 729
National and municipal hospitals	64	74	94	104	134
Health centers	189	193	198	204	204
Health center beds/units	565	919	1 459	1 459	1 459
population	1 968	1 900	1 657	1 420	1 033
Medical graduates	620	650	880 – 1 338	1 438 – 1 500	no data
Medical doctors	12 450	13 419	15 824	20 165	25 910
Ratio to population	2 344				1 500
Dentists	1 905	2 147	2 770	4 262	7 356
Ratio to population	15 440				5 600
Nurses	11 699	17 018	26 749	34 770	45 124
Ratio to population	2 510				900
Auxiliaries	3 805	6 799	9 799	12 799	16 799
Ratio to population	7 676				2 461

becoming unbalanced, thus impeding the resolution of the policy to reduce the number of doctorless villages. Another handicap is the high cost of obtaining Board qualifications. This may give rise to increase medical costs in spite of efforts by the Ministry of Health and Social Affairs to prevent such increases.

The governmental network for medical care begins at the top with the National Medical Center. Below this apex are 47 municipal or provincial hospitals, 197 health centers at county level, and 1343 branch clinics at the village level. This system defines the branch clinics as the places for primary care; complicated cases are referred to a higher level all the way up to the National Medical Center. A provincial hospital is also designed as a teaching and training institution for lower-level medical professions in the administrative realm. There are also several medical institutions such as the National Mental Hospital, the National Tuberculosis Hospital, and the National Leprosy Hospital for special types of treatment. As a part of social security, a medical insurance law was promulgated in 1963 with the aim of equalizing benefits for all citizens. However, it is still in the experimental stage due to financial problems. The government is proceeding with the plan step by step beginning on a small scale. According to the plan, about one million households, 14% of the total population, will be included in the insurance plan by 1976. Employees in factories or offices of 500 employees and over, civil servants, military men, and some self-supporting people are also to participate in this plan. By increasing and extending the number, 100% participation is envisioned by 1986. Nevertheless, success will be largely dependent on financial support from the national budget.

b) Medical relief and insurance: In 1977 the Ministry of Health and Social Affairs divided the country into 55 areas to provide medical relief to those in financial need. The number entitled to such help was estimated to be about 2.1 million or about 6% of the population. Medical Institutions in the respective areas were designated by the governor to take care of these people. There were 49 national or city and provincial hospitals and 25 private hospitals so designated. Those who need medical care may first visit a health center or its branch on the myeon level or any clinic designated by the local administrator. If the medical problem cannot be cared for at this level, the second stage of treatment will be carried out in a hospital located in the respective area.

A medical insurance system was also set up in 1975. At the beginning stage, 605 factories or companies which had 500 employees or more were associated with this system. The number of beneficiaries was 3 920 020 — insured persons 1 722 867 and dependents 2 197 153. In 1979, the system was expanded to include factories and companies with 300 employees or more. Consequently the number of insured increased to 2 587 000 and their dependents to 7 915 000.

c) Medical manpower resources: Medical manpower is not seriously lacking in numbers, but about 3000 physicians have gone to foreign countries, including the United States, and a great number of physicians are crowded into large city areas, thus creating a shortage of medical doctors especially in rural areas. To halt this tendency, the government has already transacted a countermeasure in which the residents in specialist courses at the General hospitals should have compulsory service for a fixed period in the countryside where no physicians are practicing. The establishment of four more colleges has been newly approved in addition to ten existing medical colleges, thus enabling about 1000 new doctors to be graduated every year.

Of the total 17 colleges of medicine, 6 are government run and 11 are privately administered. Eight colleges are located in Seoul, and the rest in the provinces.

5. Use of Folk Medicine in Rural Areas

To clarify the current practice or status of folk medicine for diseases and injuries in Korea, surveys were conducted of 180 households covering 1009 persons in seven townships in Kimpo-gun (county), Gyeong-gi Do (province) during a 5-day period from 1st to 5th August 1973 by Huh [9]. The seven townships were Kimpo-myeon, Yangchon-myeon, Kochon-myeon, Hasong-myeong, Taegon-myeon, Wolgon-myeong, and Kumdan-myeon. The report may be representative of the situation of folk medicine in present-day Korea.

In general the male-female ratio of the population in the study area was nearly 50 to 50. The number of persons per household was approximately 5.6. The age group of 40 – 49 years constituted the largest segment or 20.8% of this population; the age group of 30 – 39 years, 19.7%, the second; and the age group of 80 years and above, 0.7%, the smallest segment.

The interviewees were classified according to religion: 69% were nonreligious persons, 12% were Confucians, 10% were Protestants, 5% were Buddhists, and 4% were Catholics.

As regards the appeal of medical facilities for the 1-year period from early August 1972 to late July 1973, the greatest proportion or 32.4% of the interviewees received initial medical care from pharmacies, 27.0% underwent folk medicine therapy, 20.4% visited hospitals or clinics, and 1.1% (the smallest proportion) were treated by doctors who visited the patient's home.

On the whole, the greatest proportion 31.5% applied to pharmacies, 26.4% received folk therapy, and 15.6% visited hospitals or clinics.

Folk therapy for the 1-year period from early August 1972 to late July 1973 was outlined by Huh [9]. Out of the 1009 persons surveyed through interviews, 744 persons (73.7%) had been sick. The ratio of folk medicine therapy cases practiced on sick cases was 100.0% for congenital deformity, 88.8% for complications caused by pregnancy, child delivery and puerperium, and 71.1% for sickness resulting from accidents, poisoning, and violence. No folk medicine therapy was administered to treat neoplasms. The average number of sick cases per population of 100 persons was 4.7 for diseases of the nervous system and sense organs, 4.2 for digestive organs, and 4.2 for senility.

A total of 149 kinds of folk medicine therapy were practiced on 190 occasions, most of which were unscientific and empirical. Many animal products, herbs, and ordinary food items were eaten, some raw and some cooked.

The average number of cases of folk medicine therapy practiced per population of 100 persons was 13.9 for males and 23.5 for females. Its practice for females was thus approximately 1.8 times as frequent as that for males. In the case of males, the age group of 80 years and above practiced folk medicine therapy in 25.0 cases per population of 100 persons, the age group of 10 – 19 years, 19.4 cases, and the age group of 50 – 59 years in 18.2 cases. In the case of females, the age group of 20 – 29 years appealed to it in 37.1 cases, the age group of 50 – 59 years, in 33.8 cases, and the age group of 4 years and below did not refer to it at all.

Unemployed persons practiced folk medicine therapy in 78.7 cases per population of 100 persons, whereas those engaged in unclassifiable occupations turned to it in 50.0 cases, and senile and crippled persons, in 37.0 cases.

Uneducated persons sought its help in 22.0 cases per population of 100 persons, those barely able to read the Korean alphabet in 20.0 cases, and primary school graduates in 19.6 cases. Thus, most of its users were in the low educational bracket.

According to marital status, divorced persons used it in 40.0 cases per population of 100 persons, and separated persons in 38.3 cases. Those born in rural areas utilized it in 18.9 cases per population of 100 persons, and those born in urban areas in 17.0 cases, thus showing no significant difference. Buddhists utilized therapy in 21.6 cases per population of 100 persons, Confucians in 19.1 cases, and nonreligious persons in 18.9 cases. As for expenses spent for folk medicine therapy, it costs nothing for 74.7% of all cases, less than 1000 wons was spent for 14.7%, and the average cost was approximately 766 wons. In 62.1% of the cases, the utilization of folk medicine therapy was suggested by folklore sayings; in 22.6%, it was recommended by neighbors; and in 10.0%, as a matter of habit. As for the perceived results of folk medicine therapy, 32.1% of all cases were considered completely cured, 25.3% improving, 16.3% unchanged, and 14.2% still undergoing treatment.

6. Acupuncture

Acupuncture bears the same relation to native medicine as surgery does to modern-day scientific medicine. The essentials of this art and practice are taken from the *Whang-Jai-Yung chookyung,* one of the two books originating in the Whang-Jai Dynasty (2697 – 2597 BC) in China. The authorship of this book and its companion is ascribed to the Emperor. Whether or not the ruler was the real author or whether the title was given honorifically and the real author remained unknown, is unknown.

It may be noted that other such manuscripts appeared in this dynasty, and the ones preceding were ascribed to the founder of the dynasty [56].

The fundamental principle underlying acupuncture is based on the assumption that in illness the blood becomes stagnant and will not flow properly through the natural channels of the body. Acupuncture is also believed to hasten relief, over and above what might be expected from the use of drugs.

The Korean name for this art is *Ch'im,* which is a term applied to any kind of an instrument used for piercing the flesh of the body. However, the original term, and the one adhered to in this discussion, applies to needles only. At some subsequent time, however, the word *Jim* came to be used, but it signified a different form of treatment, namely the application of heat with or without medicine. Mugwort (*Artemisia*), a weed growing in all parts of Korea, is used quite extensively for this purpose under the name *Ssook Jim,* which is employed in two forms, the poultice and the fire ball. The poultice is prepared by boiling a quantity of leaves and the stalk, then placing it in a cloth, and wringing it until the water is fully removed. The hot pulp remaining is then used as a poultice. The fire ball called *Ddum* or *Ssook-Ddum* is made by crushing a small quantity of the stalk and rolling it between the palms of the hands. Afterwards the ball, varying in size from a pea to a walnut, is set on fire and placed over different portions of the body. The *Pillow Jim* is made by heating the wooden block (*pillow*) on which Koreans rest their heads while sleeping and applying it to different portions of the body for various ailments.

In 59 AD during the reign of Choong Myung, the *Wee-hak-eep-moon, A Medical Primary,* was written which included a few more rules for the application of the Ch'im and an elaboration on the rules of the *Whang-Chai-yungchoo-kyung,* the original source of the Ch'im practice. After this a succession of books appeared but none of them were of any special significance until 420 AD, when the *Tong-een-kyung* appeared, written by Wang-you-il (Fig. 11). This author modeled a man from copper, which is signified by the title of the book, meaning the copper-man book. He elaborated on all the previous teachings of his predecessors and constructed a chart illustrating the copper man's anatomy, which to this day is the accepted standard of anatomy among the practitioners of native medicine. In connection with this anatomic scheme, it was believed that there were (1) blood vessels, (2) nerves, and (3) channels. Five kinds of channels are illustrated: (a) the spleen and stomach, (b) the liver, (c) the lungs (d) the pericardium, and (e) the gall. The physiologic implications of the channels were outlined as follows. The spleen and stomach: the channels given off from the spleen and stomach convey nutritious materials for final distribution to the different parts of the body.

The liver: this organ gives off channels for the distribution of gall, and it is believed, now as it was then, that the eye is directly connected with the liver by means of a gall channel, which accounts for the yellowish discoloration of the eye in jaundice.

The lung: the channels originating from this organ are supposed to contain air during fetal life, but after birth when respiration is established, blood is supposed to enter a state that continues through life.

The pericardium: the oil channels are connected with an oil sac remotely situated in the region below the diaphragm. This probably corresponds to the omentum, and possibly is at the base of the mesentery.

The gall: these channels are confined to the upperhalf of the body, and they have not connection with the liver. They are the receptacles for a complimentary fluid, supposed to be the seat of courage.

All the above named channels are supposed to contain blood but in a modified form due, of course, to the presence of the respective substances, which they receive and convey.

The chart of anatomy consists of three parts called *Sam-cho:* regional, visceral, aud surgical.

The regional part consists of three divisions: (a) upper-third of the trunk-thorax, *Sang Cho;* (b) middle-third of the trunk-abdomen, *Jung Cho;* (c) lower-third of the trunk-lower abdomen, *Ha Cho.*

The visceral part: in this chart there are 32 anatomic structures named. The surgical part: the blood vessels, nerves, and channels represent the chief items of consideration.

In spite of lack of scientific approval, acupuncture is still empirically appreciated by Koreans, especially by the older generation and women. It has been announced several times that even surgical operations have been performed at the Kyung-Hee University Hospital, the site of the college of Oriental Medicine by anesthetizing patients with acupuncture. However, its merit will be assessed when and if scientific proof is disclosed.

One other problem is the illegal practitioners of acupuncture. Their number is estimated at 50 – 60 thousand in the country, thus far exceeding the number of licensed acupuncturists, 280. Establishment of regular courses is urgently required to prevent the unexpected health hazards posed by such illegal acupuncturists.

7. Maternal and Child Health (MCH)

a) Historical review: Until 1945 there was no organized MCH service as such although various missionary health agencies were independently engaged in modest MCH work. In September 1945, with the establishment of the U.S. Military Government in Korea, organized MCH service was started as a part of overall health care by the Department of Health and Welfare. A "City clinic" was established in Seoul which offered MCH services. Two mothers' classes were organized, and systematic health education (20 h) was given to several groups. As other health centers were subsequently established, rudimentary MCH service gradually spread to the rural areas. Moreover, various university hospitals and

other general hospitals developed their own MCH clinics and contributed to overall MCH development.

b) Population composition: According to a survey in 1959, 51.9% of the total population were below 20 years of age, while 32.7% of the female population were 20 – 45 years old [6] (Korea Statistical Yearbook, 1975).

c) Childhood problems: Lee and his associates [50] carried out an investigation of vital statistics of the inhabitants in Rural Health Demonstration Area (Kaejong, Jeonra Bug Do). They determined that the ratio of age groups 0 – 4 year olds and 5 – 9 year olds had decreased relative to the whole population since 1964:

Age (years)	1962	1963	1964	1965	1968
0 – 4	16.4%	14.5%	14.4%	13.5%	11.7%
5 – 9	14.2%	15.8%	15.6%	15.1%	14.6%

The birth rate (number of live births per 1000 population per year), which in the myeon showed a constant rate of 35 before and after the Korean War, had decreased since 1964. The birth rates in 1967 and 1968 were 21.8 and 21.2, respectively. These results were assumed to be due to the family planning program initiated in 1962. The death rate (number of deaths per 1000 population), averaging 16.1 during the period 1944 to 1953, decreased to 8.8 during the period 1956 – 1960 and to 6.2 in the period 1961 – 1968. This is assumed to be due to decreased infant mortality, which in turn may be due to availability of antibiotics and improvement of public health work.

The natural population increase rate in the studied area was about 20 (the increase per 1000 population per year) before the Korean War. During the years 1956 to 1963, it showed an average of 28 because of the decreased death rate. But it had decreased since 1964 to 15.9 in 1967 and 13.8 in 1968 due to decreased birth rate.

Infant mortality (number of infant deaths per 1000 live births per year) during 1944 – 1948 was 171.4, and 98.6 in 1949 – 1953. The rate dropped to 63.6 during the years 1956 – 1960 and decreased further to 30.2 in the period 1961 – 1968 due to better food and living conditions. These results are attributed to the effect of public health work and to the general improvement in solving the health problems of rural society.

The neonatal mortality rate (number of deaths of newborns, 0 – 4 weeks of age, per 1000 live births per year) ranged from 20 to 41 in 1958 – 1968, but decreased to 3.9 – 7.7 during the years 1963 – 1965. Neonatal deaths in 1924 – 1953 before the inception of public health work were 39.9% of all infant deaths; this proportion increased to 62.8% in 1956 – 1968. Thus the incidence of infant deaths from infectious diseases can be reduced by improving environmental hygiene and health education and giving preventive vaccinations, but it might not be easy to control deaths from such causes as prematurity, congenital anomalies, birth trauma, solely by public health measures. The increased neonatal death rate in recent years indicates a relative rather than a real increase in actual deaths when compared with the decrease in infant deaths per thousand cases.

The child mortality rate in the 0 – 4 age group expressed as deaths per 10 000 children, was 93.1 in 1960, 52.2 in 1964, and 19.6 in 1968. In 1924 – 1953 before public health work was begun, child deaths made up 30.9% of all deaths. Now they number only 9.9% of all deaths, a distinct contrast.

In Kaejong-Myeon the group 0 – 4 years accounted for 52.0% – 64.7% of all deaths in 1944 – 1953 before the start of public health work, but this decreased to an average of 17.0% of all deaths in 1966 – 1968. Deaths of persons over 5 years of age ranged from 35.3% – 47.9% in 1944 – 1953, but subsequently increased to 82.9% of all deaths. The group over 50 years of age accounted for 56.8% of all deaths in the period 1966 – 1968.

The principal cause of all deaths during the years 1924 – 1953 were diseases of infants and children from 1 – 7 years of age in descending order of frequency. In contrast during the years, 1956 – 1968, adult deaths were more frequent, the chief killers being cerebral hemorrhage, tuberculosis, diarrhea, malignancy, and accidents, in descending order of frequency.

Of the causes of infant death, prematurity ranked first both before and after the inception of active public health work. At least 70% of infant deaths were attributed to infectious diseases before public health work was begun. However, birth injuries and congenital anomalies have more recently become the main causes of infant death.

If was noticed that infectious diseases were still the main cause of childhood deaths, and thus more active preventive measures and treatment of these diseases are desired. Measles, diarrhea, and pneumonia were the three main causes of childhood deaths, with other acute infectious diseases contributing to about 80% of all cases of death in children. During the years 1956 – 1969, smallpox and dysentery disappeared as causes of childhood death, and Japanese encephalitis and accidents replaced them as the main causes of death.

The abortion rate (number of abortions per 1000 total births per year) was 79.6 during the years 1957 to 1961 and 50.0 in the period 1962 – 1968.

Natural abortion showed 28.0% at 3 months fetal age, 20.0% at 2 months, 10.4% at 10 months and 7.2% at 5 and 9 months. The number of induced abortions in recent years has been rapidly increasing and now exceeds that of natural abortions. Induced abortion analyzed according to mother's age and year, showed only a 0.9% to 2.6% increase during the years 1961 – 1963, but this has risen gradually year by year since then to reach 10.1% in 1968. This study showed a correlation between mother's age and the rate of induced abortions: 10.8% in the age group 31 – 35, 20.5% in ages 36 – 40, and 48.9% in ages 41 – 45.

d) Prenatal care: In the early stage of MCH activities, i.e., during the period 1951 – 1960, an increasing number of health centers, hospitals, and clinics were engaged in prenatal care service. The cumulative number of health agencies offering prenatal care was 423; the respective

institutions gave a total of 233 124 consultations during this 10-year period.

e) Delivery services: It should be noted that although it is not the proper function of health centers to offer delivery services, a number of such services were offered. These were mostly performed by public health nurses, who are also qualified midwives. In general, one-third of the deliveries occurred at home and the rest, at hospitals. Certain national and private hospitals offer delivery service without charge or at a greatly discounted rate. Approximately 20% of the total deliveries in Seoul are done by midwives. In view of the important role played by midwives in MCH, the National Institute for Public Health Training is offering them regular training courses. The ratios of attendance by professionals differ by area. In Seoul delivery by professionals ranges from 30.5% to 77.4%, while it is 19.3% – 49.2% in urban and 12.2% – 11.0% in rural areas.

f) Infant and preschool care services: Although a number of infants are brought to health centers, hospitals, and clinics to have their health checked, it is extremely difficult to estimate their number accurately. According to a survey conducted by a fact-feeding and coordinating committee of the ROK-WHO-AID Health Evaluation and Planning Program in 1961, a total of 531 590 consultations took place and treatment was given to 2 046 035 patients by various institutions during the period 1951 to 1960. However, it is estimated that the actual number far exceeds the reported cases. The above figures represent those younger than 1 year of age. In general, the health care of preschool children is rather neglected.

g) Premature infants, congenital malformations, and birth injuries: An increasing number of premature births have been reported every year by various health institutes. Health centers, hospital clinics, and practicing midwives reported 4702 premature infants from 1951 to 1960. On the average, premature births made up 7.7% of the total deliveries. At present, incubators are only available at large hospitals. During the period 1951 – 1960, 525 babies with congenital malformations were reported, averaging one case of congenital malformation in 100 deliveries. The frequency of birth injury was 3.9% during this same period.

h) Crippled children: The number of crippled children who received proper treatment through pediatric service was 12 349 in 1960. However, this does not represent those treated at orthopedic clinics. Approximately two-thirds of those treated in 1959 and 1960 were reported from the Department of Physiotherapy at Severance Hospital, and they were mostly poliomyelitis patients.

i) Nutritional services: Public health nurses working at various health centers are engaged in nutritional services with the main emphasis on public education. The intake of an adequate amount of protein is emphasized to pregnant women, and artificial feeding methods as well as the selection of food for babies at the weaning stage are also taught. In addition, various relief agencies such as UNICEF, CARE, and Church World Service supplied a considerable amount of skim milk powder to these health centers for distribution to those in need from the time of the Korean War until the early part of 1970.

j) Immunization: The total number of immunizations given in 1960 reached 1 393 790. Approximately 90% of the immunizations have actually been given by health centers and the rest by other hospitals and clinics. During 1960 one health center gave nearly 20 000 immunizations. The immunizations, mainly DPT, polio vaccine, or BCG, are being continued.

k) Periodic physical examination: Although various health care facilities attempt to examine babies and children periodically, this is not being carried out satisfactorily because of poor cooperation among those concerned.

l) Health education: Both the Korean Pediatric Society and the Korean Gynecologic and Obstetric Society meet regularly to discuss various problems, including MCH. In addition, the Korean Public Health Association, the Korean Family Planning Society, and the Korean Maternal Society also participate in various activities in this respect.

m) Public health nursing services: Public health nurses assigned to health centers are organizing a number of small mothers' classes for the purpose of promoting the concept of public health, especially MCH. In addition, they also visit pregnant women and babies in their homes to render appropriate service.

n) MCH status summary: To obtain an outline for figures on MCH status in Korea, Jung and Hong [59] attempted to compile data from various studies made since 1945. The health index for mothers and children selected by the author with the above-mentioned objectives were infant death rate, outcome of pregnancy, and conditions at birth of infant. In the review of 76 original papers, trends in infant death rates in all areas showed an annual reduction.

However, there are big differences between the rates surveyed in metropolitan areas and those in rural areas. Great variations in infant death rates were also observed in different socioeconomic sectors of the population. The current infant death rates in metropolitan areas were estimated at 35 per 1000 live births; higher rates up to 60 were found in rural areas.

The rate of pregnancy arrest by abortion and still births was very high and has increased steadily according to recent data, especially in rural areas. The above rates shown by a recent survey in a metropolitan population were calculated to be more than 250/1000 pregnancies and around 120 in rural areas. These trends in pregnancy arrest rates were thought to originate from the difference in induced abortion practices. More than 70% of these arrests were incurred by induced abortion in a metropolitan and urban population.

The condition of infants at birth includes two rates, that of professionally supervised deliveries at a medical facility and that of home deliveries. The rate of deliveries attended by professionals is very low, varying from 50% in some metropolitan areas to 10% in rural areas. These rates are thought to be affected by socioeconomic status.

8. School Health

a) Government program: The school health program is implemented and supervised by both the central and local government in agreement with laws and regulations pertaining to school health problems.

The Ministry of Education has a section of school physical education under the Physical Education Bureau, and its functions extend to controlling and supervising the status of school health and physical education throughout the country. The government especially emphasizes the importance of the nation's health and has initiated a strong policy to improve and promote the activities of the school health program. The Ministry of Health and Social Affairs is also concerned with the school health program as well as with health education.

According to regulations of the Ministry of Education, each school must provide general health examinations every year to determine the health status of each child and to detect the presence of physical defects or impaired health. However, in many schools these regulations have been not only as a means for gathering data without any special consideration or follow-up examination.

b) Tuberculosis and parasitosis control: The most important program in the field of school health is the control of parasite infections and tuberculosis. For early detection of active cases of tuberculosis, massive tuberculin tests and X-ray examination programs are the primary control means. These programs are planned to be extended to all schools, but at present this is very difficult and not feasible. Of all school children given tuberculin tests, approximately 50% were negative; these were immunized with BCG. About 2% of all active cases are discovered by X-ray examination of school children. For parasite control, mass examination and mass treatment have been practiced twice annually since 1967. The number of school children examined each time was about 6 million, amounting to almost 85% of the total school registration.

c) Immunization: An immunization program at the elementary school level is another important part of the school health program at present. Since many children have not received any preschool immunization against various communicable diseases, the school has been delegated responsibility for providing an effective immunization program. The instructions of the Ministry of Health and Social Affairs specify the following required immunizations:

BCG: For the first and sixth grade children of the elementary school who have a negative tuberculin test reaction.

DPT: Booster injection as well as initial immunization for children who are not immunized. Only 30% of children in urban and 20% in rural areas receive DPT immunization during the preschool period.

Smallpox: Booster vaccination in the first grade of elementary school.

Vaccines of typhoid, typhus, and paratyphoid: Performed selectively at the time of epidemics.

Influenza and Japanese encephalitis: Given to some schools on a trial basis at present.

d) Other services: School nurses are assigned to dispensaries in elementary and high schools to provide medical consultations for the school children. In addition, school physicians assist in this service on a part-time basis. At present most elementary and high schools, except for a few smaller rural schools, have school nurses, and a few colleges and universities have established their own school health centers for student health service.

Dental health services in the schools has become very active recently with the enthusiastic cooperation of dentists. The Seoul National University Dental College reports that more than 90% of elementary school children in the Seoul area are in need of one or more dental services at present. As a preventive procedure, fluoride applications to the teeth and the public water supplies are being planned and massive dental health education is being performed in the community as well as at school.

e) Health status of school personnel: To protect school children from various diseases, the health status of school personnel is very important. The Ministry of Education requires routine physical examination, including X-ray examination of the teachers in all schools. Seoul city school health centers performed 557 pre-employment health examinations in 1960 resulting in 1% of the applicants being classed as nonqualified. Of 5960 school teachers in Seoul, 1.8% had pulmonary tuberculosis.

9. Voluntary Organizations

a) Korean Medical Association: The Korean Medical Association was established on 15 November 1908 under the name, Korean Society of Medical Doctors (*Hankook Euisa Yonkoo-Hoi*). However, its activity was subdued until World War II. During the Japanese regime, another body called the Chosen Medical Association was formed in 1911 by Japanese medical doctors and was operated mainly by Japanese physicians. On 15 August 1948, the day of the Republic of Korea was established, the association was renamed the Korean (*Dae-Han*) Medical Association. It joined the World Medical Association in 1949 and the Confederation of Medical Associations in Asia and Oceania (CMAAO) in 1961. A convention is held each year. By 1976, 35 sectional academic societies were established under the association. Academic meetings are held preferably at the time of the annual convention. The association has issued a monthly journal, *The Journal of the Korean Medical Association,* since March 1967.

The main objectives of the association are to promote social and national benefits, ethical standards, the development and spread of medical sciences and their technical application, and the protection of physicians' rights. In addition to a central office, branch offices exist in each special city and province. The central office has five divisions: medical affairs, laws and regulations, academic affairs, financial affairs, and public information.

b) Korean Dentist Association: In 1918 the Chosen Dentist Association was set up with about 50 Japanese and 2 Korean members. However, in 1925 the Korean group organized a separate body, *Hanseong* (Seoul) Dentists' Society, which developed into a nationwide body, the Korean Dentist Society, in 1945, but again its name was changed to the Korean Dentist Association. The association operates seven academic societies and holds scientific meetings once or twice a year.

c) Korean Herb-Doctors' Association: Organized in December 1959 and approved by the Minister of Health and Social Affairs, this association still limits its activity to supplementary education for herbalists.

d) Korean Pharmacist Association: It originated in 1928 as the Korean Pharmacist Society. In January 1955 it was authorized as a corporate juridical person.

e) Korean Nurses' Association: The Korean Nursing Association founded in 1923 was reorganized after World War II and resumed its activities, establishing a central office in Seoul and 11 branches in the respective provinces and special cities.

The association holds seminars twice a year to enhance the standards of the members. *Dae-Han Kan-Ho* (Korea Nurse) is issued bimonthly and distributed to members. A related association, the Korean Midwife Association, was established in April 1946. To become a midwife, a registered nurse is required to train for 1 year at a hospital designated by the Minister of Health and Social Affairs.

f) Korean Hospital Association: Established in July 1959 as a corporate juridical person, this association seeks to improve hospital systems and their installment by holding seminars and academic meetings annually. Its purposes are to enhance the quality and scope of each hospital as a satisfactory institution for clinical training as well as enforcing confidentiality in hospital affairs.

g) Korean Association for Parasite Eradication (KAPE): On 12 November 1958 the Korean Association for Harmful Animals was organized as a corporate juridical person. On 11 June 1964 it was reorganized as the Korean Association for Parasite Eradication. Soon it became the center of the control movement for parasites in Korea as well as the official body for supervising parasite control. A law for the prevention of parasites was promulgated on 19 April 1966, thus providing the historical background. Branch offices in nine provinces and two special cities were established in June 1965. The government set up a Subsection of Parasite Control in the Section of Chronic Diseases, Ministry of Health and Social Affairs in February 1967. Then the Association was designated as the authorized body by the Ministry of Education for parasite examination and treatment of primary, middle, and high school students.

The ultimate objective of the association is to improve the health and living standards of the Korean people by successfully controlling and eradicating parasitic diseases. To achieve this objective, mass diagnosis and chemotherapy of parasitic infections have been carried out, focusing on school children at the outset. In accordance with the rapid development of the *Saemaul* movement, the activity program has expanded beyond the schools since 1975. The program is also set up to promote the enlightenment of the people, personnel training programs, and improvement projects for community sanitation as well as to develop research programs for effective control. Mass fecal examination and treatment have been conducted twice a year for schools (excluding universities) and also for the general public on a limited scale. In spring of 1975, 6 638 258 students out of a total of 7 736 077 were examined. Those with ascaris egg positives (2 956 722) detected in the spring examination were treated with appropriate anthelmintics. A health education program is also included in the publishing of about 15 000 copies of the booklet, *Geon-gang,* a monthly periodical on health. Pamphlets and leaflets are also distributed, particularly during "Parasite Prevention Week". Mass media, lectures, and consultations have been utilized regularly during these yearly campaigns.

h) Korean Tuberculosis Association: This organization was founded in 1953 and joined the International Union of Anti-Tuberculosis (IUAT) in 1954. Christmas seals have provided one of the main sources of income. Branch offices are situated in each of the nine provinces and in two special cities. Personnel for tuberculosis control are also allocated to each of the 197 government health centers. In 1966 a board examination for tuberculosis specialists was legalized, and a law for tuberculosis prevention and control was promulgated in January 1967. An institute for tuberculosis, Tuberculosis Research Institute, was established in 1969. The main activities of the association include literature publication, mobile services for X-ray and sputum examination, BCG production and vaccination service, case-finding activity in designated areas. *Bo-Keun-Sege,* a monthly magazine, and *Tuberculosis and Thoracic Diseases,* an academic periodical, are its main publications. An average of 0.7 million are examined annually by the mobile X-ray teams. Sputum examinations numbered 176 150 in 1975.

i) Korean Leprosy Association: The association was established as a corporate juridical person in September 1947. After a period of inactivity during the Korean War, resumed business in 1955 and set up branch offices first in the individual provinces and then in the two special cities of Seoul and Busan. Of the contributions made by the association, the following should be noted: the settlement program for leprosy cases, either bacillus positive or negative; control of and admittance of vagabond cases to the leprosarium; and educational programs through publications and lectures.

j) Korean Red Cross: Established and inaugurated on 13 February 1960 the Korean Red Cross (KRC) carries out activities such as enlightenment and publicity, expedition of legislative action, education, research activities, establishment of facilities, and public relations. The health section of the association supervises the Red Cross Hospital in Seoul, Red Cross TB sanatorium in Incheon, and some other hospitals in the individual provinces.

k) Korean Family Planning Association (KFPA): Established on April 1961, the KFPA movement's aim was adopted as one of the national policies. Its main objectives include curbing the population increase, training of personnel for the movement, improvement of maternal health, medical and clinical services, and education programs through the mass media and publications like *Happy Home,* a monthly magazine. Medical and clinical services carry out an important role in promoting successful activities. A total of 190 000 persons visited the 14 family planning (FP) clinics in 1965, 175 000 of them for FP-related services. As a result of the FP-IEC (information education communication) emphasis on vasectomy and special programs aimed at low-income persons and young men in the active duty and reserved military forces, the number of people seeking FP consultation increased 200%, and the number of vasectomies in 1975 rose by 400%.

l) Korean Industrial Health Association: This association was established as a corporate juridical person in July 1964. Several institutions have been established to carry out its functions, e.g., the Occupational Diseases Clinic in 1965 and a Training Center for Industrial Health in May 1966. A Center for Health Services was set up in Yongdongpo, one of the industrial sections of Seoul, in 1968. The official magazine, *Industrial Medicine in Korea,* is published quarterly.

m) Korean Public Health Association: Established in September 1953 at a meeting attended by public health physicians, nurses, sanitarians, and others interested in public health, this association had as its objective the development of the public health program. Not only does it deal with the investigation and research of public health, but it also cultivates health ideals among the public, and publishes academic research.

C. The Diseases of the Country

Introduction:

1. Causes of Illness

Due to the previous absence of a satisfactory recording system, especially in the rural regions, reliable statistics on disease and causes of death which may cover the whole land are not available. However, statistics from hospital records collected by the Ministry of Health and Social Affairs [18, 19] and reports by several researchers [50, 55, 284] may help to outline the situation.

Prevalence of disease may depend upon sex, age, and geographic area. Data from Ganghwa Gun [55] may

Table 15. The Most Common Diseases Seen at the 25 Health Subcenters in Ganghwa Gun (May 1975 – April 1976) {284}

Order	C-list[a] code	Disease Category	Number	%
1.	39	Acute respiratory infections	492	25.5
2.	59	Infections of skin and subcutaneous tissue	134	6.9
3.	3	Enteritis and other diarrhoeal diseases	132	6.8
4.	60	Other diseases of skin and subcutaneous tissue	111	5.8
5.	72	Neuralgia, myalgia and back pain	95	4.9
6.	70	All other injuries	76	3.9
7.	55	Other disease of genito-urinary system	72	3.7
8.	30	Otitis media and mastoidits	68	3.5
9.	51	Other diseases of digestive system	65	3.4
10.	71	Acute gastritis	63	3.3
11.	47	Peptic ulcer	42	2.2
12.	65	Other specified and ill-defined diseases	42	2.2
13.	73	Family planning	37	1.9
14.	74	Headache (simple)	37	1.9
15.	4	Tuberculosis of respiratory system	34	1.8
16.	75	Fever	34	1.8
17.	42	Bronchitis, emphysema and asthma	32	1.7
18.	27	Psychosis and non psychotic mental disorders	28	1.5
19.	34	Hypertensive disease	28	1.5
20.	31	Other diseases of nervous system and sense organs	26	1.3
21.	28	Inflammatory diseases of eye	25	1.3
22.	77	Pregnancy	20	1.0
23.	19	All other infective and parasitic diseases	16	0.8
24.	45	Other diseases of respiratory system	16	0.8
25.	52	Nephritis and nephrosis	16	0.8
26.		Miscellaneous	189	9.8
			1930	100.0%

[a] International Disease Classification C-list, 71 – 77 are added to adapt Korean rural situation

Table 16. The Most Common Diseases According to Age Group in Ganghwa Gun (May 1975 – April 1976) {284}

0		1 – 4		25 – 44		65 +	
Disease category	%	Disease category	%	Disease category	%	Disease category	%
U.R.I. [a]	48.0	U.R.I.	39.5	U.R.I.	10.4	U.R.I.	12.5
Enteritis	20.6	Enteritis	14.3	G-U infection	9.0	Neuralgia & back pain	10.9
Skin infection	5.5	Skin infection	11.7	Neuralgia & back pain	7.8	Chr. obst. resp. dis.	6.3
Otitis media	5.1	Skin disease	8.8	Family planning	7.1	Dis. sensory & nerv.	6.3
Fever	4.7	Fever	4.2	Injuries	5.3	Tbc. of resp.	5.5
Skin disease	4.0	Otitis media	3.1	Skin infection	5.1	Skin disease	5.5
G-I trouble	2.4	Gastritis	2.3	Peptic ulcer	4.6	Ill-defined dis.	5.5
Eye inflammation	2.0	Injuries	1.8	G-I trouble	4.4	C-V dis.	4.7
Injuries	1.6	Parasite	1.8	Otitis media	3.7	Skin infection	4.7
Gastritis	0.8	G-I trouble	1.3			Peptic ulcer	3.9
Inte. obst. & hernia	0.8	Eye inflammation	1.3			G-I trouble	3.9
Other Tbc.	0.8	Ill-defined dis.	1.3				
All others	3.7	All others	8.6	All pthers	42.6	All others	30.3
	100.0		100.0		100.0		100.0

[a] Tbc., tuberculosis; U.R.I., upper respiratory infections; C-V dis., cardiovascular disease; G-I trouble, gastrointestinal trouble.

help to understand the situation of disease problems in present day Korea. Ganghwa Gun is actually an island connected to the inland by a bridge. It is about 50 miles west from Seoul, and the cultural level stands more or less midpoint between urban and rural.

Due to ginseng production, the economic standard is above the national level. Of 1930 patients randomly surveyed, acute respiratory infections, infections of skin and subcutaneous tissues, enteritis, and other diarrheal dis-

eases have the highest incidence of 25 diseases known to occur on the island (Table 15). The first three types of disease show the same order of incidence when analyzed according to sex, but that of the fourth rank showed divergences according to sex: injuries in the male and neuralgia and back pain in the female. The former may be more related to farming activities and the latter with gestation and difficult household chores. In all age groups, disease of the respiratory system occupies the first rank. But from the second rank on the pattern shows difference according to age. Under 4 years of age enteritis and skin infection were second and third in frequency, but in the 25 – 44 age group genitourinary infection and neuralgia and back pain were second and third, whereas the 65-year-and-over group showed neuralgia and back pain in the second and chronic obstetric and respiratory diseases in the third rank [55] (Table 16).

2. Causes of Death

Leading causes of death among Koreans show no unique pattern according to year. From 1932 – 1942 pneumonia, diarrhea, enteritis, and meningitis were the most common diseases causing death, but in 1959 tuberculosis became second in order and diarrhea dropped to third place. But this order varies according to the area. In the rural villages of the Gunsan area, the causes of death in 1924 – 1953 were measles, diarrhea, and enteritis, premature birth, convulsion, smallpox. Recently Kim et al. [55] examined the ten leading causes of death and disease-specific death rates with the object of identifying mortality patterns of rural communities with the total 125 deaths that occurred in two townships (Naega and Sunwon Myeons) in Ganghwa county during the period 1 April 1975 – 31 March 1976. The most common cause of death was cerebrovascular disease (221.8 deaths per 100 000) followed by malignant neoplasms (158.4 deaths per 100 000). Death from pulmonary tuberculosis ranked fifth with a death rate of 55.5 per 100 000, and death from pneumonia and other respiratory diseases ranked eighth with a death rate of 47.5 per 100 000. The overall statistics imply that recently the cause of death in rural Korea has rapidly changed from infectious to non-infectious diseases.

As for causes of death in infants under 1 year of age, prematurity (16.2%), convulsion, tetanus (14.1%), diarrhea, and enteritis in 1924 – 1953, and premature birth, tetanus, and pneumonia in 1956 – 1968 were the most common of 13 diseases. As for causes of child death, measles (23.2%), diarrhea, enteritis (13.2%), and pneumonia (11.4%) were the most common of ten diseases in 1924 – 1953, and pneumonia (19.3%), encephalitis (11.5%), diarrhea and enteritis (11.5%) in 1956 – 1968. Among the causes of infant death, prematurity was the first (16.2%), both before and after the inception of active public health work. At least 70% of the causes of infant death were attributed to infectious diseases before public health work was begun.

According to a survey [50], congenital anomalies have become the main causes of infant death since public health work was instituted in the Gaejeong Rural Health Institute. Of the causes of childhood death, measles, diarrhea, and pneumonia were the three main causes, contributing with other acute infectious diseases about 80% of child deaths before the active public health work program was begun. During the years 1956 – 1968, after public health work was started, smallpox and dysentery disappeared as causes of childhood death, whereas

Japanese encephalitis and accidents assumed a leading role. This suggests that communicable diseases are still the main cause of childhood death, and thus more active preventive measures and treatment of these diseases are desirable.

3. A History of the Occurrence of Communicable Diseases in Korea

Communicable diseases in this chapter include viral, bacterial, fungal, and spirochetel infections. Of the microbial diseases, 17 diseases were designated notifiable in Korea in 1976 by the Communicable Diseases Control Law.

Class 1 Communicable Diseases includes cholera, typhoid fever, paratyphoid fever, dysentery, diphtheria, meningococcal meningitis, smallpox, yellow fever, plague.

Class 2 Communicable Diseases are the whooping cough, poliomyelitis, measles, Japanese encephalitis, rabies, mumps, epidemic hemorrhagic fever, and malaria.

Class 3 Communicable Diseases list venereal diseases, tuberculosis, and leprosy. Class 2 and 3 diseases are diseases for which only a report is obligatory for doctors and medical institutions. The subsequent diseases in the following chapters are not given according to this system, but in using the detailed three-digit categories of the Manual of the International Statistical Classification of Diseases, Injuries and Causes of Death Ninth Revision, 1975.

The first recording of infectious diseases appeared in 15 BC (King, On-Jo in the dynasty of Baek-Je) as Baek-Je Ki Byeon (Ki-Byeon may denote an epidemic). Since then many epidemics have been recorded up to the emancipation of the land in 1945 from 36 years of Japanese domination, although the records were not scientific in the strict sense. For this reason more details shall be allocated to chronologic topics beginning in the year 1945 [3].

Before 1945, typhus fever, relapsing fever, and meningococcal meningitis prevailed all over the country. After the emancipation of Korea in August 1945, the Bacteriologic Laboratory under the Health Department of the Japanese Governor-General to Korea was renovated and replenished to establish the National Institute of Health. By producing vaccines and serums it has been contributing to the prevention of various communicable diseases since then.

During and following 1946, there was a nationwide spread of smallpox, which had been introduced into the peninsula through the repatriation of people from Manchuria. Cholera was also imported on 5 May of the same year through Korean nationals repatriating from Canton and Shanghai aboard the American ship U-036. The disease spread rapidly throughout the country, resulting in the occurrence of 15 644 cases with 10 181 deaths in South Korea. Beginning in October 1946, typhus fever and smallpox swept over the country, yielding 20 810 cases of smallpox in the course of 1 year in South

Korea alone. In August and September of the same year, the Japanese B encephalitis virus was isolated by the American Army then stationed in Gunsan, Korea. From June through July 1948 the first scientific work on mumps was carried out on Yunpyung Island. From June through August of the same year a minor spread of poliomyelitis occurred in the Seoul area, and from November through the next spring there followed another spread of smallpox in various parts of South Korea.

1949: An epidemic of smallpox in the countryside resulted in the death of 2068 people out of 10 085 cases of illness. Japanese encephalitis, occurring from August through October of the same year, spread through the country, resulting in 5661 cases and 2729 deaths. This event brought about the declaration of this disease as a legal communicable disease. Indeed, the Japanese encephalitis virus was first isolated in Korean patients that same year.

1950: The outbreak of the Korean War in June 1950 prevented the reporting of communicable diseases for the year.

1951: In January typhus fever occurred both in the army and among the civilian population, especially in Gyeong-gi Do, Chungcheong Bug Do, and Gangweon Do, numbering 32 211 cases among civilians alone, an all-time high in Korean statistics. Since many of the reported cases of typhoid fever were assumed to have actually been typhus fever, the latter undoubtedly reached its height at that time. In May of the same year, for the first time in our history, the 8th U.S. Army reported an occurrence of epidemic hemorrhagic fever in the frontline areas and especially in the "Iron Triangle", including Cheolweon, Geumhwa, and Pyeong-weon. The disease seriously hindered the military operation. Since then, this disease has occurred almost periodically in late spring and autumn among the residents of that area and in the army stationed there. In June, Seoul city and Gangweon Do were attacked by relapsing fever, and from June through September of the same year, dysentery was prevalent throughout the country. Isolated bacteria showed more Flexner type-2 and type-5 than Shigella dysentery; the former two were found in over 95% of the cases. From February through July of the same year, an epidemic of diphtheria concentrated mainly in Gyeong-gi Do, where 1603 cases occurred; a total of 2534 cases occurred in the entire country. Although 81 575 cases of typhoid fever were reported, an exceedingly high number, again many cases of typhus fever were suspected of being erroneously included. From January through July 1951 an extensive epidemic of smallpox raged. Notwithstanding the obvious underreporting in wartime, 43 213 cases of smallpox with 11 530 deaths were reported. The diseases centered mainly in Gyeong-gi Do, Chungcheong Bug Do, and Gangweon Do.

In November of the same year an outbreak of influenza B occurred mainly in and around the refugee capitaol, Busan. Also an epidemic of paratyphoid fever A

occurred in Busan and on Jeju island, continuing throughout the following year. In 1951 domestic production of BCG was first attempted.

1952: On April 16, the U.S. Army established their Hemorrhagic Fever Center in Kwangjang-ri on the outskirts of Seoul. From August through September of the same year there was a nationwide outbreak of Japanese encephalitis. Domestic BCG was first used for the massive prevention of tuberculosis in Korea. In December, influenca A, which prevailed in the Far East, invaded Korea, lasting through January 1953.

1953: The old "scratch" method of smallpox vaccination was replaced by the "multiple pressure" method. From January through May of the same year, smallpox was registered in the southern provinces with 3349 cases resulting in 571 deaths.

1954: Plague, smallpox, typhus fever, relapsing fever and yellow fever were designated quarantinable diseases. On 2 February 1954 the Communicable Diseases Control Law was promulgated (Law No. 308). In this law, 20 diseases were designated as notifiable diseases in Korea. The details, even though the items were modified several times, are as follows:

Class 1 Communicable Diseases cover 13 in number: cholera, dysentery, typhoid fever, paratyphoid fever, smallpox, typhus fever, relapsing fever, scarlet fever, diphtheria, meningococcal meningitis, epidemic encephalitis, plague, and typhus murine.

Class 2 Communicable Diseases include measles, whooping cough, poliomyelitis, and mumps.

Class 3 Communicable Diseases are listed venereal diseases, tuberculosis, and leprosy.

The 2nd and 3rd classes were diseases for which only a report was obligatory for doctors and medical institutions. According to the Communicable Diseases Control Law, smallpox, diphtheria, whooping cough, typhoid fever, typhus fever, and paratyphoid fever were to be dealt with by periodic preventive immunizations.

1955: Foreign-made DPT was first available for use in Korea. From August through October of the same year, Japanese B encephalitis occurred throughout the country with 2056 cases and 837 deaths.

1956: On 13 December massive import of DPT made possible extensive preventive vaccination, and the promulgation of the Health Center Law (Law No. 406) contributed to the onset of a new era for prevention of epidemics and the improvement of public health.

1957: On 28 February the enforcement of regulations of the Communicable Diseases Control Law was issued as Executive Order No. 1257. In June of the same year influenza type A, which had spread pandemically at that time, beginning in the Far East, swept the peninsula. In South Korea alone, 5 – 6 million cases were estimated to have occurred, but fortunately the mortality rate was negligibly low.

1958: For the first time DPT was produced in Korea. On June 30 of the same year, Executive Order No. 1378 was issued to enforce the Health Center Law. From

August through October of the same year, an extensive spread of Japanese encephalitis took the lives of 2177 people of the 6897 cases reported as occurring. On 31 December the amendment to the enforcement details of the National Medicine Law was issued as Act of the Ministry of Health No. 33. On 13 January the School of Public Health was established in Seoul National University in accordance with Executive Order No. 1430.

1959: The Ministry of Health initiated a prevention and eradication program against malaria in Korea with the technical support of the World Health Organization. Malaria had been added to the Class 2 Communicable Diseases on 25 May. From August through October of the same year, 2093 cases of Japanese encephalitis resulted in 742 deaths. In October the presence of typhus murine in Korea was proven by the isolation of *Rickettsia* from the fleas of human subjects and mice and was also confirmed by serologic procedures such as complement-fixation tests and hemagglutination-inhibition tests.

No report of smallpox was made for the first time in the history of medicine in Korea. However, 2139 cases of typhoid fever and 1134 cases of diphtheria occurred, which was the highest incidence for both diseases during the previous five years. In Seoul 693 cases of diphtheria were reported.

1960: In January the National Institute for Public Health Training was established under the supervision of the Ministry of Health and Social Affairs. It was to be in charge of the training of health personnel, research, and evaluation of the technical fields of health administration. From June to October of the same year the incidence of typhoid fever showed a tendency to increase. From August through September, an epidemic of Japanese encephalitis occurred in Seoul, Jeonra Bug Do, Jeonra Nam Do, Gyeongsang Bug Do, and Gyeongsang Nam Do. From September to November, a minor spread of Japanese encephalitis centered in Seoul.

1961: An outbreak of typhoid fever affected the whole country: 4982 cases with 186 deaths, especially concentrating in the districts of Honam and Yeongnam. Japanese encephalitis was also prevalent in the same year, registering 1056 cases and 375 deaths (fatality 35.4%). The geographic distribution was mostly in Yeongnam and Honam. In November the Korean Society for Infectious Diseases was established [3].

1962: Sporadic outbreaks of influenza A_2 type infection occurred from February to April. The incidence of typhoid fever continued at the same level as in previous years. Sabin oral vaccine for poliomyelitis was imported in the autumn and distributed to 250 000 children (6 months – 4 years) in Seoul.

Whooping cough occurred in Seoul during May to July on a small scale. In Hanam Gun, Gyeongsang Nam Do, 20 cases of anthrax were reported in cattle. Local people who carelessly ate the meat became infected and died.

1963: In May polio vaccine (Sabin type) was distributed to 130 000 children (6 months – 1½ years); consequently the number of patients was reduced. In September – October, cholera El Tor was introduced into Busan port probably from Thailand. The infection was especially prevalent in the coastal areas of Gyeongsang Bug Do and Gangweon Do. Two strains, Inaba and Ogawa, were isolated in the same area, so that the infection source was considered non-unique. The total number of suspected cases was 1073; 493 verified cases resulted in 46 deaths. In May to June epidemic hemorrhagic fever occurred in the frontline areas of Gyeong-Gi Do and Gangweon Do. Even among the soldiers, 257 cases occurred with 31 deaths (mortality 12.0%). Virus croup occurred in the winter months in the Seoul area. In May – July paralytic type of poliomyelitis reoccurred even among recipients of the vaccine. Typhoid fever still showed a high incidence, reported cases reaching 3857 with 101 deaths. The drought prevailed in the summer months in Yeong Nam and Honam, a Japanese encephalitis predominated in August – October with 2955 cases and 971 deaths (fatality 32.8%). In contrast very few cases were reported in Gyeong-Gi Do and Gangweon Do where precipitation was normal.

1964: During 8 October – 13 November, cholera El Tor occurred in Inchon port: 12 cases with one death and 71 carriers were reported. The infection route could not be determined. Epidemic hemorrhagic fever was prevalent during October – November in Gyeong-Gi Do and Gangweon Do and spread down to Gyeongsang Bug Do and Chungcheong Nam Do. During November – December 37 cases of anthrax were reported among cattle in the Daegu area. Local people ate the meat thoughtlessly, and 59 were infected with 3 dying as a result.

1965: A_2 influenza was prevalent during March – May throughout the country. Cardiopulmonary symptoms were severe compared with the epidemics of 1957 and 1962. Typhoid fever was still prevalent according to the report of 3760 cases, but the rate (2.5%) was relatively low. During the summer months, considerable numbers of cases of bacillary dysentery also occurred.

1966: Japanese encephalitis occurred during August to September, mostly in Jeonra Bug Do, Jeonra Nam Do, and Gyeongsang Nam Do. Cases numbered 3563 (morbidity per 10 thousands was 12.1) with 965 deaths (mortality 27.1%). In July – November a mass outbreak of typhoid fever occurred (600 – 700) in Danyang, Chungcheong Bug Do due to the pollution of well water and streams. In September to November there was a small outbreak of infectious hepatitis.

1967: Epidemic hemorrhagic fever cases were found in Jangseong Gun, Jeonra Nam Do in October. It has been suggested that the disease was not limited to the northern part of the land, but extended even to the southernmost province of the peninsula. In September – November, an outbreak of typhoid fever centered in Samchon-Po port: 638 cases and 1 death were reported. A polluted well was suspected to be the source of infection, and the pollution was proven to have originated in a carrier who contracted the disease 1 year previously. In the course of 1 year 4230 cases (morbidity 14.3 per 10 thousand) and 53 deaths were reported, but the fatality rate was only 1.2% because of the effective application of antibodies.

1968: Rubella was prevalent in Seoul from the late winter of 1967 until the beginning of the summer season, 1968. Small epidemics of poliomyelitis in May to July and whooping cough in May – August were reported in Seoul. Typhoid fever was prevalent throughout the country. Ten cases of anthrax with two deaths occurred in the Daegu area.

1969: This year was marked by several epidemics: influenza A_2 variant (A/Hong Kong 68 (H_2N_2)) during December 1968 – January 1969 and poliomyelitis and whooping cough in May – August in the Seoul area. On 26 August a cholera case was reported in a refugee colony in Gunsan, and the disease spread to Chungcheong Nam Do and Gangweon Do: total cases were 1396 with 129 deaths (fatality 9.2%). The agent was proven as El Tor type (nonhemolytic Vibrio cholera El Tor). Bacillary dysentery infection due to *Sh. flexneri* occurred in Seoul during August to September. Typhoid fever was also still prevalent. Polluted well-water was suspected to be the source of infection.

1970: In February Jeonra Bug Do provincial laboratory isolated El Tor vibrio from two people afflicted with cholera in September of the previous year. This fact suggested that cholera El Tor might possibly persist over the winter. During June – September it was said that a number of dysentery cases were treated in clinics all over the country, although many of them were not officially reported. The isolated strains were *Sh. flexneri, Sh. sonnei,* and *Sh. dysenteriae* in order of incidence. Reported typhoid fever cases numbered 4221 with 42 deaths. In the same period, poliomyelitis cases were also prevalent all over the country. Fewer cases of Japanese encephalitis were reported compared with previous years: 27 cases with 2 deaths. Heavy precipitation during the epidemic season and mass use of insecticides were suspected of causing the decrease. During August – October cholera El Tor broke out in Chang-Yeong, Gyeongsang Nam Do and spread to Gangweon Do and Jeonra Nam Do along the coastal regions. Masan port was suspected of being the initial place of the epidemic.

1971: Diphtheria was again rampant in this year, especially in Seoul, Gyeong-gi Do, Chungcheong Nam Do, and Gyeongsang Bug Do. Reported cases numbered 556 with 28 deaths (fatality 5.0%). The favored seasons were January – March and October – December. During March – May an epidemic of influenza A_2 variant spread throughout the land. In spite of

the absence of official reports on measles many cases were observed in clinics in Seoul in the same period. Typhoid fever was still prevalent all over the land and two outbreaks occurred. In May 20 cases occurred in Busan among 120 lumber workers. The source was polluted stream water, and five carriers were detected among the healthy workers. Another outbreak occurred in Geumreung, Gyeongsang Bug Do. Villagers (43 out of 191 villagers) who took their meals at a festival complained of typhoid symptoms. During October – December, epidemic hemorrhagic fever was prevalent in Chungcheong Nam Do, and Chungcheong Bug Do, and the Yeong Nam Area among 393 civilians and 295 soldiers.

1972: During March – May influenza A_2 variant (A/England/42/72 (H_2N_2) = A: Hong Kong 5/72 (H_2N_2)) was prevalent all over the country, and the virulence seemed more severe than influenza (A/Hong Kong) in previous years. Occurrence of typhoid fever still continued without showing any decrease. In September Getah complex virus belonging to arbovirus was isolated from swine serum. The isolated strains of dysentery bacilli in health laboratories totaled 88 that year: 77 *Sh. flexneri* and 11 *Sh. sonnei.* Among domestic animals, 2 cases of anthrax (Jeju Do and Gyeongsang Bug Do, each), 38 cases of bovine tuberculosis, and 25 cases of rabies were reported.

1973: Epidemic hemorrhagic fever was reported in Jangsu Gun, Jeonra Bug Do in November 1972, and another case was found in Jinan Gun in the same province in May. During June – August, whooping cough was again prevalent in Seoul. Less use of DPT in recent years due to undesirable side effects was considered one of the reasons. In August six milk cows in Jeju Do died due to brucellosis, the first incidence of its kind in the country. No human case was found. In August – October serologically proven Japanese encephalitis cases numbered 333. The official report was 286 cases with 18 deaths. This year the weather was relatively dry during that period.

1974: During March – May influenza A_2 variant strain (A/port Chalmers/72 (H_2N_2)) was prevalent but less virulent compared with influenza in 1972. In August – September, Japanese encephalitis cases reached 330 in Seoul and the southern part of the land with a 25% mortality rate. During October to December, epidemic hemorrhagic fever occurred in Gangweon Do, Chungcheong Bug Do, and Gyeongsang Bug Do — 251 soldiers and 170 civilians [3].

1975: The reported number of Typhoid fever was 534 with 8 deaths. The cases were more frequent in urban area than in rural, and seasonaly from June to August. Crowding and unsanitary condition in slum areas of urban cities might contribute to the factors. Diphtheria was prevalent more in Seoul than rural provinces. The total number reported were 337 with 27 deaths and the peak months were from November to early March. Japanese encephalitis was prevalent in summer months as usual. The laboratory confirmed cases were 117 with 7 deaths, but it is the general opinion that clinical cases would far exceed the number. The first case and the first detection of *Culex tritaeniorrhynchus* were from Gyeongsang Nam Do, southeastern part of the peninsala. Before 1969 the first case was usually from Jeonra Bug Do, southwestern part of the land. It was the yearly trend that more numbers were reported from southeastern part of the country before 1969. Korean hemorrhagic fever cases were 819 (civilian 448, army 370, U.N. forces 1), and distributed extending to southern provinces (Sources from 1975 to 1978 by The National Institute of Health, ROK).

1976: Typhoid fever cases were 672 with 6 deaths. Diphtheria was still prevalent, numbering 493 with 33 deaths. Measles had a comparatively higher prevalence in this year. The reported cases were 7328. Influenza A/Port Chalmers/'73 (H_3N_2) virus strain variant was specially prevalent in this year with the record of 8631 cases. Japanese encephalitis cases were reported only 30 cases with no deaths. Korean hemorrhagic fever cases were 823 (civilian 515, army 304, UN-forces 4). According to records maintained at hospitals and private clinics, case fatality showed 9.97 of the total of 515 in 1976. The distribution covered the whole country except islands, although the area close to 38° parallel still showed higher endemicity.

1977: Thyphoid fever was still prevalent among urban population, especially in the peripheral regions. Even though the reported cases were 304 with 1 death, actual cases are estimated far more than the figure. People usually get antibiotics from drug-stores, rather than visiting a clinic. Whooping cough was specially prevalent, numbering 3162, cases, which were more in Seoul than in local provinces. Evasion of vaccination among urban children due to the undesirable side-effects might result such adverse phenomenon; crowded conditions of the urban children will also not to be excluded from the factor. The laboratory confirmed cases of Japanese encephalitis were 101 and 7 deaths during July and August. The first case was also reported from Jeju Do area. Korean hemorrhagic fever cases were 394 (civilian 176, army 212, UN-forces 6).

1978: Measles was again prevalent during April and May, the total reported cases were 6149. Mumps were reported 1950 cases, mostly during from June to July. In both of these there were more cases in rural provinces than urban cities including Seoul. Laboratory confirmed Japanese encephalitis were 41 with 2 deaths, and more numbers from southeastern part. The recorded cases of typhoid fever was 427. Record of Korean hemorrhagic fever among army groups was not available in this year, only 100 cases among civilians were reported.

I. Intestinal Infectious Diseases

1. Cholera[1]
(ICD 001)

Cholera is not an endemic disease in Korea, but has frequently been imported into the country. Dr. C. H. Chun, Professor of Internal Medicine, Catholic Medical College, estimates through study of the literature over 35 epidemics of cholera in the past [3]. The route of its invasion has usually corresponded to established transportation routes from other parts of Asia, e.g., occurrences in the downstream areas of the Yalu River could usually be traced to Manchuria; occurrences in Hwanghae Do and Gyeong-Gi Do, especially Incheon, might be traced to China; and those of Gyeongsang Nam Do and Jeonra Nam Do, from Japan, China, and southeast Asia (Map 3c).

In the big epidemic of 1946 the classic type cholera, sero type Ogawa, was imported by the American Military Transport Ship U.O. 36, which repatriated Korean residents from the Kwangtung areas of China. The disease spread to all parts of the country, causing 15 644 cases and 10 181 deaths. In that epidemic the fatality rate reached 65.1%.

The disease usually begins to appear in May and extends through June and July. It is most prevalent in the summer months of July, August, and September, but continues to occur in October and November. However, cholera has never passed the winter in Korea, and no carrier is reported to have passed the winter season. A number of investigators concluded that this disease does not pass from one year to another in the carrier. Since 1964 cholera invasion has occurred five times, but none of them showed overwintering. With respect to age, the occurrence is high in the young adult age group of 20 – 40. Cholera occurs frequently in unsanitary areas, especially among the working classes. Usually the disease first occurs in the region where it has entered, leaving many victims. The areas of initial attack in the past were Busan, Gunsan, Incheon, Nampo, Uiju, and the areas along the Duman river.

An epidemic cholera El Tor in 1969 resulted in 1396 cases and 129 deaths with a fatality rate of 9.2%. Again in 1970 another epidemic occurred: 206 cases with 12 deaths (fatality rate 5.8%). But improved quarantine inspections and services have reduced the outbreak very effectively. Adequate treatment in recent years has also drastically reduced mortality. Many carriers between the ages of 21 – 30 have been found during epidemics. The excreting period of bacteria is mostly 1 – 2 weeks, but a case of cholera El Tor bacteria lasting 1 month has also been reported.

---

[1] The subsequent diseases in the following chapters are given the detailed three-digit categories according to the Manual of the International Statistical Classification of Diseases, Injuries, and Causes of Death (ICD) 1975, Ninth Revision, Vol. 1; World Health Organization, Geneva 1977.

2. Vibriosis Parahaemolyticus
(ICD 008)

This recently discovered food poisoning, caused by *Vibrio parahaemolyticus,* is an enteric infection contracted by eating contaminated sea products, such as fish and shellfish.

Vibrio parahaemolyticus has been known as an important pathogen of food poisoning outbreaks in Japan, but it was not well studied in Korea in the past. As Koreans have a custom of eating raw fish, it is quite reasonable to suppose that there are many infections of *V. parahaemolyticus* in summer. In the literature, Chang et al. [79] isolated the organism from a diarrheal patient. Chun [80] investigated an outbreak of acute gastroenteritis due to *V. parahaemolyticus* in a small army unit in Daegu city and isolated the organism from 22 freshly evacuated stool samples in the early stage of diarrhea or abdominal discomfort. The biologic characteristics and antibiotic sensitivity were enough to confirm its identity. It also showed positive Kanagawa Phenomenon, and its occurrence is more common in the summer and autumn months.

3. Typhoid Fever
(ICD 002)

Some evidence of typhoid fever was prevalent even in the Three Kingdom Era (18 BC – 668 AD). It is, no doubt, an important endemic disease in Korea. As widespread epidemics of this disease have occurred nearly every year in the past, clinicians have suspected typhoid fever in all obscure cases with symptoms of long-standing fever [3].

a) Outbreaks: After 1940, it suddenly soared up to the level of 10 000 cases annually. In 1951 following the beginning of the Korean War, an unprecedented epidemic occurred with 81 575 cases and 14 051 deaths. This epidemic was eventually controlled by extensive preventive measures such as mass immunizations, well-water disinfection, and other sanitary measures with the assistance of UN organizations. Since 1958, this endemic disease has again been on the increase, and about 2800 cases with 125 deaths were reported from the entire country in 1960. The 1974 statistics indicate 656 cases with 8 deaths, a fatality rate of 1.2%. 672 cases with 6 deaths in 1976, 304 cases with 1 death in 1977, and 427 cases with 2 deaths in 1978 were reported (Government Statistics 1979). It is most prevalent in the young from the age of 15 to 25, when they are most active. However, when there is an explosive epidemic, the age group from 6 to 10 has the largest number of cases with infants, adults, and the aged showing fewer cases. Incidence is higher in males than in females. In statistics from 1912 to 1941 the fatality rate among 109 437 cases was 14.6%. The rate was lower in infants, being proportional to the age. The seasonal mortality variation showed a slight increase in the cold season. No significant sex difference was found. However, in the present situation, in which only advanced cases are hospitalized, it is difficult to

estimate the fatality rate as a whole, which is higher by far than in other countries, including Japan. This is mainly due to the fact that people visit a physician only when the disease has gone too far. In the past typhoid fever was prevalent mainly in densely populated cities with no other specific characteristic in the geographic distribution. Today, probably due to the development of communications, transportation, etc., more cases are reported in villages and rural areas.

b) Epidemiologic characteristics: Little difference has been found to depend on sex, age distribution, or seasonal variations in comparison with other countries.

The high incidence of *Salmonella typhi* carriers has to a certain extent been attributed to the abuse and inadequate use of antibiotics in this country, since antibiotics may be purchased freely at any drugstore without prescription. Kim [78] reported that the asymptomatic carrier rate of *S. typhi* was 0.54% of 1855 food handlers in Daegu, where typhoid fever incidences were frequently noted.

According to statistics for the 10 year period from 1932 to 1941, 1146 (0.5%) were positive from 229 718 healthy persons examined, and 671 carriers (2.8%) were found in 23 692 recovered persons examined. As investigated in Busan during the Korean War, 0.5% of restaurant employees examined were found to be carriers of typhoid and paratyphoid salmonellosis. In 1958 four carriers with strains of *Salmonella* (1.7%) were found [3].

Water may be responsible when there is an explosive outbreak of this disease anywhere. For example, in an outbreak in Danyang [3] during July – November 600 to 700 cases were found in the same region, and it was proven that polluted well-water was the cause.

c) Problems of antibiotic therapy: Chloramphenicol has been in use in Korea for typhoid fever therapy since 1952. Appearing during the war and as there was no legislative regulation for the use of antibiotics in Korea the drug could readily be purchased by anyone at the market or in a pharmacy. Accordingly, there has been a tendency for people to abuse the antibiotic on typhoid-like disease without reliable diagnosis, thus producing undesirable deviations in the symptoms and course of the disease. When a patient uses antibiotics, he feels completely cured when the fever subsides and is less careful about his activities, food, and sanitation, thus providing an opportunity for an intestinal perforation, which has happened in 16.7% of typhoid cases.

4. Paratyphoid Fever and Salmonellosis
(ICD 002, 003)

Until recently the Korean medical profession has not paid much attention to salmonellosis. Data on paratyphoid fever appearing in immunologic studies during the period 1912 – 1941 are generally similar to those of typhoid fever. Occurrence by month is more prevalent during the summer months.

Numerous cases of salmonellosis were observed among persons with undefined fevers during the Korea

War, and various salmonella bacteria were isolated from the blood and feces of the patients.

a) Salmonella types: Cases of ulcer formation in the lower part of the ileum were relatively common in cases with *Salmonella paratyphi A.*

Various kinds of Salmonella were isolated in Koreans, particularly during the Korean War: *S. paratyphi A, S. enteritidis, S. typhi murium, S. blegdam,* and *S. paratyphi B.*

The frequency of occurrence of typhoid fever is much higher than that of the paratyphoid fevers. Hospitalized typhoid-paratyphoid patients in Seoul National University Hospital in 1959 showed 30 typhoid and 6 paratyphoid A, but no paratyphoid B. In 1960 – 1961 only typhoid fever cases were hospitalized, suggesting that the once common paratyphoid A had disappeared [3]. However, Government Statistics (1974) indicate that paratyphoid fever still exists in Korea, indeed 42 cases were reported in 1970, 5 in 1971, 9 in 1972, 2 in 1973, but none in 1974.

b) Food-borne salmonellosis: Cases of massive food poisoning (theoretically, food infection is a better term) have been frequently reported. An example of such epidemics is that occurring in Busan in May 1952 in which more than 138 persons were infected and 4 died after eating the meat of an ox which died of a disease caused by *S. typhi murium.* Another accident occurred in Goyang Gun, Gyeong-gi Do, in September 1955 in which more than 80 persons were infected, and 7 died after eating the meat of an ox that had died of disease. It was later proven to be a *S. typhi murium* infection.

5. Shigellosis
(ICD 004)

Data collected before Liberation in 1945 are not meaningful, since the figures might to some extent represent the numbers of advanced cases or charity cases. During the Korean War in 1951 dysentery must have reached its peak occurrence along with all other communicable diseases. Since the ready availability of antibiotics in 1956 only 100 cases were reported to the proper authorities, thus probably indicating a general negligence in reporting.

The incidence of dysentery in Korea is believed to surpass the rate of population increase. However, it is certain that the mortality rate has visibly decreased due to the mitigation of symptoms and efficiency of modern therapy. The government statistics of 1974 report only 72 cases with the morbidity rate of 0.2%, and 1 death (fatality rate 1.3%). It may occur at any time of the year; beginning in June, it becomes epidemic until August, dropping again in September and October. However, during the war in the prisoner-of-war and refugee camps where the water supply and other surroundings were unsanitary, more cases of dysentery occurred in winter. It is most prevalent among infants and children from 1 to 9, and the rate is inversely proportional to age. It is generally accepted that the incidence rate is higher in males than in females.

Before 1940 the case fatality rate was reported to be from 10.4% – 19.4%. This figure represents only severe cases in which the case fatality rate may be unduly elevated. In more recent years, the fewer deaths, if any, are now attributed to effective drugs readily available. However, in spite of the convenient availability of antibiotics, the fatality rate is still not low, even in urban areas: 1.9% in 1969, 4.1% in 1970, 4.3% in 1971. Mass outbreaks are known to have occurred in school dormitories and army camps without written reports being made.

Epidemiologic characteristics: The reasons for dysentery's prevalence in Korea in the past are considered to be ignorance of sanitation in the general population, abundance of flies, insufficient waterworks including wells, the contamination of drinking water, use of stream water (brooks and rivers) for cooking, bathing, drinking, and laundering, unsanitary toilets and sewage disposal facilities, drinking of untreated water, poor personal hygiene, and no adequate control of carriers. In the past, a general misconception reigned that Koreans were highly resistant to dysentery. This has been proven wrong by the fact that the case fatality rate in Koreans was much higher than in Japanese (Koreans 25.1% as compared to Japanese, 8.8%). Of course, the reason for the high fatality rate in Koreans is attributable to the hospitalization of only the more severe cases and even then after much delay. However, the claim that Koreans have a stronger resistance to dysentery is unfounded.

The statistics of an investigation from 1929 to 1930 revealed 72 (1.81%) positive carriers among apparently healthy individuals examined, 304 (0.22%) transient carriers among 135 333 recovered patients, and another 157 (1.9%) transient carriers among 9845 recovered cases. In 1952 an investigation by the National Institute of Health revealed 0.52% of carriers among restaurant employees and still 0.3% among Seoulites in 1961. In the past, the isolated bacteria showed a relatively large proportion of *Shigella dysenteriae,* whereas recently *Sh. flexneri* is the major organism found in Korea, followed by *Sh. sonnei* and *Sh. boydi.* Interestingly, almost no *Sh. dysenteriae* was isolated during 1971 – 1972. Some strains isolated in 1960 showed resistance to sulfadrugs and later to antibiotics. One report [87] described 70% – 80% resistance to streptomycin, chloramphenicol, and tetracycline derivatives. However, ampicillin, nitrofuratin, and gentamycin are still susceptibile to the *Shigella* group.

6. Amoebiasis
Entamoeba histolytica Infection
(ICD 006)

Dysentery in Korea is caused mainly by infection of *Entamoeba histolytica. E-jil* has been the local term for the disease since olden times, and the name was already in use in 1202 AD in the era of King, Shin-jong (1197 to 1204 AD) Korea Dynasty. However, there was no scientific identification of the infection from other origins.

Because of the insiduous nature of the clinical manifestation, most Koreans treat amoebiasis by themselves without visiting clinics. Otherwise, it is presumed that there would be a larger number of amoebiasis cases than those reported.

The incidence of *E. histolytica* has been examined by several investigators. Kessel [131] found 41% incidence by four successive examinations with the direct smear method in 208 examinees, and Choi [132] reported 1.5% positives among 2000 fecal samples examined singly by direct smear, compared with 30.2% upon six repeated examinations in the Seoul area. Soh et al [133] reported 4.3% of *E. histolytica* carriers out of 10 320 fecal samples in Seoul, but found no special age or seasonal differences among them. Nevertheless, the clinical cases are predominant throughout the summer. Recently, Kim et al. [134] reported 6.4% incidence by direct smear, zinc sulfate floatation technique, and formalin – ether concentration technique with 2250 fecal specimens collected from ten localities in different provinces. The incidence was seen to predominate in areas with poorly controlled pollution.

Amoebiasis may occur throughout the year, but more so in June, July, and August. The report on Korean Health-ROK-WHO-AID Health Evaluation and Planning Program [18] indicates that the majority of cases involved infants (76.6%), the youngest age was 3 months. No sex-specific differences were determined. Two deaths resulted from 73 cases; and these were caused by undernourishment during the Korean War. Among 53 confirmed cases, 22 had a past history of dysentery (41.5%). Brooke et al. [135] reported 87% of *E. histolytica* carriers in the prisoner-of-war camp on Gojeo Do island, where the sanitary conditions were very inadequate. However, K. M. Cho found 2.1% of positives among 665 on the same island in 1974. He also reported 32.4% of positives among 105 children in an orphanage located in the suburbs of Gunsan city, Jeonra Bug Do. However, he found 7.4% of positives among 149 inhabitants in the same area (unpublished data). The above figures endorse the impression that the incidences vary considerably by reporters and years.

Cho et al. [136] made an extensive study in Jeju Do where pigs are raised with human night soil and stagnant water is used for drinking and laundry. Two repeated direct fecal examination were carried out, 24.3% of the 738 fecal samples were positive for *E. histolytica* cyst. With this in mind, the occurrence of hepatomegaly in these islanders was examined. Two villages were sampled at random for the study. Sinum Ri (houses, 100; population, 630) is located on the coastal area of the northern part, and Yeongheung Ri (houses, 97; population, 398) is about 2 miles inland from the coast. In Sinum Ri 11.3% of the 150 examined showed hepatomegaly and over half of them were positive for *E. histolytica* cysts in the feces. In Yeongheung Ri 37.1% of 213 showed hepatomegaly, and 59.7% of the hepatomegaly cases were cyst positive. Moreover, 54% of the hepatomegaly cases had histories of diarrhea from a year before, and 10% had experienced a certain kind of hepatic disease. They complained of various general symptoms: fatigue, headache, and shoulder pain. The main gastrointestinal symptoms were nausea, capricious appetite, diarrhea, and abdominal pain in decreasing order of occurrence. Soft hepatic margin was found in 83%, and 44% complained of tenderness of the liver. To elucidate the hepatic involvement of *E. histolytica* on the island, 238 patients were investigated at the local clinic in Jeju City. In 64 cases of liver abscess, males outnumbered females. The largest group distribution was in ages 30 – 39, and both an increase of WBC and a decrease of RBC were noted. Six cases presented *E. histolytica* in the liver, 13 cases revealed protozoa only in the stool, and amoeba was not demonstrated in 45 cases, neither in abscess nor in stool. In microbial cultures of aspirated liver abscesses 42.1% revealed positive results, testifying to the presence of *Escherichia coli, Alkaligenes fecalis, Aerobacter aerogenes,* and paracolon group, in decreasing order. In 105 hepatomegaly cases, only 14 showed cyst positive in the stool.

There have been few reports concerning epidemiology of amoebic infection in Korea. Water pollution from the unsanitary disposal of human night soil as well as paratenic carriers such as flies are considered means of transmission of *E. histolytica,* as is the case in foreign countries. Soh et al. [137] examined the roadside soils in Seoul area and found cysts of *E. histolytica* 1, *E. coli* 3, and *Iodamoeba butschlii* 1 among 19 samples. Kim [138] reported that *E. histolytica* cysts were found in 24.8% of hog feces, 6.8% of drinking water, 4.1% of house rats, 6.2% of compost, and 0.01% of house flies in Jeju Do.

In conclusion, amoebiasis is one of the common contagious diseases in Korea. Although many studies have been reported in the past, a number of unsolved problems still remain, including poorly controlled sanitary conditions.

7. Intestinal Infections due to Other Specific Organisms

a) Balantidiasis coli (ICD 007.0)

Three cases were reported in Korea [133, 141]. Choi and Kang [141] found one case, a 58-year-old female admitted to Ehwa Women's University Hospital with complaints of an abdominal tumor, distension, flatulence, and abdominal pain.

b) Giardiasis lamblia (ICD 007.1)

This is one of the known pathogenic flagellates in Korea. Soh et al. [133] reported 489 cases (4.7%) of 10 320 specimens in Severance Hospital Laboratory. However, reports vary. Chung et al. [152] reported incidences of 11.1% and 13.1%, whereas Choi [132] reported 0.2% and 0.9%. It is more common in younger ages. Soh et al. (1961) analyzed the age distribution: 10.7% – 11.7% in 1 – 10 years of age, 5.1% – 8.7% in 11 – 20 years of age, but decreasing to less than 1.0% gradually in 60 years of age and above.

c) Coccidiosis, Isopora hominis infection (ICD 007.2)

Several cases of this infection were reported up to 1961 [133, 147] with no clinical manifestation. But Chyu [144] found three cases who complained of mucous stool, diarrhea, and abdominal pain.

d) Entamoeba gingivalis infection and Trichomonas tenax infection (ICD 007.3)

Of patients with periodontal diseases at the Dental Clinic of Severance Hospital, infected cases with *Entamoeba gingivalis* and *Trichomonas tenax* occurred in 38.1% and 12.5%, respectively, of the cases [129]. Kim und Cho [130] examined 254 persons in the Korean Air Force, males of on average 25 years of age (range: 19 – 32). The survey set up six categories: periodontal disease, caries, pericoronitis, periapical abscess, other diseases, and normal. The positives made up 54.3%, *E. gingivalis,* 47.2%, *T. tenax,* 15.4%, and simultaneous infection with both protozoa was 8.3%. As regards the prevalences according to oral hygiene status, of 139 persons with periodontal disease 60.4% showed positive for *E. gingivalis* and 20.9%, positive for *T. tenax.* Of nine persons with pericoronitis 30.3% showed *E. gingivalis* and 22.2%, *T. tenax. E. gingivalis* and *T. tenax* positives were observed in 36.7% and 8.9%, respectively, of 90 healthy patients. Caries, periapical abscess, and other disease groups showed negative for both. The overall findings indicate that positive rates of both oral protozoa increase proportionally to the periodontal index and the calculus index.

II. Bacterial Diseases

1. Tuberculosis
(ICD 010 – 018)

The third tuberculosis prevalence survey was carried out in 1975 by the Korean National Tuberculosis Association (KNTA) in cooperation with the Government. The KNTA selected 82 sample areas from the list of sample areas prepared by the Economic Planning Board, 36 of which were urban and 46 rural. During March to September workers of the KNTA examined 30 509 persons over 2 months of age, using an examination form composed of 33 items. Results indicated a prevalence rate of radiologic pulmonary tuberculosis of 3.3%, a marked decrease in comparison with the 4.2% in 1970 when the second survey was carried out. The following is an outline of the situation based on three nationwide surveys in 1965, 1970, and 1975 and also some available data compiled in 1965 [88, 89, 90].

a) Infection rate: Tuberculin (PPD) method was applied in systematically sampled areas in Korea to determine the prevalence rate. Individuals with BCG vaccination scars were excluded from the survey. Among the 10 – 14 year age group, rates of 69.5% in 1965, 54.1% in 1970, and 49.6% in 1975 were determined. The 5 – 9 year

age group showed 33.7% in 1965, 26.1% in 1970, and 15.9% in 1975, indicating that the rate has declined significantly during the past 10 years. However, no decline was observed among the group above 15 years of age. A WHO report [91] emphasized that the infection rate should be slowed down to less than 1% at 14 years of age so as not to be considered a public health problem. For this reason, Korea still needs to continue its strenuous efforts to control the disease.

b) Prevalence rate: Tuberculosis is observed as active shadows on X-ray film or as Tubercle bacterium-positive sputum in laboratory examination (Fig. 12).

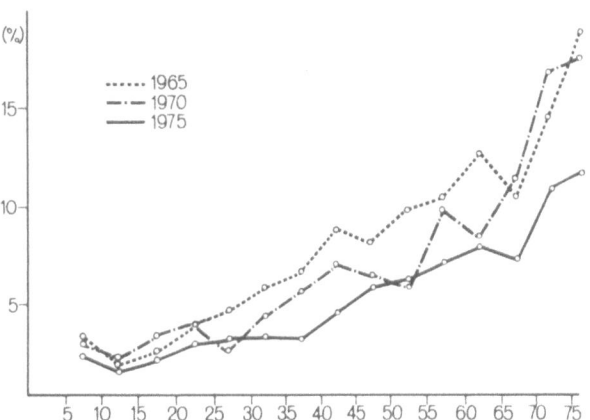

Figure 12. Prevalence of radiological active pulmonary tuberculosis by age (1965, 1970, 1975)
(Source: S. J. Kim [90])

Morbidity rates for ages above 5 were determined by X-ray findings to be 5.1% in 1965, 4.2% in 1970, and 3.3% in 1975. At the same age levels, Tubercle bacterium-positive sputum tests showed 0.94% in 1965, 0.74% in 1970, and 0.76% in 1975. However, these figures are still high when compared with morbidity rates in Japan in 1973, 0.94% by X-ray findings and 0.08% by sputum examination. Sales workers, miners, and farmers, all the most affected occupational groups in the past, showed remarkable decreases. Geographically, however, no difference was observed between rural and urban. But a sex difference was noticeable. In 1970 morbidity rates were 5.1% in males and 3.1% in females. The male group may have more chances to contract the disease while working outside. The grade of intensity of the disease in 1965 was mild in 62.8% of the cases, moderate in 24.0%, and severe in 13.2%, but 1970 showed 69.7% mild, and 20.2% moderate, and 9.2% severe; 1975 continued this trend with 65.8% mild, 24.5% moderate, and 9.7% severe.

c) Incidence rate: The incidence rate was examined during 1970 and 1971. New tuberculosis cases reported during this period comprised 0.8% of the total population above 5 years of age. Of them, 0.4% were shown normal by X-ray film 1 year before, 7.8% previously had had nontuberculoid X-ray shadow, and 12.5% had had a nonactive tuberculosis X-ray shadow the previous year.

Sputum examination showed an incidence of new cases at a 0.25% rate. Of them, 0.12% had a normal X-ray picture, 0.79% showed nontuberculous or nonactive tuberculosis, and 2.93% were bacillus negative but had been tuberculosis patients the previous year.

d) Factors of prevalence: It is difficult to determine, a direct connection between climate and season and the occurrence of tuberculosis, in spite of the popular belief that a cold, wet climate favors, and a dry, warm climate protects against it. Tuberculosis does not occur any more frequently in one season than another, but it is generally accepted that there are more frequent pulmonary hemorrhages in the changeable weather of spring and fall than at other times.

Mycobacterium tuberculosis is excreted with sputum, urine, stools, etc., and again infects new persons by inhalation or ingestion. However, the establishment of the infection depends upon the susceptibility and resistance of the human being. With this in mind, several factors are encountered in Korea which offer favorable conditions, such as nutritional imbalance, overwork, and living environment of poor economic communities. Of these factors, the crowding of a family in a small room may provide more chances for spray contagion. Yun and Yang [92] surveyed tuberculosis infection rate and some environmental conditions among the inhabitants of Gangwhal Myeon, an artificially reclaimed land area located at the Yellow Sea side of Jeonra Bug Do. Even though the survey dates back many years, the data may still be in conformity with the present situation in many parts of the rural and slum areas of cities where living conditions have not changed much. The number of examined totaled 881 people from 153 families with an average 5.7 persons per family. These families possessed about 3.5 acres of rice fields, thus more than the standard area of rural tenants. In general, traditional rural houses are built of earth and wood. Owing to its simple structure, the houses serve only to protect from wind and are not built to admit sunshine. For instance, the room doors face south, and the entrance gate is set up facing south a short distance away. The small paper window of the rooms faces north. Consequently, no sunrays enter the living room throughout the year. The New Community Movement in recent years exerted influence on the rural inhabitants to change the roof from straw to tile, but the essential structure inside has not been altered. The living space is small for the family size (Table 17).

Table 17. Living Space in Rural Villages of Ganghwal Myeong, Jeonra Bug Do {92}.

[a] No. of rooms	No. of families (%)
1	55 (35.9)
2	90 (58.8)
3	8 (5.3)

[a] 1) area of living space (m²)
   per person, 2.28 0.01 m²;
2) space of living room (m³)
   per person, 6.24 0.03 m³;
3) number of person per
   room: 3.4.

It was a serious problem that 35.9% of the families had only one room in the village. The average number of 3.4 persons in a room indicates overcrowded living when compared with the room area. Such conditions explain the high prevalence of tuberculosis which will continue as long as the room conditions are not improved. In the villages surveyed, tuberculin-positive rates showed 57.9% in males, 61.3% in females or an average of 59.7%. Even children under 5 years of age showed 23.2%, indicating that many active tuberculosis cases of infectious origin exist in this area and elsewhere in Korea. Thus, the poverty and unsanitary living environment may lead to strong infection with tuberculosis. Protein-deficient diet, indifferent care for patients in the family, unsanitary house construction, overcrowded living, poor ventilation, and other environmental conditions seem to act as major factors in high prevalence of tuberculosis in Korea. Long-time confined living during cold seasons, especially in the northern part, is apparently an additional reason for the high prevalences. However, the strenuous efforts made to control the disease by voluntary organization and government following the Korean War are reasons for optimism.

e) Mortality: The tuberculosis mortality rate in Korea is unknown because of the lack of diagnostic standards for cause of death. However, the mortality rate is estimated at 70 per 100 000 of the population.

f) Treatment program: In 1959, altogether 233 treatment centers including 5 sanatoria with only 4200 hospital beds existed throughout the country. At both hospital and ambulatory treatment centers, over 20 000 new patients were annually given free chemotherapy with government-supplied drugs for a period of 12 months or longer. However, over 100 000 patients are estimated to be under modern chemotherapy by private physicians either on an outpatient or inpatient basis.

g) Socioeconomic loss due to tuberculosis: Through the cost – benefit analysis of the Tuberculosis Control Program in Korea and on the basis of available statistical data, the following resultant figures concerning the economic loss incurred by tuberculosis morbidity and mortality have been estimated [9]. These economic losses reached a total of 44.31 billion won in the year 1970. This amount broke down into 3.31 billion in medical care expenditures, 25.88 billion in decrease in income incurred by infection of the disease, and 15.12 billion in losses resulting from tuberculosis mortalities. The yearly ratio of cost to benefit was 24.1 to 36.1 in 1965, 27.1 to 95.8 in 1966, 39.5 to 173.2 in 1967, 67.7 to 296.0 in 1968, and 87.9 to 456.8 in 1969. The benefit from the tuberculosis control program amounted to a total of 105.79 billion wons, whereas the total cost was 429.5 billion wons. The cost (X) and benefit (Y) of the tuberculosis control program for the 5-year period 1965 – 1969 were indicated by means of a linear regression equation $Y = -82.5 + 5.97 X$. In other words, benefit amounted to, so to speak, 5.97 times the cost [102].

h) BCG vaccination: Inoculations are recommended to be given to all newborn infants from 1 month after birth to 1 year and to school children having a negative tuberculin test. In 1975 statistics showed that 41.3% of the total population had already received the inoculation, including 80% of all school children. KNTA conducted an evaluation of BCG vaccination for the quality control of BCG vaccine in 1975. In the survey, 77% tuberculosis prevention effectiveness was seen in the 5 – 19 year age group.

2. Plague
(ICD 020)

The occurrence or epidemics of plague in Korea has not been recorded. However, the recent trend toward an intricate communication and transportation network between Korea and southeast Asian countries has created concern regarding the importation of communicable diseases, especially plague, from endemic foci in other countries. Anticipating an unexpected outbreak, Ahn and Soh [81] carried out a fauna study of rats and fleas along coastal regions, focusing on main harbors and some inland places. *Xenopsylla cheopis* was identified in 16.4% to 83.6% of 1448 rats caught, and the flea index was 1.72 on the average (Table 18).

The investigators proceeded to examine the flea indexes from areas surrounding main ports in Korea. An insecticide susceptibility test was also performed. The applied method was the standard test and techniques recommended by World Health Organization. The results obtained are summarized as follows:

Total number of rats caught was 1448; these included the

Rattus norvegicus, 1208 (83.4%);
Rattus rattus alexandrinus, 238 (16.4%);
Rattus norvegicus var. hibernicus, 1; and
Apodemus agrarius, 1.

Total number of fleas collected was 2480, including

Xenopsylla cheopis 1470 (59.3%);
Monopsyllus anisus 774 (31.2%);

Nosopsyllus fasciatus 217 (8.8%);
Leptopsylla segnis 10 (0.4%);
Ctenophthalmus congerner congeneroides, 6 (0.3%);
Ctenocephalides canis, 1;
Nosopsyllus nicanus, 1;
Rhadinopsylla attenuata, 1.

3. Leprosy
(ICD 030)

a) History: Leprosy is known to have first been reported in Korea in Jeju Do about 1300 AD. Now it is most prevalent in the southern provinces of Korea; Gyeongsang Nam Do, Gyeongsang Bug Do, Jeonra Nam Do, and Jeonra Bug Do. Leprosy control in the early era of modern medicine in Korea was directed mainly toward isolation due to the lack of proper specific medicine. For this purpose several leprosaria had been constructed before 1945: Ai-yang-won in Yeosu established by Dr. Wilson in 1949, Ai-rak-won in Daegu established by Dr. Fletcher in 1913, Sorog Do leprosarium in Sorog Do island established by the Japanese Government General in 1916. However, the advent of the sulfone drug after the World War II converted the control movement from isolation to treatment and rehabilitation. In addition to the leprosaria, some clinics offered early detection, effective treatment, and appropriate guidance of the infected cases. An example is the Special Skin Clinic, built in downtown Seoul in 1957 with funds from World Vision, a Christian Crusade Mission, managed by Professor J. Lew [93, 94, 97]. The Research Institute for Leprosy in St. Lazarus village was funded by Sasakawa Foundation of Japan in 1976. Its aim is to strengthen Leprosy control.

b) Leprosaria: At present there are one national leprosarium and five private leprosaria (Map 4 a).

c) Number of leprosy patients: Although the exact number of Leprosy patients is difficult to obtain, experts are able to estimate the number according to their profes-

Table 18. *Geographic Distribution of the Rat-Flea in Several Regions of Korea {81}*

Region	City	Xenopsylla cheopis (%)	Monopsyllus anisus (%)	Nosopsyllus fasciatus (%)	Flea index
North-west	Inchon and	16.4	68.5	13.7	0.50
	Islets, western	–	99.6	0.4	1.35
	Gyeonggi Do	60.2	34.6	–	2.04
	Seoul	95.6	3.7	–	3.75
South-west	Gunsan and Janghang	81.4	9.6	9.0	2.69
South	Yosu	0.4	45.9	53.0	1.16
Southeast	Busan	83.6	15.4	1.0	1.39
Total					1.71

Monthly flea indexes were:

0.35 – 0.72 in January, February and March,
2.02 – 2.59 in April, May and June,
0.65 in July, 1.58 – 2.84 in August – November.
In general, the index showed a decline in the summer months, but an increase in the fall season. *M. anisus* increased in the spring season, and *X. cheopis* in the fall season.

sional knowledge of the disease and past experiences in this field. It is estimated there are several times the number of registered cases (17 000) or 70 000 – 100 000 patients. The total number of leprosy patients in Korea in 1968 was estimated at 60 157. This figure was the sum of the total number of known, registered leprosy cases up to 1968 (37 571), the total number of estimated unknown cases (20 232), the number of leprosy cases not included in the estimation of the total number of unknown cases because of failure in identifying their birth place (1454), and the number of possible leprosy cases among the household contacts of lepromatous leprosy patients (900) [95].

Until 1967 five national and three private leprosy hospitals existed, but four of the national leprosaria were transformed into resettlement villages in 1968, and the number of inpatients was also greatly reduced. At the end of 1961, 18 307 patients were reported in leprosaria and colonies, but only 5000 patients remained in the two national leprosy hospitals, and 2000 cases were in five hospitals for the disabled by the end of 1973 [93, 94]. A total of 13 126 ex-leprosy cases had been resettled in 87 resettlement villages under the self-support program by the end of 1973.

In addition there are 13 stationary outpatient clinics, 13 mobile units, and 102 paramedical leprosy case workers attached to the local health centers. These units have been actively engaged in case finding and treatment of domiciliary patients (14 619 cases by 1973). Reconstructive surgery has been performed for deformities since 1962, and also the program for prevention of deformity is being greatly emphasized.

Since 1960 gradual and significant increases have been observed in the number of new and early leprosy patients, and these increases in the number of new cases indicated that the public understanding of leprosy itself has been greatly promoted year by year [93].

d) Geographic Distribution: Among the registered cases of leprosy in 1955, about 90% of the patients were from Gyeongsang Nam Do, Gyeongsang Bug Do, Jeonra Nam Do, and Jeonra Bug Do.

e) Epidemiology [93, 94, 95, 96, 97, 98]: The classification of the disease type revealed that the lepromatous forms, including borderline groups, comprised 54% and the tuberculoid forms, including indeterminate groups, 46% in 1973. Bacteriologic examination of leprosy in patients showed that 15.2% were bacteriologic positive in 1968, and 12.3% in 1972. These findings are interpreted to indicate that the epidemicity of leprosy in Korea has been decreasing, and chemotherapy of leprosy has been effective.

A survey of the age at onset of the disease established that the peak group was in the 15 – 19 age range. The peak of present age had increased to 50 – 54 years of age in 1968 from the 30 – 35 years of age group in the past. This shift in the peak of present age clearly indicates progress in controlling leprosy.

Of the patients registered in 1955 and 1957 the ratio of male to female was 1.8 : 1, revealing that the rate in males is a little less than twice that of females, almost in accord with the worldwide ratio.

Investigation in 1957 revealed that approximately 14% (2509 patients) had previous contact with known patients and were able to identify the contacts as the possible source of their illness, while approximately 59% were unable to remember any contacts. In the former group, 63% recalled extrafamilial contacts as compared with 37% familial contacts. Of the familial transmissions 28% shared the same room. Investigation revealed that about 2 years and 8 months had elapsed since the onset of the disease until accurate diagnosis. First diagnosis was made by the leprous family or by a layman in about 60% of all cases, and 40% were diagnosed by doctors, herb doctors, or medical attendants. In the latter group about 25% received relatively early diagnosis by doctors. It was found that about 2 and ½ years usually passed before proper treatment and that the period between onset and isolation was 4 years and 5 months.

Of the patients surveyed in 1955, 78.7% were engaged in agriculture at the onset of the disease, a rate somewhat similar to the occupational proportion in the entire population. More patients were found among physical labor groups. Approximately 61% of the patients were able to pay for their medical care, whereas 39% could not afford medical attention. The patients receiving higher than primary school education numbered 58% in 1955 and 47% in 1960.

f) Control: The stationary outpatient clinics, mobile units, and the integration of leprosy campaign in the general public health network through health centers for early case finding and domiciliary treatment proved to be an effective system for controlling leprosy in Korea. Also the resettlement program for the recovered ex-leprosy patients, including self-support schemes and institutions for the far-advanced, disabled cases, significantly facilitated their social rehabilitation, giving great relief from the age-old accumulated leprosy problems in Korea.

In conclusion, the leprosy control campaign has been progressing well, and the prospects for overall control of leprosy in this nation appear quite bright [93].

4. Diphtheria
(ICD 032)

a) History: No chronologic record for this infection is available, suggesting that it was probably overlooked and included as one of the "throat diseases" until the scientific name was introduced. Formerly the incidence appeared to be increasing with every passing year after 1937. However, massive production of DPT vaccine since 1958 has made possible an extensive preventive vaccination program. Likewise the promulgation of the Health Center Law No. 406 contributed to the establishment of a new era for preventing epidemics and improving public health. Since then the incidence of diphtheria has visibly decreased [3].

An investigation of the seasonal variation over 30 years, from 1912 to 1942, revealed that the incidence began to rise in October, reaching a peak during the months of December, January, February, March, and April, and decreased in May, reaching bottom in July and August, i.e., the incidence rate is much higher in the winter and spring than in summer. The majority of cases (60%) belonged to the age group under 10, especially young children between 1 and 4. No significant difference exists between incidence in male and female. A reinfection in 6 of 106 cases (5.6%) as well as some other recognized relapses in other cases have been reported. In 18 345 cases reported in 1912, the fatality rate was 27.5% with 6668 deaths. Several other reports show a fatality rate of 5.3% – 32.8%. In 1963, 713 cases were reported with 63 deaths and 8.8% fatality rate [3]. Some investigators claim that antibiotics (especially penicillin and erythromycin) do not affect the diphtheria bacteria at all. Although they do noticeably prevent bacterial complications, there is the risk that the patient having taken antibiotics, will delay coming to the hospital until long after the onset of the disease.

b) Carriers: One investigator [84] reported a carrier rate of 24.5% with 10.8% of the mothers of infected children being carriers. As regards sex difference, the carrier rate is slightly higher in males though not significantly. No noticeable difference was observed in age distribution. April, March, January, and November showed higher rates than July and August, e.g., more carriers were noticed when the disease was prevalent than when it was not. As for the sites of bacteria in carriers, the nasal cavity and pharynx were most common. Healthy carriers seem to play an important role in the spread of diphtheria, as is the case in other communicable diseases. A high carrier rate results in a high incidence rate. A lack of thoroughness in the isolation and treatment of patients may also produce the same result.

c) Epidemiologic characteristics: No epidemiologic characteristics have been found other than that the case fatality rate is much higher in Korean children than in other countries. Nationwide statistics show the fatality rate of diphtheria to be 9.8% (1957), 10.2% (1958), 7.6% (1958/1959), 4.2% (1959 – 1960) and 4.5% – 5.0% (1972 to 1974). Sporadically diphteria shows a gradual increase among adults in Korea. The diphtheria bacteria found in the Seoul area [86] in 1938 – 1939 and in 1948 – 1949 had more T. gravis, 54.2% and 42.0% respectively, than T. mitis, 34.8% and 38.0%, respectively.

5. Whooping Cough
(ICD 033)

When this disease became endemic is not known. However, whooping cough is constantly being called to the attention of the public.

Since 1955 it has been designated a Class 2 Communicable-Disease, and 7 000 – 20 000 cases per year have appeared in the Annual Health Statistics. This amount is assumed to be far less than the actual occurrence of cases due to the lack of thoroughness in reporting [3].

According to an investigation in 1948, 12.1% of the population studied had previously had whooping cough. The age distribution was 13.8% in the ages from 1 to 5, 14.0% from 6 to 10, and 9.3% from 16 to 20. The disease was probably very prevalent in the past [3].

In recent years, however, it has occurred very infrequently in families above the middle class in urban areas, but is still prevalent in the poorer quarters of cities and rural areas. Therefore, there is no evidence of any decrease at the present. In densely populated cities, cases occur incessantly throughout the year, especially in April, May, June, and July. A large number of cases occur in infants under 1 year and from 1 to 2; the youngest age was 20 days old. The case mortality rate was 2.4% – 21.5% before antibiotics came into use. A reliable report in 1948 investigating various localities gave a fatality rate of 6.7%.

During the years 1955 – 1964, the reported cases totaled 90 982, and 1093 deaths (1.2%) were recorded. Most of the deaths were caused by complications such as bronchitis, pneumonia, dyspepsia, encephalitis, and tuberculosis. But adequate application of antibiotics in recent years has diminished the mortality rate drastically. According to the Annual Health Statistics, the case fatality rate of whooping cough was nil in 7348 cases reported during the years 1970 – 1974. Another report records no death among 914 cases treated with chemotherapeutics [82].

Due to the incompleteness of reports, an overall picture of geographic distribution has been difficult to acquire. In view of the situation in recent years in Seoul city, whooping cough is still prevalent in the poorer quarters on the outskirts of the city. Also a trend of mass outbreaks in orphanages and institutions for the destitute is developing but there are no written reports.

The epidemiologic characteristics are a relatively low age of occurrence and prevalence mainly in the early spring and summer. At present primary schools, kindergartens, and sunday schools provide the sources of infection from which the disease finds its way into families to spread to the younger children and infants. Difficulties recognized in preventing and treating the disease are inadequate immunity, carriers even among adults, high infectivity in the catarrhal stage when the typical symptoms have not yet appeared, and no effective drug for the infection. DPT has been used routinely since 1955 to prevent the disease. But one report shows that even after the complete regimen of vaccination, there was 1.9% infection, with only 9.6% infection among the group with incomplete vaccination.

6. Scarlet Fever
(ICD 034)

The historical background of this infection is obscure; indeed, until modern medicine was introduced in the last part of nineteenth century, it was often confused with measles. Showing little fluctuation from year to year, it did not assume epidemic dimensions until the cases began increasing among Japanese residents in

Korea. After reaching a peak in 1931 and 1932, it has been on the wane ever since. The clinical manifestation is rather mild in Korea, so that it has come to be of little importance clinically. However, when Korea was under Japanese occupation, the statistics showed a visible difference between Korean nationals and Japanese residents in Korea as regards scarlet fever incidence. Such differences between natives were also usually found in other communicable diseases, demonstrating an interesting immunologic and epidemiologic phenomenon. The reasons given for the low incidence of scarlet fever in Korean nationals were (1) natural immunity of the Korean people, (2) failure to diagnose scarlet fever as such, and (3) national customs and economic conditions preventing patients from seeking medical care. However, there is no doubt the incidence rate was lower in Koreans than in Japanese residents in view of the results of the Dick test and in many other respects [3]. After World War II as symptoms of scarlet fever become milder and the differentiation of this disease from other eruptive diseases became obscure, reports tended to be neglected. Only two cases in 1972 and one case each in 1973 and 1974 were reported. There were no deaths. Even though no statistical data is available, newspapers reported occurrences in 1976, and a government report of 1979 indicates that the number of cases increase again; 9 cases in 1977 and 107 cases in 1978.

No significant difference was found between males and females in incidence and mortality rates. In the past more than 40% of all cases occurred in the ages 1 – 4, while above 4 years of age incidence was inversely proportional to age. Above 35, few cases occurred. During the period 1912 – 1941, 4392 cases were reported (Korean) with 1486 deaths (fatality rate 34%). The high fatality rate was believed to result from reporting only the severe cases. The fatality rate was higher in children under 9 and in the aged above 50. Many reports indicate that scarlet fever cases begin to appear from the onset of the cold season and increase up to May of the next year [3].

The Dick test is the assay procedure available for studying the toxin filtrates of streptococcal cultures. The results of the Dick test made prior to 1945 on 3142 people (429 Koreans and 2713 Japanese) in Seoul and Pyeongyang revealed differences. The Dick test positives among age groups 1 – 16 numbered 15.7% – 33.4% in Koreans and 22.5% – 68.2% in Japanese. The rate of positivity in Koreans turned out to be 23.9% in males, 29.9% in females, and 24.9% average, lower than that of Japanese residents (44.4% males, 43.4% females, and 44.0% average) [3].

7. Meningococcal Meningitis
(ICD 036)

In 1911 this disease was mentioned in a scientific report for the first time in Korea, and in 1924 meningococcal meningitis was designated a notifiable communicable disease. Since the epidemic in 1934, it seemed to be on the wane until 1944 and 1945, when nationwide epidemics took place. Since 1950, when antibiotics began to be extensively used, an epidemic has not occurred, but sporadic cases and the shocklike state of collapse known as the Waterhouse-Friderichsen syndrome have at times draw attention to the clinical aspects [3]. Even though the occurrence of the disease has decreased since antibiotics became available, the statistics still indicate that about 20 cases are reported every year. But only two cases are recorded in the government statistics of 1976, and 3 in 1977, and one in 1978.

The investigation from 1924 to 1941 showed one peak in September and a second in March and April, with the lowest points in July and December. The first peak was in September because the reported incidence rate in September was exceedingly elevated in 1924, 1927, 1937, 1938, and 1939. However, the view of the fact that encephalitis was prevalent in Japan and spread to Korea in the years 1924, 1927, 1935, and 1939, the exceedingly elevated incidence rate reported in those years is assumed to be a result of erroneous diagnosis of Japanese encephalitis as meningitis. It has been recognized in various scientific reports that meningococcal meningitis cases occur in Korea mainly in the spring months; March, April, and May.

A report revealed that 69.4% of the cases belonged to the age group under 15, and 41.4% were children under 5. An investigation conducted from 1922 to 1941 on 2016 cases (Korean) revealed the age distribution from 1 to 4 (24.1%), from 5 to 9 (25.7%), and from 10 to 14 (18.3%), i.e., 68% of the total are less than 14. Incidence was higher in males than in females [83].

8. Tetanus
(ICD 037)

Very recently the occurrence of this infection in neonatal babies has tended to increase. Since considered fatal, there was a great resistance to hospitalization, and thus it was not recommended in the past.

According to M.G.L. Annual Report (U.S.A.) 2.4% (1951) – 2.7% (1950) of the Clostridia isolated from persons wounded in the Korean War were *Cl. tetani*. It is astonishing that 12 of the 25 tetanus patients hospitalized in the Seoul National University Hospital during the period from early 1960 to the end of July 1961 were neonatal and umbilical tetanus cases.

There is no remarkable difference according to age since perhaps most of the cases are caused by traffic accidents. Since the number of patients is small, it is difficult to find a definite tendency in the monthly distribution, but apparently, the occurrence rises in the warm months. As for sex distribution in nonneonatal patients; the male to female ratio is 11 : 2, and in neonatal patients, 10 : 2. Fatality was 58.5% – 66.6% in neonatal patients, and 38.4% in nonneonatal patients who were admitted to Seoul National University Hospital during the Period 1960 – 1961 [3].

In 1951 early in the Korean War as many as 165 cases of tetanus occurred, of which 152 died (92.1%). But, recently as anti-tetanus inoculations are given periodically, not even one case has been reported since 1959. In conclusion, the high morbidity of neonatal infants suggests that health education for pregnant women is important.

III. Viral Diseases

1. Poliomyelitis
(ICD 045)

Poliomyelitis has been considered a less important disease for a long time in Korea because only paralytic cases were included in the statistics, and few occurred in the past. Nevertheless, it is hard to believe that the polio virus existed in our surroundings to such a negligible degree. On the contrary, the virus must have been extensively distributed in the unsanitary quarters and had frequent contact with the people. People who were infected and died from the disease were readily recognized as such, and a majority of the survivor were naturally immunized through contact with the infection. This is a reliable scientific explanation for the strong immunity of Korean people to the disease, and the few cases of the paralytic type. As proof of the above explanation, the incidence rate in the American army, stationed in Korea since the end of World War II, was by far the highest in the American army all over the world during the same period. Moreover, serologic tests on Koreans in 1951 showed the presence of the antibodies in 70% of those aged 2 – 3, 97% in those 4 – 5, and 100% in those above 18 years of age.

An investigation of densely populated areas in subtropic zones (Korea included) showed 80% of newborns to possess the antibody through transplacental immunity. After losing a large part of it by the 6th month after birth, they regain the immunity with an increasing intensity, so that by the 5th year more than 95% of them acquire sufficient antibodies. The possession of transplacental passive immunity at birth is recognizable, but by 1 – 2 years the antibody decreases to its lowest level. After the 2nd year the antibodies are acquired for one or two types, whereas in the 5th year the antibodies for all three types become demonstrable in the body.

As stated above, at the time when only paralytic cases were considered polio, the occurrence of this disease appeared infrequent. However, written records prove that it was prevalent in 1940 in the Daegu area, in 1948 in the Seoul area, and in 1957, 1959, and 1960 again in the Seoul area [3]. The government statistics of 1979 shows the incidence in recent years: 23 in 1975, 77 in 1976, 35 in 1977 and 2 in 1978 in Korea, although the actual number may exceed far more than the figures. It may occur at any time of the year, but especially in May, June, July, and August.

Ninety per cent of the cases were under 3 years of age; 3.78% were from 3 to 4 years; thus, about 99% of the cases occurred under 5, and only 1% of the cases were above 5. Some reports showed that male cases were slightly higher in number, whereas other reports showed equal numbers in both sexes.

Status and written records so far available can be of little help in gaining a clear understanding. To the investigators who traveled to various localities in Korea to acquire the data, the epidemic areas of this disease appeared nationwide, since even in isolated rural villages crippling by polio was found. Fatality rate is relatively low. However, one report recorded only 1 death in 61 cases in 1971 (statistics, 1974). On the average, 40% – 50% of cases in Korea show mild manifestation without lasting complication, 20% – 30% leave a slight paralytic symptom, and 25% end in limb paralysis.

Both killed vaccine (Salk's vaccine) and live vaccine (Sabin's vaccine) are used in Korea. Since 1965 live vaccine has been given orally to infants and children in the cities; Seoul, Busan, Daegu, and Gwangju. The incidence in the cities decreased following 1965, but in 1968, when both vaccines were unavailable for export to Korea, the incidence again increased. One report from the pediatric clinic in Severance Hospital described the incidence in 1966 as 12 cases, but this greatly increased by year, 52 in 1968 and 284 in 1968.

Another report from the clinic recommends combined vaccination with both vaccines. It is reported that 20% – 28% of paralytic cases occurred among children who received only Salk's vaccine, but only 1.7% – 3.3% among those who received both vaccines. However, a stronger suggestion has been that the vaccines be administered to all newborns.

2. Smallpox
(ICD 050)

An epidemic of smallpox (*Son-nym*) is believed to have occurring in Korea in the period of the Three Kingdoms (18 BC – 668 AD or even earlier). History indicates that frequent occurrences of this disease took the lives of a large number of people during the Yi Dynasty when the nomenclature of disease was relatively correct. In 1946, one year after the liberation of Korea from Japanese rule, there was an unprecedented spread with 20 810 cases, introduced into the peninsula, through the repatriation of people from Manchuria, from where the virus was most likely to enter the peninsula and spread southward. An extensive epidemic of smallpox raged from January to July 1951 during the Korean War. Notwithstanding the obvious underreporting in war time, 43 231 cases with 11 530 deaths were recorded, mainly in Gyeong-gi Do, Chungcheong Bug Do, and Gangweon Do.

In the year prior to the liberation of Korea (1945), Pyeongan Bug Do and Hamgyeong Bug Do provinces geographically adjacent to Manchuria served as starting points for the disease. The occurrence of smallpox reaches its peak during the months of February, March, April, and May. It was especially high in March and April and lowest in August, September, and October, e.g., the spread was greatest during the warm spring and generally subsides in the hot summer to rise again slightly around November.

Since 1962 the well-planned vaccination program by the government has been effective in preventing the disease, and for the first time in the history of preventive medicine in Korea it became completely controllable in 1959. WHO declared in 1960 that smallpox had been eradicated from the Republic of Korea.

3. Varicella – Chickenpox
(ICD 052)

Varicella has existed since olden times; however, on many occasions, it was overlooked due to its mild symptoms or else, sometimes diagnosed as smallpox. Through a series of available data, no clear-cut difference was proven in sex distribution, although generally more boys were affected than girls. The patients were distributed between 3 to 10 years of age in 1958, 1959, and 1960. The seasonal distribution for the 3 years of high occurrence was in April, May, and June, when measles was also prevalent [3].

4. Measles
(ICD 055)

In the past large epidemics occasionally occurred in Korea. Although the term *Hong-Yeuk* (red epidemic) was not confined to measles, long-lasting epidemics of measles were recorded in 1668 and 1707. Some of them might have included scarlet fever, rubella, and some other exanthematous diseases. Over 10 thousand deaths occurred in the Pyeongan Do area in the 1707 epidemic due to *Hong-Yeuk*. Since the large epidemics repeat every year, almost all people have been infected once during their lifetime, so much so that measles was recognized as the most prevalent diseases of Korea [3].

Of the few reports on epidemiologic studies, one indicates that the morbidity rate in 1962 was 162.4 per 1000 population, and 167.7 in 1963. Consequently the probable number of infected cases is estimated at 1 012 609 (males, 528 872 and females, 483 737) in 1962 and 1 041 834 (males, 55 831 and females, 486 003) in 1963.

a) Infection rate: Records show that 58.3% of the 1 – 5 age group is infected, 93.4% of the 6 – 10 age group, and 96.7% of the 11 – 20 age group. This indicates that almost all inhabitants experience the infection during their lifetime. Even the remaining portion is regarded as consisting of subclinical or latent cases.

b) Epidemiology: Measles are much commoner in infants than in the advanced age group. Statistics show that 1-year-olds (13 – 24 months) make up the largest group followed by 2- and 3-year-olds.

Higher prevalence occur in the spring months, March through June, and the lowest prevalence is in

October and November. The mortality rate is high in the younger ages, but almost no deaths occur in adults. The rate in 1948 was 5.9% on the average, reaching 25.7% of all infantile deaths. In the same year the mortality rate due to measles in New York was 0.2%. Complications occurring in fatal cases were pneumonia, tuberculosis, and others. However, since the mass application of antibiotics, the fatality rate has shown a rapid decline. Statistics in 1961 – 1963 indicate that the number of measles cases was 56 370 with 506 (0.9%) deaths. Government statistics for 1974 report only five deaths of 4867 cases. One of the prevailing superstitions is that crayfish juice given to a child with measles or scarlet fever will cure the disease. Thus, many children are infected with *Paragonimus,* the details of which will be described in the corresponding chapter on *Paragonimus.*

c) Vaccination: Death due to measles is almost nil among persons over 5 years of age in Korea. For this reason, it is important to prevent infection in the under-five age group. Inoculation with the convalescent serum or gamma globulin may prolong the incubation period and make the clinical manifestation mild. Vaccines with Edmonton strain (Rubeovax) and Schwartz strain (Lirugen) were introduced in 1965. Since the start of their massive application, the intensity and pattern of measles prevalence has changed drastically. The annual large outbreak in the spring season has almost disappeared. But government statistics of 1979 indicates that measles is still prevalent in Korea; 4973 in 1975, 7328 in 1976, 5064 in 1977 and 6149 in 1978.

5. Rubella
(ICD 056)

This disease has been confused in this country with other exanthematous diseases such as measles and scarlet fever. The first scientific report was made by Ikeda in 1912 [3]. According to various relevant reports, distribution is mostly concentrated in the 2 – 13 year age group. Seasonally, April and May are the months of high occurrence, coinciding with measles outbreaks. Hemagglutination inhibition titers among maternal and infant groups in 1974 were 58.3% (7/12) and 58.8% (10/17), respectively [7]. The positive titers (1 : 10 and above), among females over 16 of age appeared in 70% – 75%.

6. Epidemic Parotitis – Mumps
(ICD 072)

Very little attention has been paid to mumps due to its good prognosis and the short duration of its clinical manifestation. An investigation conducted on outpatients visiting a children's hospital in Seoul from January 1956 to July 1961 showed a total of 185 cases (male 119, female 66). Most cases were in the age group 2 – 10, whereas cases under 1 year were met once in 7 – 11 month intervals. The oldest age was 12 years and 11 months. Another survey conducted on Yonpyeongdo island at a primary school on the west coast reported 56 cases (18.0%) of 239 students examined [71]: 53 appeared in June and 3 in July, the majority were 8 – 11 years old (44% or 78.5%), and no sex-specific differences were noted (male, 18.0%; female, 20.7%). The cases were more common among the children in lower grades, with no case occurring in the 6th grade. Incidences in recent years show 813 in 1970, 1431 in 1971, 1655 in 1972, 590 in 1973, and 2039 in 1974 (*Yearbook of Public Health and Social Affairs,* 1974). Family infection was recognized in ten cases in four families. Crowding and frequent contact may provide the probable source of infection. It is interesting that no adult case was observed during the survey. Serologic tests in the immunologic sense will be needed to ascertain the epidemiologic factor.

7. Japanese Encephalitis
(ICD 062.0)

Japanese encephalitis (JE), which has occurred almost yearly in epidemic dimensions was proven by Sabin [60] for the first time by isolating the virus of the disease among U.S. Army soldiers in the Gunsan area. He also proved the presence of the antibody of JE in Korean residents from several areas, including Gunsan. According to several reports on Encephalitis lethargica, the disease was generally assumed to occur even before 1935, but was apparently misdiagnosed as epidemic meningitis. Since the occurrence of a large nationwide epidemic of JE in 1949, it has been recognized as one of the major subjects to be investigated from the epidemiologic point of view [3].

a) Seasonal occurrence: Beginning in July and reaching a peak at the end of August, the disease tends to disappear in October. In general, the rainy season continues from June to the middle of July, changing into the hot and dry season in August. The weather becomes cooler until October. Such climatic conditions in Korea seem favorable for the propagation of the mosquito vector of JE, *Culex tritaeniorhynchus.*

b) Geographic distribution: Although many reports revealed that the occurrence or latency of JE had been considered in the whole Western Pacific region as far north as the 39th parallel, the regular seasonal epidemic of JE was reported in Korea, Japan, Okinawa, Taiwan, mainland China, and Guam. The JE epidemic in Korea took place later in the season than in the other areas mentioned above; also the epidemic duration was the shortest with the highest morbidity. It is more prevalent in the southwestern part of Korea along the borders of the Korean straits and the Yellow Sea. While JE can be found in all areas of Korea during the epidemic, higher morbidity is usually found in Honam and Youngnam districts. Located on the Honam Plain, Jeonra Bug Do province especially has been considered the endemic focus of JE in Korea. This province is mostly covered with rice paddies, and strangely enough the average air temperature during winter is little higher than that of the Youngnam District. In the central parts of the province, the Mangyeoung river runs for more than 50 km

from the reservoir, forming a network with the irrigation ditches. But Youngnam District tends show a heavier endemicity pattern than in Honam District in recent years.

c) Age distribution: In Korea approximately 90% of all JE cases have been found among the nonimmune group under 15 years of age. The morbidity was highest among these 4 – 6 years of age, and lowest among those 2 – 3 years and 7 – 10 years of age.

d) Investigation of JE in Korea: Even though scientific reports were initiated in 1935 by Shiba and Chun, active investigations and preventive measures against JE have been conducted by the ROK Government since 1951, and the public has been interested in the disease since then [3]. Thereafter, preventive measures have been improved with close cooperation between clinicians and the government. Rodents, some birds, swine or other domestic animals, frogs, and snakes have been identified as natural reservoirs for the disease.

Although the mortality rate has been lowered lately as compared with the epidemic in 1949, the actual number of patients is not declining by virtue of improved diagnosis.

e) The vector: The wintering of *Culex tritaeniorhynchus* Giles (=*C.t.*) is a main cause for the distribution. In temperate countries, like Korea, Japan, and maritime Siberia, the primary vector of JE is believed to overwinter as the adult and to transmit the virus during the warm part of the year. The reverse is the case in the tropics and to a lesser degree in the subtropics, where *C.t.* is present and active throughout the year.

Knowledge regarding the manner whereby *C.t.* overwinters is essential for an understanding of the fate of JE virus or at least the way *C.t.* repopulates in the Korean peninsula early spring and in summer. It is well known that *C.t.* hibernates at the adult stage in Japan, and it has also been proven by the WHO Vector Ecology and Control Research Unit (VECRU) that this mosquito does not overwinter in Korea in the larval stage. Searches for overwintering adult mosquitoes have unfortunately been negative so far in the Korean peninsula [61]. Despite continuous efforts for 5 years, the VECRU team was unable to collect any overwintering specimen of *C.t.* The earliest capture of *C.t.* was reported on the 7th of April in Busan. Other species of mosquitoes were collected, often in appreciable numbers [61]. It is, however, assumed that *C.t.* overwinters in the adult stage and that oviposition commences in late spring. A severe or mild winter would influence the number of females that successfully overwinter. Such geographic and climatic conditions might be favorable for the propagation and overwintering of the mosquito vector of JE. However, the importance of the geographic factor has not yet been firmly established since the 1974 statistics indicate yet another geographic pattern (Table 19).

The number of cases and deaths by year shows a chronologic fluctuation (Table 20).

However, it is a moot point whether this factor or a wet mild spring is the most conducive to initiating the

Table 19. Number of Japanese Encephalitis Cases and Deaths by Province in 1974 {2}.

City or province	Cases	Deaths
City of Seoul	47	4
Busan city	1	–
Gyeong gi Do	35	2
Gangweon Do	3	1
Chungchong Bug Do	6	–
Chungchong Nam Do	12	–
Jeonra Bug Do	6	–
Jeonra Nam Do	5	2
Gyeongsang Bug Do	5	–
Gyeongsang Nam Do	1	1
Jeju Do	–	–

Table 20. Yearly Incidence and Death from Japanese Encephalitis {3}.

Year	Cases	Death	Morbidity per 100 000	Fatality
1965	752	284	2.6	37.7
1966	3597	957	12.3	26.6
1967	2691	810	8.9	30.0
1968	1226	396	4.1	32.3
1969	67	13	0.24	17.1
1970	254	54	0.80	21.2
1971	372	43	1.16	11.5
1972	337	83	1.03	24.6
1973	769	157	2.32	20.4
1975	117	3	0.3	2.5
1976	30	0	0.1	0.0
1977	101	7	0.3	6.9
1978	41	2	(?) [a]	4.8

[a] no census

summer population build-up. Temperatures must therefore be considered when regarding overwintering and the spring cycle.

It would certainly be easier to find hibernating females on Kyushu Island than in Korea because of the much higher density of the overwintering population. It would also be interesting to know if *C.t.*, a species known to fly considerable distances, could migrate every year into Korea from southern countries and reintroduce the virus. Although improbable, this hypothesis should not be fully rejected, since *C.t.* hibernation has not been proven in cold continental Asia. Both factors could be combined, hibernation and migration without contradictory reasons. However, a problem arises whether or not the females of *C.t.* keep the JE virus during winter. Not all the nulliparous females hibernate even if the majority does. The chance of thus preserving the virus seems very small, although under experimental conditions, *C.t.* females infected with the virus in autumn can overwinter successfully and transmit the virus to susceptible pigs in the spring [62]. Over 60 000 overwintered females were examined for the virus in 1965 – 1973 in the Nagasaki area without success [63]. The JE virus isolated in February 1973 from *C.t.* females in Amami and Oshima Is-

lands, situated on the fringe of the subtropics, does not mean feeding activity of the mosquito ceased.

From observations and theoretical calculations [64], the number of generations has been estimated to vary in temperate climates from 7.7 in Kagoshima, 7.4 in Nagasaki, and 2 in Vladivostock. In Korea, the number of generations has been estimated about five in Sintaien, Jeonra Bug Do, and probably only four in Seoul, but that number can vary slightly from year to year [65]. The question arises whether the mosquitoes from northern places like Seoul or Manchuria really hibernate on the spot or do they migrate from the subtropical zone or neighboring warmer countries, such as Japan or China, each spring? It is extremely difficult to answer this question. Southwest Korea is characterized by a mediterranean type of climate, often called subtropical, but more properly classified as warm temperate because of its evergreen trees (*Lauretum*), the presence of ivy (*Hedera rhombea*) and shrubs like *Camellia*, and a year average of 14 °C.

If there is a southern overwintering, there is nothing to oppose the hypothesis of a quick northern repopulation and rapid building up of the population. It would not exactly be a migration north, as in the case of the numerous butterflies and moths which every year repopulate Great Britain from breeding grounds in the mediterranean region, but a spot-to-spot reconquest of ecologic niches left unoccupied by the destruction of the species in November – December. This is purely hypothetical and only a marking of mosquitoes would provide adequate proof. However, the building up of an enormous population of certain species occurs much quicker in the arctic zone. Finally, overwintering of *C.t.* in the northern areas is highly probable, but it has not been proven.

On a practical note, whether it is hibernation or repopulation south-north, there is no point using larvicides at spring time to stop the future rapid population buildup. Experiments made in Japan on a small island have shown that only summer treatment is efficient, and spring larviciding does not stop a quick recovery of the mosquito numbers during the transmission season. Now research should be done in Korea in the extreme south in areas with heavy breeding, e.g., Kohung, and particularly in the southwestern islands, using perhaps new techniques (such as sweeping nets or D vac) during warmer weather than was the case in the survey in March 1974 [61].

8. Korean Epidemic Hemorrhagic Fever (ICD 065)

a) Characteristics: Epidemic hemorrhagic fever has not been reported previously in Korea, but its epidemicity and pathogenesis suggested the possibility of endemic disease in Korea even before 1950. It most likely was overlooked because of both lack of knowledge and interest and its rare occurrence depending on a special ecology [66].

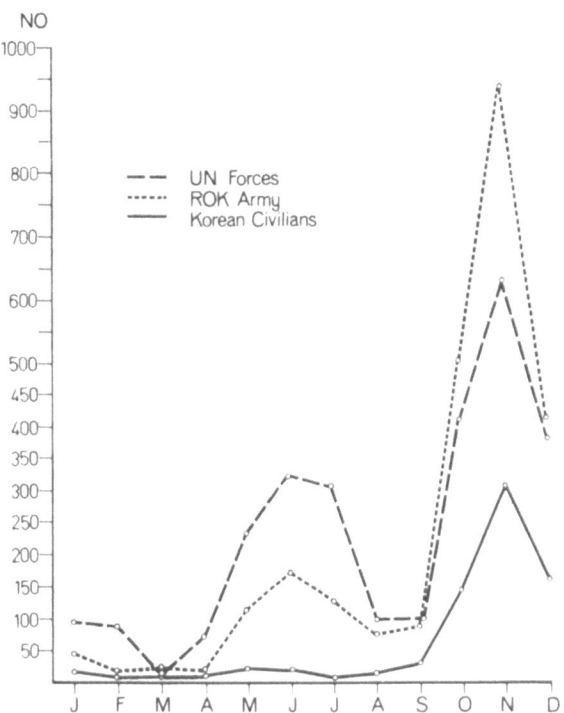

Figure 13. Incidence curve of Korean hemorrhagic fever (1951 – 1971) (Source: S. W. Kim and C. H. Chun [66])

From the outbreak of the Korean War in 1950 until the armistice in July 1953, Korean hemorrhagic fever became one of the military diseases of major concern in Korea (Fig. 13). Since 1965 hemorrhagic fever has been reported to occur among civilians residing in the southern part far below the cease-fire line. In 1971 the number of civilian patients exceeded that of military personnel. The endemic area has gradually extended southward, at present time including almost all parts of South Korea except the islands.

This disease has now created a sensation because of its high incidence, relatively high fatality rate, unknown causative agent, and transmission route; it is of great sociomedical concern.

Korean hemorrhagic fever is an acute infectious disease characterized by functional and morphologic vascular disorders similar to other Far Eastern types of hemorrhagic fever. Plasma transudation, proteinuria, petechia, hemorrhage, and hypotensive shock are the main pathologic features of this particular disease. Around 70% of all cases fortunately take a mild course, reaching the convalescent stage within 2 weeks of onset. The remaining 30% suffer several complications, such as shock, bleeding, renal failure, electrolyte imbalance, pulmonary edema, and bacterial infection, all of which require therapeutic management. Fatality was measured from 15% to 33% in the past periods (1951 – 1954) when severe cases were mainly managed at clinics, but it has decreased to 5.1% to 8.6% nowadays (1975) with more logical treatment based on the pathophysiology of this particular disease.

In spite of intensive research, the causative agent has not been successfully isolated nor has the natural reservoir been completely confirmed. As regards the mode of

transmission, different epidemiologic viewpoints presume the etiologic agent to probably be a virus carried by wild rats [67]. But the disease can be transmitted from man through blood and urine in the early stage. An ample elaboration of the disease pattern is unavailable because of the lack of confirmative serologic tests. A few hypothesis were proposed in the past based on unestablished serologic epidemiology.

b) Endemic areas and occurrence status: In 1951 – 1953 before armistice the endemic area was limited to the Demilitarized Zone between South and North Korea. This area is 50 miles in width; the endemic foci was localized in the central part of the frontline. At present the endemic area extends over the whole of South Korea. The northern central part of South Korea (Gyeong-gi Do and Gangweon Do) has been recognized as the endemic focus.

During the Korean War (1951 – 1953) the incidence of epidemic hemorrhagic fever was high in the United Nations' allied forces residing in the central part of the frontline of the Korean conflict. However, its incidence among the personnel of the Allied Forces has declined since the armistice in July 1954 and has gradually increased in the Korean army, which gradually assumed the combat position of UN forces. Thus, hemorrhagic fever has become a disease of major military concern in Korea.

Since 1966 the incidence of hemorrhagic fever has been reported even in civilians living in the southern part far below the frontline. In 1971 the epidemic incidence of hemorrhagic fever was reported among civilians in almost every part of South Korea. That this incidence exceeded that of military personnel has caused considerable sociomedical concern.

Through 1951 to 1968, the areas of heavy occurrence among military personnel were confined to Yoencheon, Cheolweon, and Pocheon. In 1970 when the Korean army took over the whole defense line the area was extended to Paju, Hwacheon, and now patients are reported nationwide.

Although almost all of the cases occur sporadically (90%), epidemic outbreaks may occur when a number of susceptible persons move into the endemic areas. An outbreak shows a simple curve without secondary cases. In a limited endemic area the entire course of illness usually ranges between 8 and 16 days from the first case to the last, but may last as long as 27 days in special cases. The incubation period is believed to last 23 – 27 days, as was observed in a civilian who went to an endemic area to visit an ancient tomb. Person-to-person transmission does not occur.

c) Seasonal incidence: The incidence of hemorrhagic fever has been reported year round, although two peaks seem to occur in the incidence curve, namely late spring (May – July) and late fall (October – December). In military personnel stationed in highly infectious areas (endemic focus area), the main peak was the second peak of "the late fall." However, analysis of the incidence of the civilian group residing in highly infectious areas (newly established endemic areas), shows that there is only one peak (late fall) in the incidence curve. This is similar to the incidence curve observed in northern Manchuria and Asian Russia.

The present overall incidence curve in Korean hemorrhagic fever shows a monophasic figure, beginning in October and ending the following January with a peak in November, thus differing from that seen during the Korean War or early postwar period (Fig. 13).

The late spring peak began to appear in 1975 and 1976 (from 47 and 54 cases, respectively, in June and July in the Korean army) and was presumed due to heavy ranger exercises necessitated by recent tension in the DMZ.

d) Sex, age, and racial incidences: It is commonly accepted that no sex-age difference as regards susceptability to the disease exists among civilians; nor does racial difference play a role among UN forces.

In neighboring areas of the frontline of the cease-fire zone, considered highly infectious and populated mainly by both males and females who farm vigorously together, the incidence of the disease was the same among male and female civilians. However, in those areas far from the frontline, lightly infectious areas (newly established endemic areas), the incidence of male to female ratio was 2.6 – 4.2 : 1. The increased incidence in males was ascribed to the fact that the male had more chance to be exposed to the infectious agent from the viewpoint of human behavior. This is analogous to the decreased incidence of younger and older age groups in whom physical activity is relatively limited and chances of exposure to the infectious agent seem to be fewer.

In the civilian group performing vigorous activities 70% – 80% of the patients fell in the 21 – 50 years age group, who were also ascribed an increased chance of exposure to infection.

No racial difference (among UN forces) was determined as regards susceptibility to the disease. Also the morbidity rate is similar without being influenced by specialty, duty, or arm of the army among the UN troops. Paul and McClure [68] found no difference in morbidity rate between various groups or races in the UN troops. If a difference were to occur, it would result from a difference in location (heavily or lightly infectious area).

e) Occupational incidence: The jobs of 312 civilian patients [239] were analyzed: 227 farmers, 28 teachers and students, 19 businessmen, 13 public officers and clerks, 8 laborers and carpenters, 6 dressers and technicians (factory workers), 2 miners, 1 policeman and 1 fireman, 1 driver, and 6 unemployed.

Two facts deserve emphasis. First, the incidence rate of citizens with jobs such as businessman, public officer, clerk, dresser, technician, driver, policeman, and fireman is now relatively high although it was not in the past. Second, the increasing number of patients among citizens living in urban areas such as Daegu, Cheongju Gangreung, Weonju, and Incheon calls for attention.

f) Prognostic factors: The following factors contribute to the severity of the disease and influence the prognosis

unfavorably: delayed medical care, rough transportation to the medical facility, prolonged high fever, severe dehydration with metabolic acidosis, early excessive fluid administration, prolonged or recurrent hypotension, persistent hemoconcentration, prolonged anuria, progressive severe electrolyte disturbances (particularly hyperkalemia), deep coma, prolonged severe vomiting and abdominal pain, marked face flushing with mental confusion, and irritability even in the early phase.

In the past the main causes of death from Korean hemorrhagic fever have been acute renal failure, shock, and pulmonary edema with bleeding. Intensive care and better management for shock and acute renal failure could reduce the fatality rate from 15% to 6% among military personnel and from 15% to 10% among civilians. The major problems at present in managing these patients are how to treat pulmonary edema with bleeding systemic infection, gastrointestinal bleeding, and CNS involvement. Therefore, these manifestations are of great prognostic significance. The causes of death were analyzed in 19 fatal cases of 370 patients with Korean hemorrhagic fever who were admitted to the KHF research center during the 12-month period from January to December 1975 [66]. The case fatality showed 5.1%. In the febrile phase, 1 (5.3%) of 19 fatalities was dead on arrival from a rapid clinical course, and the direct cause of death was not obtainable clinically. In the hypotensive phase, the direct causes of death in 7 (36.8%) of 19 fatal cases were shock and acute pulmonary edema. In the oliguric phase, 7 (36.8%) of 19 fatal cases died.

g) Epidemic characteristics: In 1954 Gauld and Craig [69] proposed certain features to be characteristic of epidemic hemorrhagic fever although the causative agent remained to be found and the transmission route was still hypothetical. Now in 1976 it seems reasonable to revise these as follows. The suggested revisions are in parentheses [66]:

1. To produce the distribution of cases found in epidemic hemorrhagic fever, the vector or vehicle of transmission must possess the following characteristics:

   a) Wide distribution throughout the rural sections of the endemic area (although the main outbreak is in rural areas, the disease occasionally develops within the urban areas).

   b) Greatest activity during the dry spring and autumn months with some activity throughout the year (activity during the spring could be excluded now that conditions differ from those during the Korea War).

   c) Limited range of movement (rapid extension of endemic area suggest that the movement of vehicles is rapid and its range may be wide).

   d) Infectiousness only in localized foci (infectiousness is not localized but expanding).

   e) Ability to infect human beings for only a short interval of time from a single focus (ability to infect human beings increases for a short time, and disease activity remains in the area and tends to reside there).

f) Ability to infect human beings without leaving evidence to indicate bite and mode of transmission.
2. Person-to-person transmission does not occur.
3. Company mess and water supplies are not involved.
4. No specific type of terrain can be implicated.

9. Viral Hepatitis
(ICD 070)

Only after the Korean War this disease was recognized as being infectious. Previously it had occurred in every part of the country in sporadic or small epidemics and had been called catarrhal jaundice or simple jaundice. Since this disease had not been a notifiable disease up to now, there is no way of finding out how many cases occurred annually. According to an investigation in 1960 by Chun and Lee [3] eight cases (2.8%) broke out among 285 persons including doctors, nurses, clerks, and other adults working at Seoul National University Hospital. According to other data obtained from two

the first case among Koreans was recorded in 1965 [72]. A 12-year-old boy complained of throat pain and enlarged lymph nodes. Heterophile antibody was demonstrated. Since then one case of a 31-year-old female [73] and another of a 10-month-old male baby [74] were reported. The number of cases is expected to increase even though no further scientific achievement regarding the etiology and epidemiology concerning Epstein Barr virus is forthcoming.

11. Trachoma
(ICD 076)

Approximately 11% of patients visiting eye clinics before 1945 were trachoma patients, but this percentage has been gradually diminishing with improving living conditions and effective antibiotic treatment for trachoma. The monthly distribution of patients on the average during the period 1938 – 1948 was as follows (Annual Statistics, Ministry of Health and Soc. Affairs, 1961):

	Jan.	Feb.	Mar.	Apr.	May	Jun.	Jul.	Aug.	Sept.	Oct.	Nov.	Dec.
% of Trachoma patients	7.6	6.3	8.7	9.9	7.1	5.2	8.3	7.3	7.5	6.4	4.8	6.6

university hospitals and other clinics, the percentage of patients suffering from this disease was 0.3% – 0.8% (1959 – 1960) of infant outpatients and 0.36% – 1.11% in adults. Also, according to this report, a small epidemic of this disease most probably occurred in Seoul in 1960. The only data with which we can estimate the monthly occurrences were those supplied by Chun and Lee [3]. These show that viral hepatitis occurs any time of the year, the occurrence being a little higher during the autumn and early winter. The incidence is more remarkable among small children. The age distribution of patients shows that there are very few cases under 1 year of age (the youngest was a girl of 8 months). More cases occur between 2 – 10 years and are remarkably numerous in the 2 – 5 age groups. In adults, the occurrence was high in the 21 – 40 age group, and the highest was the 31 – 40 group. The occurrence was generally greater in males than in females. However, in 1959 the male to female ratio of hospitalized patients in a university hospital was 6 : 7.

The fatality rate among hospitalized patients was 1.5% – 2.0% in infants and 1.8% in adults. Since we may assume that only severe cases were hospitalized, the actual fatality rate of this disease my be considered actually below 1%, the value reported in other countries. Though the susceptibility to this disease is greater in small children, thus more cases occurring among them, these cases are usually milder than those of adults and have a quicker recovery.

10. Infectious Mononucleosis
(ICD 075)

Sporadic occurrences were reported among foreigners during the Korean War in 1950 and after. Nevertheless,

The occurrence was higher and usually more serious in females. As regards age, the occurrence was highest in the 11 – 15 age group and tended to decrease subsequently, only to rise again in old age.

Epidemiologic peculiarities are observed in Korea. For example, among orphans, who live together in the most unsanitary conditions, mass occurrences of trachoma showed a high rate of 10.1%, in striking contrast to the 0.6%, 0.3%, and 0.2% of the primary, middle, and high schools in the cities where sanitary and living conditions were favorable.

One of the major reasons for the relatively low occurrence of trachoma in this country as compared to other countries in the Far East is the favorable natural conditions of weather and climate. The high rate of occurrence in the Manchurian and Mongolian regions before 1945 was probably due to the unique yellow dust caused by the continental climate, although the human factors of living conditions in those regions must have undoubtedly played a role.

12. Influenza and Common Colds
(ICD 487)

Influenza has been recognized as *Dok-Gam* (toxic cold) in Korea, and in the majority of cases is differentiated from the common cold without isolation and identification of the virus from patients.

Annual occurrence of cases and deaths: Since influenza and common-cold-like diseases have not been notifiable diseases in Korea, there is no way of finding out the actual number of occurrences. The estimated number of cases of epidemic Asian flu (A2 type), which prevailed during June – August 1957, was 5 – 6 million, while the occurrence of 3 – 4 million cases was estimated for the B

(B1) type influenza epidemic prevailing during April to June 1961. During these two epidemics the number of fatal cases was considered extremely small. The 1957 epidemic showed a high incidence rate among the younger population (the highest being between 21 to 25 years) and a relatively low rate in the age group above 60. In the 1961 epidemic, however, no remarkable difference in susceptibility among different age groups was observed. The difference of incidence by sex was not remarkable [3].

Geographic distribution: In recent years big epidemics with A2 variant have occurred: 2 – 3 million in 1965 and 3 – 4 million in 1972 have been estimated. It is difficult to find any regional differences in occurrence in the above two influenza epidemics. The occurrence seemed to depend very little on the density of population.

Epidemiologic peculiarities: Influenza is not believed to have epidemiologic characteristics. There were epidemics in this country in the past by A, A1, A2, and B, and the swine-type or other similar antigenic influenza viruses, the periods of prevalence were similar to those of other countries. In other words, the antibody-carrying population for the swine type of influenza was most numerous among the age groups above 60 who spent their infancy or childhood during the period of 1918 – 1920; for the A type (especially the A FMIO), among the 10 – 20 year age group who were very young during the period of 1946 – 1951, when that type of infection was prevalent; for the A PR8 type, among the 21 – 30 year age group, who spent their childhood when that type was prevalent, namely 1934 – 1946; and for the A2 type (Asian type), which began to prevail in 1957 and is still prevalent, among the 1 – 9 year age group. According to season, epidemics through A2 and A2 variant occurred in the winter and early spring. No clear-cut difference in incidence according to sex was noted in Korea, but the disease is more common in ages under 20 years, especially among young children. Fatality is very low unless complications such as pneumonia or heart trouble are present [3].

IV. Rickettsioses

1. Louse-Borne Epidemic Typhus
(ICD 080)

The first scientific record of an outbreak of rickettsiosis in Korea was described in 1911 in the Health Statistics [3]. Subsequently one big epidemic occurred in 1918 with 8123 cases, and this was followed by outbreaks of varying intensities. Another vast epidemic raged in 1951 during the Korean War, tallying-up 32 211 cases.

However, for several reasons, such as improvement in sanitary practices, effective insecticides, vaccination, and medicine, the occurrences decreased rapidly, and large outbreaks have almost disappeared, although a few cases have been reported yearly for several years since 1952.

Another outbreak occurred in an orphanage in Yeong-dong-po, Seoul in April 1957, in which 37 of 239 orphans were infected [76]. According to the literature, this may have been the last reported outbreak in Korea. Serologic tests [77] reported that 45% or more of 15-year-olds and older showed positive titers indicating that a fairly large number of latent and subclinical cases were probably involved. High occurrences were observed during March – June; and the lowest were in August to October.

Although the massive application of DDT during the Korean War altered seasonal characteristics, the fact, nevertheless, remains that lice ultimately became DDT resistant. The youth group (20 – 24 years of age) showed higher morbidity than other groups. The occurrence was more frequent in slum areas where beggars, laborers, and other poor people were living in an unsanitary environment. However, fatality was higher among the clerical occupations and medical personnel [3].

During the period 1912 – 1941 fatality was 11% – 30% of 20 605 cases. Antibiotics have greatly reduced lethality; for example, 306 cases occurred in an army camp during the Korean War, but due to application of chloramphenicol and Terramycin only two patients died.

2. Scrub Typhus
(ICD 081.2)

Even though scrub typhus was reported among U.N. Army soldiers stationed in Korea, this disease has not yet been observed among Koreans. Nevertheless, the following facts may support the possibility of its occurrence in Korea. *Rickettsia tsutsugamushi* was isolated from rodents (*Apodemus agrarius*) near the 38° N parallel zone. The pathogenic agent was isolated from trombicular mites (*Trombicula pallida*), which are considered the vector of the disease, and the presence of another trombicular mite, *T. scutellaris,* was confirmed [3]. In reports from U.N. Army soldiers, scrub typhus cases were found in districts near the 38° N parallel zone and also in Masan, Gyeongsang Nam Do. It is presumed that the disease may occur anywhere in the country due to the nationwide distribution of a possible vector, *Trombicula scutellaris.*

3. Murine Typhus
(ICD 081.0)

About 31 cases were verified as murine typhus [75] by the isolation of *Rickettsia typhi.* This disease is rather sporadic, and its mild symptoms may very often be confused with epidemic typhus or typhoid fever. It is more prevalent in males than in females; of these 31 cases 23 were male. At the present stage, no epidemiologic study has been conducted in Korea. However, the species of rat clearly acts as a definite reservoir of murine typhus, and at the same time *Xenopsylla cheopis,* which infects them likewise plays a role as a vector of the mode of disease transmission.

V. Other Arthropod-Borne Diseases

1. Malaria
(ICD 084)

History records that the mother of King Sejong died from malaria in 1420. Although no scientific statistics are available, malaria, especially vivax malaria, was prevalent in Korea from olden times up to World War II. It is popularly known as Haru-geori (every-otherday fever attack) or *Chare-Bae* (enlarged spleen). Until World War II, malaria was recognized as one of the unavoidable diseases in the summer months. However, with the advent of effective insecticides, its prevalence has drastically diminished. Now it is limited almost entirely to some areas in the Taebaeg mountains. In the meantime, a pre-eradication survey was conducted in a joint ROK-WHO action in July – October 1959. The survey covered 37 *Guns* (counties) in all nine provinces of the country. A total of 22 005 children, 1 – 14 years of age, were found to have an average spleen rate of 1.1% in a range of 0.0% – 3.8%. Spleen enlargements were found scattered over the whole country; no distinct areas with high spleen rates could be determined. During the survey, 16 498 blood samples were taken. A total 17 cases of vivax malaria were confirmed microscopically. All 17 positive smears were detected from 531 smears collected in Yeongju Gun of Gyeongsang Bug Do. Five of nine villages surveyed in this Gun were found to be malarious, with a parasite rate of 3.2%. In addition, four positive falciparum malaria infections were detected among 560 narcotic addicts at the Narcotic Hospital, Seoul, and two cases of vivax malaria among the Turkish troops stationed in Korea.

Plasmodium vivax is the most common species of malarial parasite found in Korea. It is a subtropical type according to Gill's pattern and endemic in nature. During 1961 a total of 13 000 reports with corresponding blood smears were examined at the Central Malaria Eradication Service of the Ministry of Health and Social Affairs; of these 5206 positive *Pl. vivax* cases were confirmed [145]. Throughout 1962 to the end of October, a total of 6369 blood smears were examined, and 2575 positive cases were confirmed. During these 2 years, the number of positive cases found in the northeastern area including the four provinces of Gyeongsang Bug Do, Gangweon Do, Gyeong-gi Do, and Chungcheong Bug Do comprised 97.3% of all positives. This fact suggests that almost all localities in the southwestern (plain) area are nonactive and retrogressive in malaria occurrence, where malaria transmission is low. In contrast, the northeastern (mountainous) area has a rather large number of residual foci where malaria transmission is high. According to entomologic survey findings, anopheline mosquitoes in Korea have a mainly zoophilic (*A. sinensis*) habit, and since its man-biting rate is rather low, it would cause a low infection rate of malaria. Of 7517 dissections of *Anopheles sinensis* only one was found to be

naturally infected with the sporozoites in August 1962 for the first time in Korea [276]. In view of these facts, the abundance of anopheline mosquitoes does not always seem to coincide with incidence of malaria. The case incidence obviously varies from Myeon to Myeon in a *Gun* (county), and it also differs from village to village even in one Myeon (township). A malarial focus usually comprises a village, in which there are breeding places. These are apparently persistent and localized pocket foci in the northeastern area, thus establishing a localized endemicity.

Pl. malariae were first reported in 1929 in Chungcheong Nam Do and since that time, several sporadic cases of quartan malaria were found in the same province and in Gyeong-gi Do. In the Narcotic Hospital, Seoul, 16 of 1404 narcotic patients were infected with *Pl. malariae*, but none have been reported in recent years [148]. The epidemiologic picture of the *Pl. malariae* in Korea is still an enigma.

Pl. falciparum was first reported in 1936 among typhoidlike cases in Seoul. Positive cases [18] of falciparum malaria were found among 1959 blood films from narcotics patients in the period August 1960 – December 1961. Infections of *Pl. falciparum* found among narcotics patients are recognized as mainly due to accidental infection by the *Plasmodia* through use of contaminated (incompletely sterilized) instruments, such as syringes or needles. The transmission of falciparum malaria could not be accomplished by the mosquito vector due to the bionomics of the anopheline mosquito and climatic conditions in Korea. A temperature of 30 °C, optimal for the cyclic development of *Pl. falciparum* in the mosquito, occurs only for a brief season in Korea, and the length of a cycle is 10 – 12 days at 30 °C. But the average temperature in August, the hot month, only ranges 25.0 – 27.5 °C in the southern part of the land. In 1964 – 1965 *Pl. falciparum* was found among refugees from the Vietnam War, however, with effective treatment and control measure no further spread occurred [3].

The species of Korean anopheline mosquitoes are as follows:

A *sinensis* Wedemann, 1928
A. *lesteri* Baisas and Hu, 1936
A. *sineroides* Yamada, 1925
A. *koreicus* Yamada and Watanabe, 1918
A *koreicus edwardsi* Yamada, 1925
A. *lindesayi japonicus* Yamada, 1918

Of these six species *Anopheles sinensis* has with the greatest density of nationwide distribution; A. *sineroides* is second. The former is believed to be the most likely vector of malaria in Korea. Korean anopheline mosquitoes are in principle zoophilic, but with very low human blood rate. H. J. Lee (1967) in National Institute of Health reported the anthropophilic and zoophilic rates of A. *sinensis:* human 1.7; cattle 54.8; pig 42.5; and others 1.0. Their breeding places are rice paddies and clean streams. Domestic animal sheds are favorite resting places for A. *sinensis,* but there are few mosquitoes inside homes (Table 21).

Table 21. Collection of Anopheline Mosquitos in Korea
(1960 – 1962) {276}

Locality	Place collected	A. sinensis	A. sineroides	A. koreicus
Yeongju	Stable	1514	6	287
	Room	73	0	0
	Outdoor	2	0	0
	Total	1589	6	287
Andong	Stable	689	231	66
	Room	12	2	1
	Outdoor	141	105	13
	Total	842	339	80
Yeoju	Stable	434	2	2
	Room	687 *	0	0
	Outdoor	3785	0	3
	Total	5086	2	5
Total	Stable	2637	239	355
	Room	952 *	2	1
	Outdoor	3928	106	16
	Grand total	7517	347	372

* One positive was found by salivary gland dissection

As regards seasonal prevalence of the mosquito, it begins to appear from the end of March or the beginning of April and starts its overwintering at the end of October, the maximum density occurring during June or July.

2. Relapsing Fever
(ICD 087)

In 1914 14 cases in Seoul comprised the first reliable report of this disease in Korea. In 1940 relapsing fever was proclaimed a notifiable disease in Korea; it prevailed at one point during the period 1940 – 1959 with cases totaling 8356 and 564 deaths (6.7%). However, it is now very rare even in clinics. This reduction might be due to extensive delousing measures with DDT and lindane spraying.

Morbidity was highest among children and young adults (10 – 30 years) and persons living in poor sanitary conditions. Due to effective use of antibiotics, the fatality rate is now approaching zero. As regards the seasonal prevalence of the disease in the past, it began to appear in December and reached a peak during April and May, after which it tended to decrease by October or November.

VI. Venereal Diseases

a) General outline: Venereal diseases (VD) include syphilis, gonorrhea, soft chancre, lymphogranuloma venereum, and granuloma inguinale, all of which are transmitted by sexual contacts.

The status of venereal diseases including its morbidity in Korea can hardly be estimated either in the past or present because of the absence of accurate statistics [118, 122]. Just before 1945 the morbidity was comparable to that of neighboring countries, and gonorrhea, syphilis, and chancroid were the most common venereal diseases in Korea. Lymphogranuloma inguinale was seen occasionally, but only five cases had been reported by Seoul University Hospital, and no case of granuloma venereum had been reported in all Korea. Estimated numbers of venereal disease per 1000 population at that time were 20 cases of syphilis, 35, gonorrhoea, and 15, chancroid; in 1961 these were 5.6 and 0.3, respectively. It is widely accepted that the highest prevalence of venereal diseases probably occurred around 1945, and the incidence tended to decrease subsequently until the outbreak of the Korean War in 1950. During the Korean War, the incidence of venereal diseases increased for a while due to the social confusion created by the war. Without doubt the number of venereal diseases is much less at present than in the past, possibly due to the use of new drugs such as sulfadrugs, penicillin, and other antibiotics [107, 108, 109, 110, 111, 112, 113, 114, 115, 116, 118, 119, 120, 121, 122, 123].

Since the Korean War no lymphogranuloma inguinale or granuloma venereum have been reported. Moreover, congenital syphilis, ophthalmia neonatorum, and gonorrhea in children, so prevalent in the past, are also gradually disappearing nowadays.

Although venereal diseases have been made notifiable communicable diseases, this is entirely disregarded. There are 109 VD clinics, either governmental or nongovernmental, throughout the country. Monthly reports of these clinics to the Ministry of Health and Social Affairs are not fully reliable. Moreover, reports of new cases and contact cases are not to be expected under the present situation. The reform of regulations on VD is imperative.

Nevertheless, the statistics of 1975 by MHSA/ROK indicate that of 1 052 152 patients examined VD numbered 3129 cases of syphilis, 35 997, gonorrhea, 537, chancroid, 9256, urethritis, and 16 225 others. In general, the patients visit private practitioners who are not well trained in diagnosis of treatment. Very few of the 142 dermatology specialists and 192 urology specialist are qualified in VD. The Ministry of Health and Social Affairs carries out educational activities including the distribution of posters and leaflets, but its activity is exclusively limited to the urban areas.

Although prohibited by law, prostitutes are still numerous in urban areas. Entertainers or dancers are supposed to be the most dangerous haborers of VD, and yet most of them are left without any control measures. The growing formation of resistance against penicillin is one of the problems in using antibiotics, but this does not impose any serious problem in controlling VD at present. However, it is recognized that nongonococcal urethritis is increasing gradually in spite of the reduction in the incidence of gonorrhea. Exchange of health certificates at marriage could be an effective method of VD control.

b) Socio-economics: The sociomedical study of prostitutes by Wang [122] revealed that their average age was

25.5 years and average number of years of education was 6.8, implying the completion of primary school. Living conditions were generally poor; for example, in Pyeong-Taeg 37% of the girls did not have running water and depended on well-water. Data on the toilet facilities revealed that 78% of the prostitutes living in the control area in Wanweoldong, Busan were still using traditional Korean privy style toilets, and in Pyeongtaeg also 91% were using privy toilets. Over two-thirds of the prostitutes only bathed once or twice a week. Therefore, improvement of living conditions was considered a critical problem.

Generally speaking, the girls have some knowledge of syphilis and gonorrhea, and the measures necessary to prevent these diseases. However, over half of the girls have abused antibiotics to treat as well as to prevent disease, and only 65% of them required their partner to use a condom. Therefore, much improvement is needed in this area.

A survey of knowledge about contraceptive measures showed that 96% of the girls had good knowledge about oral contraceptives, 90% knew about condoms, 88% had knowledge of the douche method, and 62% use condoms for contraception.

1. Syphilis
(ICD 090 – 097)

Syphilis was introduced into Korea from Peking, China between 1506 and 1521 [124].

In Korea the current status of incidence of syphilitic infection can hardly be assessed accurately because there are no reports of confidential statistical data. However, recent reports on syphilis indicate that syphilitic infection has been gradually increasing in Korea since around 1962 to 1963: 24.6% in 1965 [119], 25.7% in 1966 [120], and 33.5% in 1965 [121]. The data of cases reported from out-patients clinic (OPD) of hospitals [120] indicate that the recent rising incidence of syphilis patients apparently commenced around 1963. Accumulated reports show that the incidence of infections of early syphilis patients has also been increasing in Korea since 1962 [122].

The incidence of syphilitic infection will be reviewed under the following categories, prostitutes, pregnant women, reported cases from OPD, and normal healthy groups. Although the sampling groups of the researchers were different, the seroreactor rates in healthy groups showed high incidence of syphilis. In the serologic study of young, healthy Korean athletes in 1962, Kim and Lew [109] observed that 7.4% of 337 were reactive to the VDRL (Venereal Disease Research Laboratories) test, and 1.5% of 337, reactive to the RPCF test (Reiter Protein Complement Fixation test). A similar study by Kim et al. [114] showed that 1.4% of 1779 high school graduates were reactive in STS (Serologic Test for Syphilis) and 4.7% of 611 youths were reactive in 1965.

The serologic surveys of Army and Air Force personnel showed that in the Army 3.4% were reactive, and in the Air Force, 4.6% [110]. A mass survey of syphilis was carried out in 1966 in the Korean Service Corps [114], and the results showed that 10.9% of 2010 were reactive to the VDRL test, and 5.7%, reactive to the RPCF test.

In the leprosy patient group, the seroreactor rates showed a very high incidence of 18.7%, and 6.9% were reactive to the RPCF test [114]. Although the percentages are lower than those reported in other countries [105, 106], these significantly high seroreactive to RPCF percentages in the leprosy patients' group indicated a higher incidence of syphilis among them than in other groups.

From all available data on the incidence of syphilitic infection in the various groups with their variations, syphilis incidence is again assumed to have increased in recent years in Korea.

Sources of infection: It is well accepted that the spread of venereal disease is closely related to environmental factors such as social, economic behavioral. Many investigators have shown that periodic recurrence of early syphilis is associated with war and the economy [107].

The effectiveness of penicillin therapy for syphilis, together with the early use of penicillin for other infections has, on the one hand, reduced the consequences of syphilis; however, it has also resulted in indifference of the risk of venereal disease infection on the other. Penicillin has not prevented the recurrence of syphilis, present since around 1956 – 1957, even though no evidence has been found that penicillin-resistant *Treponema pallidum* has occurred.

As elsewhere prostitution is a major source of venereal diseases. However, in recent years this feature has been gradually changing. After the Korean War, the cultures of developed countries surged into Korea. The improvement of socioeconomic conditions, industrialization, female emancipation, and the breaking-up of old social patterns followed becoming increasingly important factors in the spread of venereal diseases and other communicable diseases.

Analysis made of the source of infection in female syphilis patients revealed that 90% were prostitutes and entertainers, and 5% were housewives and unmarried girls [112].

Contradicting high percentages of syphilis infection in males through contact with prostitutes and entertainers, a report by Kim et al. [113] showed that 22.2% had had contact with prostitutes 22.2%, with business girls, 27.7%, with female students, and 27.7%, with others. From these data, the source of venereal infection seems to have gradually spread into the general public. Although homosexuality has been emphasized as one of the causes of increasing syphilis in the developed countries [103], data on the extent of homosexuality and its role in the spread of syphilis infection is not available in Korea.

Pregnant women: Few reports are available on the incidence and pattern of syphilis in pregnant women. Studies made at Severance Hospital from 1959 to 1966 showed that the seroreactor rates from syphilis in pregnant women have been continuously increasing since 1963, starting at 1.4% in 1959. Compared with the seroreacter rates in developed countries [107], reactive rate of syphilis in Korean pregnant women appears very high.

Prostitutes: Information on the frequency of syphilis in the prostitute group is based on the results of the STS. These data may not present the true incidence of syphilis, but they can provide certain indicators of trends in syphilis.

Before World War II prostitutes were regarded as a major source of syphilitic infection in developed countries. The situation is now changing, although in a new sense "bread-and-butter" prostitution remains in developed countries [107]. In Korea prostitution is not only a major source of syphilis, but also of other venereal diseases. Public Health and Social Statistics indicated a persistent decrease in the number of syphilis patients among entertainers from 1948 until 1962, when a gradual increase started to appear. Several follow-up studies suggest the continuously rising rate of syphilis in Korea since 1963. With the Korean War, economic difficulty and changing moral standards brought about the development of concentrations of prostitutes around army camps known as "villes".

The problem with venereal disease among prostitutes living in the "villes" and consequently among soldiers has been one which has perplexed health authorities. Research into the problem and efforts to solve it have yielded only fragmentary results thus far.

An extensive research project has been undertaken to determine the root causes of and best methods to prevent venereal disease in the villages around military installations. The "villes" around large military installations were selected as the target of this research. In particular, the Pyeongtaeg "ville", the Gunsan "Silver Town" area, and the "Texas Town" area of Busan were studied. The study took 2 years, from April 1973 to March 1975 [122].

A total of 988 serologic tests for syphilis were carried out. The VDRL method recorded 11.5% reactive cases, Kolmer C – F showed 5.9% reactive, and RPCF test showed a 4.3% reactive rate.

2. Gonococcal Infections
(ICD 098)

Exact data on gonorrhea is not available at the moment in Korea. However, the figures collected from the health center network and some other hospital laboratories by the Ministry of Health and Social Affairs show that prevalences have not changed as expected in spite of the use of effective antibiotics. In 1965 positives numbered 37 366 of 1 438 272 tested and 35 997 (28 656 by slide smear and 7341 by culture and sensitivity test) of 1 052 157 tested in 1974. This suggests that no notable change was achieved during the 10 years.

But in Severance Hospital Clinical Pathology Laboratory, Dr. Samuel Lee (1976 personal communication),

analyzing gonococcal-positive specimens during the years of 1969 – 1973, found yearly increases: 79 (1969), 119 (1970), 125 (1971), 160 (1972), and 185 (1973).

An extensive survey was conducted by Wang et al. [122] in Pyeongtaeg, Gunsan, and Busan areas where military camps are located. The survey subjects were living around the army camps and serving as prostitutes or entertainers. An overall gonorrheal morbidity rate of 14.1% was determined among the 1282 girls studied. This shows a rather dramatic rise compared to other studies. The majority of the increase is attributed to an actual increase in the VD rate, but more precise methods of isolation and culture using Thayer-Martin media also contributed to the increase. In Pyeongtaeg, a gonorrhea morbidity rate of 5.3% (endocervical), the lowest in any of the "villes", was recorded. The Pyeongtaeg policy of enforced hospitalization until cure effected, tended to eliminate contact with infected girls and also make the girls more aware of and concerned about prevention of venereal disease. The „Texas Town" prostitutes in Busan had a 15.5% (endocervical) morbidity rate, the highest of any of the areas, probably attributable to sexual contact with many foreign sailors. In culturing for gonorrhea, rectal as well as cervical cultures were found to be necessary. Rectal cultures revealed 71 cases (39.2% of total positive cases), some of which were negative by endocervical culture, and endocervical cultures yielded 181 positives from the 1282 cases surveyed.

Sex and age distribution: Data from Severance Hospital Clinical Laboratory shows more male cases than female; in 1973 the figures were 139 male and 46 females. It is noteworthy that infants and children 9 – 12 years of age were proven gonococcus-positive from specimens from the genitourinary tract, except for one case under 1 month of age in which the specimen derived from the eyes. Most cases were observed among age groups 20 – 29 and 30 – 39 (Fig. 14).

Penicillin treatment failed to cure 40% of the gonorrhea patients in clinical tests; moreover, 16% of the strains used in the sensitivity tests showed resistance to penicillin. Therefore, penicillin was judged to be rather ineffective against gonorrhea. Ampicillin, however, is still quite an effective antibiotic against gonorrhea.

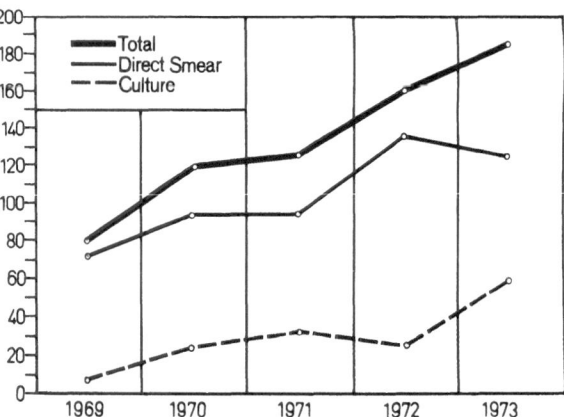

Figure 14. Number of gonococcus positive specimens examined during the years 1969 – 1973 at Severance Hospital Laboratory (Dr. S. Y. Lee) (Ref.: Personal data, not reported)

Ethics: Venereal diseases are directly related to ethical problems. The radical social change of modern industrial and technologic society has also caused radical change in the ideals, sanctions, and techniques of sex and sexual behavior of modern man. The main characteristics of this trend tend to be that of exploitation, lack of religious affiliation, and lack of regard for the sexual partner. The degree of sexual immorality seems to correspond directly to the prevalence of VD. In other words, the increase of VD cases reflects the decline of sexual morality.

Another crucial ethical aspect of VD is its epidemic nature. An innocent wife, infected through her immoral husband has to suffer from VD. Furthermore, an innocent baby has to suffer from the dreadful disease because of the moral failure of its parents. Thus, the problem of immoral sexual behavior and VD is closely related to family relationships, the responsibility for VD in infants resting with the parents. In this way, VD may infiltrate, not only among the delinquent community, but into the deep strata of the nation.

The most important factor in the VD problem is prostitution. Prostitutes may be classified into two groups: registered and nonregistered. The former is obliged to receive medical examination once a month, but the latter is almost defenseless: they practically perform the role of focus in VD. Tourists may contribute as transmitters, now that the tourist industry is booming.

The above factors, intermingling with each other, may accelerate the wide spread of VD among nations in spite of highly effective diagnostic, preventive and therapeutic measures. For this reason, the following items should be considered to eliminate VD:

1. Educate moral attitudes toward sex and sexual behavior in society,
2. develop socioeconomic stability,
3. strict control of entertainers among restaurant and hotel business,
4. detection and treatment of asymptomatic carriers,
5. regulate abuse of antibiotics.

VII. Mycotic Diseases
(ICD 110 – 118)

Mycotic infections, which are prevalent in Korea, can be grossly categorized into superficial and systemic.

1. Superficial Mycosis

Superficial mycosis commonly found among Koreans is dermatomycosis, the agents of which are *Microsporum ferrugineum* and *M. canis.* The fungus causing tinea capitis is *M. ferrugineum. Trichophyton schoenlein, T. rubrum, T. mentagrophytes,* and *T. violaceum* are also commonly found among dermatomycosis cases. Pityriasis or tinea vesicolor results from infection by *Pityrosporum orbicularae* or *P. ovale. Candida albicans* also causes cutaneous or vaginal candidiasis.

2. Systemic Mycosis

The following cases of systemic mycosis have been reported in Korea: sporotrichosis, systemic candidiasis, cryptococcosis, actinomycosis, and aspergillosis. Other systemic cases, such as chromoblastomycosis, subcutaneous phycomycosis, torulosis, geotrichosis, mucomycosis, and nocardiosis are extant in Korea. Increasing communications among nations may enhance the spread of mycotic diseases to neighboring countries. Indeed, mycosis already seems to be increasing yearly for several probable reasons, primarily due to abuse or misuse of the followings: antibiotics, corticosteroid hormones, immunosuppressants, anti-contraceptive drugs, and cytotoxic agents for malignant neoplasms. In addition, tissue or organ transplantation, increase of endocrine diseases, frequent contact with domestic animals, use of chemical fibers for clothing and shortage of anti-mycotic drugs may be reasons for the increase of mycosis in Korea [99].

Flora of the surrounding environment are not to be ignored. Suh [100] examining the aerial fungi in Daegue, collected 19 species, of which hormodendrum (25.1%), penicillium (13.6%), alternaria (10.6%), and aspergillus (7.1%) were predominant. Soil fungi, also collected, numbered 34 species. The commoner species, found even as deep as 20 cm in the earth, were penicillium (22.4%), hormodendrum (9.3%), aspergillus (5.7%), spicaria (5.6%), botrytis (3.8%), and alternaria (3.7%). Thus environmental factors, habitats of the inhabitants, and behavioral factors described above may have enhanced rising morbidity of mycotic diseases in recent years.

VIII. Parasites and Parasitic Diseases

Scientific reports on human parasites in Korea were initiated by Mills [126] and Suzuki [127]. During the past 65 years, 45 species of helminths and protozoa (11 nematodes, 7 trematodes, 7 cestodes, and 20 protozoa) have so far been recorded in the literature. The prevalence of these diseases have been covered on a nationwide scale in several publications [128].

1. Clonorchis sinensis
(ICD 121.1)

One of the main endemic parasites in Korea is *Clonorchis sinensis.* Its endemicity has been recognized since olden times, and considerable data concerning its geographic distribution have been reported from the early part of the twentieth century. Relatively recently Hunter et al. [198] performed a survey by fecal concentration method throughout the country, visiting mainly city areas at random, and found the endemicity to depend on the area. Walton and Chyu [187] estimated, after a skin test survey, that in South Korea alone some 4.5 million of 30 million people were infected by *Cl. sinensis* posing a public health problem of considerable magnitude.

Table 22. The Second Intermediate Hosts of Clonorchis sinensis in Korea {200}

[a] No	Family	Species	Author	Year
1	Cyprinidae	*Abbotina rivularis* (Basilewsky)	Kobayashi	1923
2	Cyprinidae	*Acanthorhodeus asmussi* (Dybowski)	Kim	1961
3	Cyprinidae	*Acheilognathus gracilis* (Regan)	Kobayashi	1924
4	Cyprinidae	*Acheilognathus yamatsute* (Mori)	Kim	1961
5	Cyprinidae	*Acheilognathus signifer* (Berg)	Kim	1961
6	Bagridae	*Coreobagrus brevicorpus* (C & V)	Kim	1961
7	Cyprinidae	*Culter alburnus* (Basilewski)	Nishimura	1938
8	Cyprinidae	*Culter brevicauda* (Gunther)	Kim	1961
9	Cyprinidae	*Gnathopogon koreanus* (Berg)	Kobayashi	1924
10	Cyprinidae	*Gnathopogon strigatus* (Regan)	Kobayashi	1924
11	Cyprinidae	*Gnathopogon maejimae*	Chun	1962
12	Cyprinidae	*Hemibarbus longirostris* (Regan)	Nishimura	1938
13	Cyprinidae	*Hemibarbus labeo* (Pallas)	Kim	1961
14	Clupeidae	*Ilisha elongata* (Bennet)	Lee *et al.*	1958
15	Cyprinidae	*Microphysogobio koreansis* (Mori)	Kim	1961
16	Cyprinidae	*Parapelecus eigenmanni* (Jordan & Metz)	Kobayashi	1928
17	Cyprinidae	*Pseudogobio esocinus* (T & S)	Chun	1962
18	Cyprinidae	*Paracheilognathus rhombea* (T & S)	Kobayashi	1928
19	Cyprinidae	*Pseudogerilampus notatus* (Bleeker)	Lee *et al.*	1958
20	Cyprinidae	*Pseudorascora parva* (T & S)	Lee *et al.*	1958
21	Cyprinidae	*Pungtungia herzi* (Herzenstein)	Nishimura	1938
22	Cyprinidae	*Sarcocheilichthys kobayashii* (Mori)	Kobayashi	1928
23	Cyprinidae	*Sarcocheilichthys mori* (Jordan & Hubbs)	Kobayashi	1924
24	Cyprinidae	*Sarcocheilichthys variegatus* (T & S)	Kobayashi	1924
25	Cyprinidae	*Sarcocheilichthys wakiyae* (Mori)	Kim	1961
26	Cyprinidae	*Zacco platypus* (T & S)	Chun	1962
27	Cyprinidae	*Zacco temmincki* (T & S)	Chun	1962
28	Cyprinidae	*Carassius carassius* (L)	Soh	1952
29	Cyprinidae	*Cyprinus carpio* (L)	Lee	1958

[a] 1 – 27 were cited from the chapter "Clonorchis and Clonorchiasis" (Komiya Yoshitaka) in "Advances in Parasitology Vol. 4, pages 53 – 106, edited by B. Dawes, Academic Press, London, New York, 1966. 28 and 29 have been frequently reported as intermediate hosts of *Clonorchis sinensis* in Korea by several researchers (Ref. Soh, 1952; Lee, 1958, and others).

Epidemiologically the distribution may be closely related to distribution of the intermediate hosts. Kim [199] reported about 2.5% – 5.2% of snails examined in the Gimhae district of Gyeongsang Nam Do, one of the high endemic areas, were infected by the cercariae of *Cl. sinensis* during summer months.

a) The first intermediate host: Parafossarulus striatulus var. japonicus (Pilsbry) (local name: *Waeurong*), known to be the molluskan host of *Cl. sinensis* in Korea, is widely distributed in the plain areas along the rivers of the southern part of Korea. Small in size (height 10.0 mm, diameter 6.5 mm, longest axis of aperture 5.0 mm) the host has a coiled, pale amber shell with a rather thick operculum. Another related snail, *Bulinus kiusiuensis,* was collected from only two restricted localities, Kyomwipo in North Korea and Gongju in South Korea; however, it has not yet been proven whether this species serves as an intermediate host of *Cl. sinensis* [200]. Kim [201] collected the *parafossarulus* snails from Goyang, Gyeong-gi Do, one of the endemic areas, and found that the cercarial shedding occurred mainly during May – October, but not during November – April.

b) The second intermediate host: Kobayashi [200] identified nine species of freshwater fish as the second intermediate host of *Cl. sinensis.* Subsequently additional species have been reported, and to date the officially recorded and recognized species number 29 [200] (Table 22), most belonging to the family *Cyprinidae.* Of these fish, *Pseudorasbora parva* (local name, *Chambung-O*) is the popularly known species that is strongly infected with the metacercariae of *Cl. sinensis.* Kim [201] collected 13 528 metacercariae of *Cl. sinensis* from only one *Pseudorasbora parva.*

c) Epidemiology: Domestic animals in endemic areas play an important role in spreading the eggs of *Cl. sinensis.* In Goyang, one of the endemic areas, Park [157] found eggs of *Cl. sinensis* in 3 out of 51 fecal samples from domestic animals, and Kim [276] found on the average 8.7% of animals in the same area to be egg positive. Some of the positives can be explained on the basis of simple coprophagy of human night soil containing the eggs. Walton and Chyu [187] reported that up to 53% of the males and 14% of the females in Gyeongsang Bug Do, one of the endemic provinces, were infected, whereas in Jeju Do, one of the islands, the corresponding figures were only 3% and 1%, respectively. In most of the endemic areas a higher rate is observed among the 41 – 50 year age group. Most likely eating habits play a role, for adult males are fond of rice wine with uncooked fish, but women do not usually practice such habits. The preschool age group is also strongly infected by the parasite [158]. Kim [199] examined the worm burden of *Cl. sinensis* in 469 school children in endemic areas and found the EPG (eggs per gram stool) to be 1590 (maximum,

Table 23. Clinical Features of Clonorchis sinensis

Subjective symptoms		Objective signs	
Fatigue	31.6%	Liver enlargement	29.1%
Dyspepsia	30.2%	Pain on pressure in the epigastric region	18.2%
Hypochondrial pain	24.2%	Fever	13.9%
Appetite loss and miscellaneous	18.9%	Jaundice	9.9%
		Edema	8.9%
		Abdominal flatulence	8.0%
		Ascites	5.6%

14 000) on the average, whereas in adults it was 4050 (maximum, 68 300) on the average.

The infection among the infant group may be due to maternal carelessness in handling raw fish. Generally people are accustomed to prepare freshwater fish raw by simply soaking them in vinegar or red-pepper mash and then consume them, sometimes after imperfect roasting. Small size and soft fish such as *Pseudorasbora parva,* which has higher metacercarial incidence, is more preferred because of their flavor. The repeated consumption of raw fish may effect the high incidence and endemicity.

d) Clinical features: To determine the clinical features of *Cl. sinensis* infections, Soh and Im [202] examined the patients' records of Yonsei University Severance Hospital for the years 1960 – 1967 and sampled 484 cases known to have passed eggs of *Cl. sinensis.* The charts were reviewed, in addition to checking laboratory data, clinical complaints, and physical finding of the patients. The following is the summary of these results (Table 23). The associated diseases in the infected cases of *Cl. sinensis* were mainly hepatic and gastrointestinal.

The above findings indicate that *Cl. sinensis* infection in Korea is closely related to troubles of the digestive system and causes more-or-less abnormal liver function which leads to general damage such as fatigue, dyspepsia, etc. Although a clear-cut conclusion may not be possible due to the diversity of the associated diseases, *Clonorchis* infection may play the main causative role in these symptoms.

e) Prevention: The nationwide campaign recently begun to prevent *Cl. sinensis* infection in Korea can be divided into two action spearheads, governmental and civic. In the early part of 1960 the Ministry of Health and Social Affairs established pilot centers in endemic area of clonorchiasis and paragonimiasis. Sixteen counties (*Gun*) were selected. The main activities were the administration of appropriate medicine to infected cases and education on how to handle and treat the intermediate hosts. Unfortunately, the program has become very inactive in recent years. The Korean Association for Parasite Eradication (KAPE) and the Korean Red Cross (KRC) have also participated in the control movement. Although KAPE has primarily concentrated its activities on *Ascaris* control, the *Clonorchis* problem is being linked up with it. KRC established a 25-year program for prevention of endemic diseases in Korea in 1965 – 1966 with the purpose of educating school children in each endemic area along the rivers in Korea: in Hangang,

Geumgang, Mangyeongang, Yeongsangang, and Nagdongang. Each area was divided into five parts, from the upstream to the mouth of the rivers. Educational lectures to the school children and their teachers were to be repeated five times or until the problems of those endemic diseases were eliminated from the areas.

2. Paragonimus westermani (ICD 121.2)

The lung fluke, *Paragonimus westermani,* has been the only species of its kind found in Korea. Kobayashi studied during 1918 – 1919 the morphology of lung flukes from a variety of both natural and experimental hosts in Korea and suggested that the specimens from all these hosts, including man, belonged to the same species [184]. The animal parasite, *P. iloktsuenensis* was identified in wild rats in Gyeongsang Nam Do along the Nagdong river [185]. Paragonimiasis caused by infection of *P. westermani* has been recognized as one of the medically important endemic diseases in Korea since olden times. Among Koreans it is familiarly known as *To-Jil* (earthborne disease).

The first nationwide survey was carried out in 1926 with sputum specimens collected from recognized endemic areas in all provinces of Korea. The overall egg positives in sputa were 24 907 (7.9%) of 353 729 people, ranging from the highest incidence (45.0%) in Jeonra Nam Do to the lowest (0.2%) in Hamgyeong Bug Do, the northernmost province of Korea [186].

Soh and associates (1979) found a different type of Paragonimus which is temporarily named as *P. pulmonalis* by Professor I. Miyazaki, Japan, from Jeju-Do and southern coast of the Korean peninsula (to be published in Yonsei Rep. Trop. Med. Vol. 10, p. 1, 1979).

In the past an intradermal test using purified adult antigens of *P. westermani* was administered by various investigators in Korea. Walton and Chyu [187] reported 1229 positive reactors of 9771 persons tested for skin reaction. The positives were found in all provinces, the highest percentage occurring in Jeju Do (47%) and the lowest, in Gyeong-gi Do (3%). An estimated 0.8 million people in South Korea were considered infected. Other nationwide surveys have recently been conducted separately by local health centers of the Ministry of Health and Social Affairs and the Korean Red Cross Team during the period 1965 – 1967. According to these two surveys, 17 505 (7.5%) of 237 465 inhabitants in over 100 different cities and counties tested were found positive. Various reports showed that the main endemic foci in South Korea are distributed in the southwestern part of the peninsula and Jeju Do. However, Kim [188] found several endemic foci in Taebaeg mountain villages in Gangweon Do, the eastern part of the land. In general, most of the known endemic foci are located in mountain areas, inhabited by the Cambaroides species, Ganghwa, Jeju, Goheung, Goyang, and Taebaeg mountain ranges.

a) Definitive hosts: Although paragonimiasis has been recognized for many years in Korea, epidemiologic studies were not conducted until the first quarter of the twentieth century. These showed a very wide distribution in domesticated and wild mammals that feed on crabs and crayfish. Several animals were reported as hosts in Korea. Easily infected domestic animals include the dog, hog, cattle, and rabbit; domestic fowl were found to not be hosts. Hogs usually do not consume the crustacean host, so it is presumed that infection in such animals may be due to the ingestion of contaminated garbage or drinking water [189]. Wild animals such as the tiger, cat, and weasel are infected whenever they consume infected crustaceans. Although not officially reported, Soh [1958] found *Paragonimus* eggs in the feces of a tiger in a Seoul Zoo.

Of the first intermediate hosts, *Semisulcospira libertina* (Gould) is the most important species in Korea among *Thiaridae* (*Melanidae*), although there are several varieties. Also *S. libertina extensa*, *S. libertina multicincta*, *Melania amurensis*, *M. gottschei*, *M. modiperda*, *M. modiperda var. gunaira*, and *M. paucinta* were listed as probable intermediate hosts by Kobayashi [275].

b) Second intermediate hosts: The known second intermediate hosts of *P. westermani* in Korea are crabs and crayfish. Of the crabs, the genus *Eriocheir* is the popularly known second intermediate host, of which there are three species: *E. sinensis*, *E. japonicus*, and *E. leptognathus* [12]. *E. sinensis* is distributed in the western part of the land from Pyeongan Bug Do to Honam plain; *E. japonicus* occurs the eastern part of the land from Hamgyeong Bug Do to Yeongnam and the southern coast of Jeonra Nam Do and Jeju Do; *E. leptognathus* is found only in Sineuiju, Pyeong-an Bug Do, but the species has not been proven to play a role as intermediate host. Endemic foci of paragonimiasis almost correlate with the distribution. The metacercariae in the crabs are found mostly in the gills and the muscle of the proximal part of the ambulatory legs. The infection rate with metacercariae in Eriocheir sp. reached 59% in one of the endemic areas in the southwestern part of Korea [191]. Chyu et al. [192] reported that 1023 (64%) of 1593 *Eriocheir japonicus* examined on Jeju island were infected and the average number of metacercariae per crab was 68.

Cambaroides similis, called *Ga-Ze* in Korea, is the species of crayfish commonly inhabiting mountainous streams. Metacercariae of *Paragonimus* are mainly distributed in muscles or the liver of the cephalothorax; however, the gills and muscles of the appendages also contain small numbers of metacercaria. Kobayashi [190] found 1016 metacercariae in one crayfish inhabiting an endemic area. The infection rate of crayfish in the southern endemic part of Korea ranges from 45% to 94% [194]. Although there are several other species of crab and crayfish in Korea, no definite reports have yet shown that they serve as the intermediate host of *P. westermani*. Soh et al. [193] reported that *Palaemon nipponensis* in Goheung also contained metacercariae of *P. westermani*.

In all the endemic areas in Korea, the most probable mode of human infection by *Paragonimus* is the habit of eating crabs by immersing them in soy sauce (*Ke-jang*). Crayfish are not usually eaten raw by Koreans, but the raw juice of the crushed crayfish is used as medicine for measles and has been a significant source of infection, particularly among children in rural areas. This superstitious practice is fading at present. Nevertheless, the *Ke-jang* (soaked crabs in soyabean sauce) food habit still seems to be the main cause of the endemic prevalence of paragonimiasis in Korea.

c) Habitat change: Water pollution and unusual drought give rise to unfavorable conditions for the normal ecology of the intermediate hosts which may help disrupt the persistence of the snail-mediated parasites in a certain area. Goyang in Gyeong-gi Do is an examplarly endemic area in this sense. In the past this area was famous for active paragonimiasis, but nowadays the infection is remembered almost as an episode. Establishment of factories upstream may hinder the breeding of snail, crayfish, and crawfish. Massive use of pesticides may also pollute the stream water, resulting in the same phenomenon. With this in mind, Kim [188] carried out an experiment to determine the effects of several pesticides on the life cycle of *P. westermani*. He found that some of the pesticides played a significant role destroying the eggs and metacercariae of *Paragonimus* and had destructive effects on crayfish (*Cambaroides similis*) and mollusks as well.

d) Clinical features: The clinical manifestations of pulmonary paragonimiasis are rather mild or moderate in most cases. However, sometimes an active thoracic paragonimiasis develops into an ectopic type. Over 200 cases of ectopic paragonimiasis have been reported up to date, the cerebral type showing the highest frequency. This type of paragonimiasis has often been observed particularly in children given crayfish juice for measles, as mentioned above. Cerebral paragonimiasis produces devastating neurologic sequelae and is often fatal. In the acute state of cerebral involvement, symptoms of meningitis usually predominate; hemiplegia and epilepsy are also seen in the chronic stage. According to Kim and Walker [195] convulsive seizures occurred in 29 of the 45 cerebral cases. Oh [196] also reported that the most common initial manifestation of cerebral involvement was epileptic seizures, which were seen in 40 of 62 cases. Several cases of spinal paragonimiasis have also been reported in Korea.

Possible methods of controlling this fluke disease have been often discussed and rested by various researchers: elimination of the intermediate hosts and reservoirs, the prevention of human infection through health education, and prohibiting the selling of crabs collected in endemic areas. But the fluke problem still remains unsolved. Kim [197] has subjected about 1400 cases to mass chemotherapy in Jeju with Bithionol. According to her, if well planned and well accepted by the local people, mass chemotherapy can achieve more than a 80% cure rate. This project could be the most effective control measure in areas where human beings are the main source of infection.

3. Metagonimus yokogawai
(ICD 121.5)

The distribution of Metagonimiasis almost parallels the distribution of the second intermediate host, *Plecoglossus altivelis*. According to Hunter et al. [198], Seogwipo in Jeju Do was the area with highest endemicity, 8.6%, other areas are Daegu with 6.9%, Chuncheon, 1.2%, and Busan, 1.1%. Reports in recent years have added other endemic foci: Hadong Gun in Gyeongsang Nam Do and the Seomjingang area of Jeonra Nam Do. Yeo [204] found that 221 (42.4%) of 521 inhabitants residing in Hadong-gun had discharges containing the ova of *Metagonimus sp.* Soh et al. [43] also reported a high prevalence of metagonimiasis in Jeonra Nam Do along the Seomjin river, where in the past only *Cl. sinensis* had been identified as an endemic parasite. In this area overall egg positives for *Cl. sinensis* were determined at 21.7%, but positives for *M. yokogawai* numbered 123 (41.6%) of 296 examined.

Metagonimus infection occurred at a rate of 44.0% in the downstream area (Gwangyang), at 55.0% in middle-stream area, (Gogseong), and at 29.0% in the upperstream area (Gurye Gun). Of 663 *Semisulcospira libertina* collected, 62 (9.4%) were cercarial positive for *M. yokogawai,* and *Plecoglossus altivelis* contained metacercariae of *M. yokogawai* in 100% of the fish from the districts of Gogseong, Gurye, and Gwangyang. The *Plecoglossus altivelis,* a sweet fish, usually inhabit clear, rapid, and cool running water. Its infection rate is registered at 54% in Milyang, Gyeongsang Bug Do, and at 90% in Seogwipo of Jeju island, with an average number of 23 – 280 metacercariae per fish. Distribution in the fish showed 45% in the scales and fins, 26% in subcutaneous tissues, 14% in muscle, and 14% in the head portion. Many salmonoid and cyprinoid fish have been found to harbor the metacercariae of *Metagonimus* since it was first detected in *Carassius carassius.* However, three species of fish, *Plecoglossus altivelis, Pseudorasbora parva,* and *Carassius carassius* are the known principal sources of human infection of *Metagonimus* in Korea, the first being the principal one.

4. and 5. Taenia solium and Taenia saginata
(ICD 123.0 and 123.2)

The diagnosis of *Taenia* infection by stool examination alone is not reliable because of the blind uterus of *Taenia*. Soh et al. [133] examined 3615 rural people by questionnaire and found that 129 (3.6%) complained of passing segments of Taenia. However, in Jeju Do, known as a highly endemic area, Cho et al. [213] found positive cases in 216 (38%) of 577 villagers and in 267 (16.4%) of 1631 high school students. The percentages of expelled worms in 98 cases were *T. saginata* with 86.7%, *T. solium* with 4.1%, mixed (*T. saginata* and *T. solium*), 6.2%, and unidentified tape worms due to poor preservation, 3%. Lee et al. [214] performed an epidemiologic study in a rural area of Jeonra Bug Do and found 13.3% (male, 16.2%; female, 10.3%) were positive of 289 examined by questionnaire. During the survey, they identified four cases of *Taenia solium* infection from 14 samples of persons treated with Extractum filicis: The reared worms numbered one in two cases, three in one case, and five in one case. Overall incidence of the two tapeworms in the country is estimated at about 4% [133]. Although adult worms of *T. solium* are rare, *Cysticercus cellulosae* have often been reported from various clinics. These infest the brain, orbita, subcutaneous tissues, and many other tissues. In the brain, larval tapeworms are responsible for epilepsy of the Jacksonian type. Cerebral cysticercosis is not uncommon in Korea.

The eating habits of the residents are responsible for the relatively common incidence of taeniasis; about 3.7% to 4.8% of them consume pork and beef raw or improperly cooked. Some data on meat surveys are available. A report in 1926 indicated 37.5% positive *Cysticercus bovis* infection among 508 cattle, 1 – 2 years of age [215]. In Jeju Do, 330 hogs of 17 404 in the slaughter houses were infected with *Cysticercus cellulosae* (Prof. Kim Seung Ho, private communication, 1976). Poor disposal of fecal matter from infected cases as well as unsanitary habits may play a role in the incidence of cysticercosis.

6. Brugia malayi
(ICD 125)

A considerable number of studies have treated on aspects of the filarial worm in Korea, ever since Yoon [176] found one male filaria from the left inguinal lymph gland during autopsy of a patient who had complained of elephantiasis in both legs. Although he reported the parasite as *Wucheria bancrofti* without any morphologic study, it is now generally believed that *Brugia malayi* is the only species of human filaria in Korea. The malayan filaria is distributed predominantly on Jeju island and its neighboring island, although several areas on the mainland show scattered endemicity.

Of a number of reports on incidence, Senoo and Lincicome's survey during the period 1942 – 1944 [179] covered the broadest area. They examined the night blood specimens collected from 5000 inhabitants of 25 villages on Jeju-Do and the mainland. According to the result Jeju Do showed the highest microfilaria rate (26.6%), Chungcheong Nam Do, 15.4%, Jeonra Nam Do, 8.6%, Jeonra Bug Do, 9.3%, Gyeongsang Bug Do, 6.0%, and Gyeongsang Nam Do, the lowest, 0.2%.

Lee [178] examined the night blood specimens collected from 229 children living in Sewha Ri, Taeheung Ri, Wimi Ri, Hayae Ri, Soncheon Ri, and Sinkwang Ri in Jeju Do and obtained a microfilarial rate of 11.4%. Seo et al. [177] carried out the examination for microfilaria among inhabitants in all age groups from 15 villages throughout Jeju Do. Blood films of 2139 persons were examined and 183 (8.6%) showed microfilariae, the incidence varying according to geographic sources from 0.8% to 19.5%. The above results suggest that filariasis is prevalent all over the island. The mean microfilarial density was 1.91% per mm³ of blood with a range of 0.30 to 5.23 per mm³ of blood.

Two reports have been made on filariasis in Gyeongsang Bug Do by Senoo and Lincicome [179] and Hwang et al. (1963). Senoo and Lincicome, examining 1002 inhabitants of five villages in Yeongcheon Gun, Yeong-il Gun, and the annexed Guns of Gyeongsang Bug Do, found 60 (6.0%) microfilaria-positive cases, of 524 males and 18 (3.8%) of 478 females. The microfilaria rate in males was higher than that of females. Hwang et al. [180] found 29 microfilaria-positive cases among 378 inhabitants living in Sinjeon Ri, Yeongju Gun as well as three cases of elephantiasis. These studies thus prove that filariasis is distributed in the northeastern areas of Gyeongsang Bug Do and that they are continuing to occur.

Igsan Gun, Jeonra Bug Do along the Geumgang river was known to be an endemic area. Bun [181] reported that 24 (18%) of 134 inhabitants residing in Igsan Gun were found to have microfilaria in their blood. However, he could not discover microfilaria in the blood of 37 elephantiasis patients, although most of them (about 70%) showed a positive reaction to an extract of *Dirofilaria immitis*. Seo et al. [177], examining microfilaremia cases in an army recruitment camp, found a relatively high rate of microfilaremia among the draftees from Jin Do island of Jeonra Nam Do. Some scattered endemic foci are suspected on the islands belonging to these provinces. Islands already recognized to have foci are the Heuksan islands in the southwestern part of the peninsula. Nonsan, along the Geumgang river, in Chungcheong Nam Do is also a known endemic focus. Paik [182] reported 19 (9.2%) microfilaria cases of 206 inhabitants in the county. In spite of that, almost no clinical case, including elephantiasis, has been reported from the area in recent years.

Other areas were found by Seo et al. [177], who examined 24 816 draftees (Map 4 c) from all over the country in an army recruitment camp. In the survey, the highest rate of microfilaremia caused by *Brugia malayi* was found in the group from Jeju Do (3.5%); Gyeongsang Bug Do showed 1.4%, and Jeonra Nam Do, 1.2%. But even though the number was small several positives were also noticed among draftees from other provinces, with the exception of Gyeongsang Nam Do. All 3807 draftees from Gyeongsang Nam Do showed negative findings.

It is not definitively known what kind of mosquitoes are involved as the major vectors in the main endemic areas. In Jeju Do, known as one of the main endemic areas in Korea, *Aedes togoi* is the most suspect because of its abundance and vigorous attacks on humans. In Gyeongsang Bug Do, *Anopheles sinensis* has been incriminated as the probable vector for brugiasis.

7. Ancylostoma duodenale
(ICD 126.0)

8. Necator americanus
(ICD 126.1)

The species of hookworms identified in Korea are *Ancylostoma duodenale* and *Necator americanus*. Soh et al. [165] found only one case infected with *N. americanus* from a total 63 hookworm infections in Chuncheon and Gunsan. Recent information also indicates that *A. duodenale* is still the dominant species so far in Korea.

The incidence in the 0 – 4 year age group is 2.7% to 4.0%. This increases gradually up to the 15 – 19 age group and stays at about 15% in the adult up to old age. As regards the worm burden, 95.2% of the infected cases were reported to range from 800 to 3200 by EPGF [166]. Choi [166] classified the intensity distribution by age groups. The positive cases were divided into two groups, a 6 – 12 year age group and a 13 – 46 year age group. The ranges of EPGF were 800 – 1200 in the former and 400 – 800 in the latter. This suggests that the worm burden in the younger ages is relatively higher than in older ages (Table 24).

Oral infection seems relatively common in Korea. Salted pickle is prepared with contaminated vegetables that are fertilized with human night soil. The filariform larvae survive as long as 2 days in the pickle, *Kimchi*. Early ingestion may result in hookworm infection [168, 169].

The distribution of hookworm differs according to soil type and geographic site, being sparser in heavy soil and coastal areas than in mountainous or sandy areas

Table 24. Age Distribution of the Intensity of Hookworm Infection {166}

EPGF (eggs per gramm feces)	Children (6 – 12 years)		Adult (13 – 46 years)	
	No. of cases	%	No. of cases	%
0 – 400	0	0	2	28.6
400 – 800	1	4.8	4	57.1
800 – 1200	7	33.3	1	14.3
1200 – 1600	5	23.8	0	0
1600 – 2000	4	19.0	0	0
2000 – 2400	1	4.8	0	0
2400 – 2800	2	9.5	0	0
2800 – 3200	1	4.8	0	0
	21	100	7	100

[167]. In general, 8.3% of urban and 12.0% of rural populations are estimated to be infected by hookworm [128]. Under certain circumstances the ova of hookworm which are discharged from the human body develop an infective form in the soil. Thus soil also serves as an important medium for the spread of human paasites. The development of parasite eggs is believed to not only be strongly influenced by sunshine, temperature, moisture, wind, and rainfall, but also by the character of the soil, e.g., geologic composition, and acidity. Many studies have treated the interrelationship of parasite infection and the natural environment. Beaver [170] reported the effects of soil composition on the development of ascaris and hookworm infections. Ascaris and hookworms are widely prevalent among Koreans, and these infections have become a most serious public health problem. No nationwide epidemiologic study on the relationship between soil character and parasite infection has been carried out in this country. Kim [167] attempted only to clarify the relationship between soil composition and ascaris and hookworm infection in an area in which the natural living environment was similar. A summary of the results of this study follows. The surveyed area was divided into four regions: mountainous, hilly, plain fields (in Jeong-eub Gun), and offshore islands (Gogunsan islands in Oggu Gun) in Jeonra Bug Do. Prevalences differed by area. In mountainous areas with high sand content of the soil, more hookworm infection was present than in the plain area with a heavy soil. The ratio of ascaris to hookworm infection was almost the same (1.0 : 1.0) in the mountainous area with over 70% sand content. However, in the hilly regions with about 40% sand content the ratio was 1.9 : 1.0, and in the plain fields, the ratio was 4.5 : 1.0. In this way, the prevalence of hookworm infection parallels the sand content of soil. The incidence of ascaris infection showed no significant change in the areas containing over 40% sand, but the hookworm had its peak concentration in the area of the highest sand percentage (60% – 70%).

To supplement the field findings, Kim [167] carried out in vitro experiments with different types of soil. In soil containing no sand, the hookworm ova were destroyed within 1 week, but the survival rate increased as the percentage of sand content increased, and even in dry sandy soil, the hookworm ova survived about 2 weeks. In contrast, 80% of ascaris eggs developed to infective form regardless of the soil's sand content; even in dry conditions they survived for 3 weeks. However, in soils of 100% sand or clay, the ova developed but rapidly died. Thus soil composition was suggested to be closely related to prevalence of ascaris and hookworm infections in Korea.

On the islands, Gogunsan island (Jeonra Bug Do) hookworm was identified in only one case of 86 examinees (2.5%), but there was a high prevalence of ascaris (82.3%). Supplementary experiments demonstrated that whereas salt in the island soil inhibited the development of hookworm ova and larvae, ascaris eggs were not affected. The island soil contained NaCl at the rate of 0.13 to 1.0 g/dl, none equaling 3.0 g/dl, the rate of sea water. Hookworm ova did not develop to the second stage larvae in soil containing over 1.0% NaCl. Thus, salty soil is considered to have some influence on the prevalence of hookworm infection, especially on the islands and in the coastal regions. Some previous reports on the geographic prevalences of ascaris and hookworm endorse this probable trend. In general, prevalence of hookworm infection is lower in coastal and island areas than in inland areas, although ascaris infection shows no difference of prevalence. Possibly the hatched larvae on the surface soil are affected by the continuous salty, windy atmosphere along the adjacent coast [167].

No nationwide program for hookworm control has yet been carried out, although some researchers have tried mass-treatment programs in limited areas during the winter months aiming to disrupt the cycle [171]. However, health education through mass media has led to a remarkable decrease in hookworm incidence in recent years. Application of drugs and thorough washing of vegetables has also helped to lower the incidence, reducing by 32% – 37% during the past 22 years. Massive application of herbicides may intensify this decline [172]. The overall results indicate that conventional pesticides have no significant lethal effect on the eggs of hookworm, trichostrongylus, ascaris, and trichuris. But the larval forms, especially the rhabditoid forms of hookworm and trichostrongylus, are destroyed by special pesticides.

9. Ascaris lumbricoides
(ICD 127.0)

Ascaris lumbricoides is one of the popularly known human parasites among Koreans. It is distributed widely all over Korea, from the most northern part to the most southern, regardless of temperature or other conditions. The chronologic data indicate that the incidence declines with age [128]. Incidence of 80% in 1949 continued at the same level until 1967, but practices of mass examination with mass treatment since then have led to a slowdown. In 1976 the Korean Association for Parasite Eradication (KAPE) summarized the overall incidences: 30.2% in urban areas and 48.6% in rural areas, still high figures compared with many other countries.

It is noteworthy that in rural areas the prevalence has reached 45% – 50% even in the 0 – 4 year age group. This is close to the average value of 48.6% in all age groups. One slight decrease in the 10 – 14 years age group is considered to be due to repeated mass treatment in schools. Intensity of the worm burden varies by areas. Then KAPE also examined the ascaris-worm burden both in rural and urban people and found 44.4% of 744 positives in average the EPG (eggs per gram of feces) range of 1000 – 4900; the EPG over 30 000 was 0.7%. The data may explain a part of the situation as regards worm burden in the year of 1976. Park [157] examined parasite incidence in three rural villages and reported 66.3%, 74.6%, 75.9% of ascaris. Kim [158] reported that

20.8% of rural infants were infected with ascaris, reaching 90.9% in children up to 6 years. The poor treatment of night soil, contaminated vegetables, and polluted soil surrounding houses may contribute to such high incidence. Many vegetables are contaminated with parasites at the source by fertilization with night soil.

The vegetables collected from markets in urban areas are also contaminated with an abundance of eggs of ascaris [159]. Ascaris eggs in Korean pickle survive for over 40 days in the summer season [160]. However, these vegetables are obviously not the only source of ascaris. The contamination of the soil by promiscuous defecation and fertilization with night soil should be considered an important source of ascaris infection in Korea, not only in rural but also in urban areas. Roadside soils even in cities are likely to be polluted with fecal matter, and consequently various parasite eggs and protozoan cysts are found [128]. Earthworms may perform the role of spreading parasites in the soil [157].

Emphasis in the battle against parasites in Korea has been mainly concentrated on ascaris control. Methods used include mass treatment, health education, and sanitary improvement. Systematic mass treatment has been carried out by KAPE with governmental support since 1965, on a regular basis in spring and autumn since 1969. The combined efforts of the past 22 years have resulted in a 10% – 20% decrease in parasites, although the sanitary environment has not yet appreciably improved.

During the spring survey of 1973, KAPE analyzed the ascaris-positive cases according to the fertilization or nonfertilization of eggs. The ratio of unfertilized (U) eggs of ascaris to fertilized (F) should be a good basis for control measurement. The report by KAPE in 1973 suggests optimistic prospects. Of a total of 3 558 999 ascaris egg-positive students, the ratio of U : F differed according to the level of infection. In provinces or cities with incidences below 50% the ratio was approximately 1 : 2 to 1 : 5, yet the ratio was 1 : 6 – 1 : 20 where the incidences were 50% or above. This is an indication that the ratio of unfertilized ascaris eggs increases with decreasing infection levels. The repeated mass treatment may contribute to speeding up the rapid reduction of ascaris infection in Korea. Komiya [161] in Japan emphasized an annual practice of mass treatment to a given population group. According to his example, 60% – 70% of ascaris infection resulted in a 5% – 10% annual decrease of the disease level in each year, and when the prevalence reaches 30% the curve decline becomes more rapid. Due to Komiya's theory, the problem of soil-transmitted helminthic infection in Korea is proceeding along successful lines. At the moment, no nationwide mass treatment is practiced for trichuris, hookworm, or trichostrongylus. Nevertheless, health education through various means has stimulated the people to treat and prevent cases on an individual basis. The reason for the decreasing incidences of these parasites in recent years compared to 1949 can be explained by the above trends.

10. Trichuris trichiura
(ICD 127.3)

Distributed all over the country, *Trichuris trichiura* was recently indicated to have an overall incidence of 63.1% in rural and 69.7% in urban areas. Infection in the 0 – 4 year age group already shows a 34% increase up to about 63.0% in the 5 – 9 year group and then levels off until old age. Worm burdens by EPG range from 100 to 10 000 and above, but 82.9 of 745 examined in the range of 100 – 999 [128, 156].

The source of infection may be accounted for by contaminated vegetables or polluted dooryard soils [128]. Choi [159] examining vegetables in the market, found trichurid eggs 20 (6.7%) in lettuce, 41 (8.0%) in radishes, and 57 (42.2%) in chinese cabbage of a total 188 examined. As yet no effective anthelmintic for destroying the parasite has appeared. This is one of the reasons why the progress of the control movement is so sluggish. However, a newly synthesized anthelmintic, Mebendazole, is expected to spur on the control movement.

11. Enterobius vermicularis (Oxyuris)
(ICD 127.4)

Prevalence of *Enterobius vermicularis* or pinworm in Korea during the past quarter of a century has varied between 7.9% and 56.1% [163, 164]. A survey localized the highest incidence in the children of orphanages (44.9%), next, primary school children (33.3%), and lowest of all, inhabitants on an island (7.9%). To study epidemiologically the leading cause for such a high incidence of pinworm both in rural and urban areas, Park [162] carried out an investigation of the incidence of pinworm infection in 72 families comprising 207 (male, 128, female, 79) rural inhabitants in Gyeong-gi Do and 502 (male, 297; female, 205) members of 115 upper-class families in Seoul, 58 families totaling 296 (male, 140; female, 156) suburban inhabitants in Seoul, and 799 (male, 446; female, 353) orphans accomodated in 14 orphanages in Seoul and Gyeong-gi Do. Epidemiologic factors of pinworm infection as well as some biologic characteristics of the ova of *Enterobius vermicularis* were also studied.

The incidence of infection in families based in Seoul and in rural areas was in Seoul, 48.6% on the average, 38.2% of suburban inhabitants, and 30.9% of rural inhabitants. Incidence in orphanage children was highest with 54.2%, whereas the upper class in Seoul was the lowest with 13.3%. In this way pinworm infection has become one of the nationwide problems. Sex played no noticeable role, and no relationship between family size and incidence was determined. Children infected with pinworm seemed to play an important role in the high incidence in individual families. The number of rooms in a family house and the number of persons per room seemed to have an influence on pinworm infection. The ova were found in 16.7% of 90 dust samples collected from the living rooms of Seoul suburban homes; moreover 4.5% of them appeared in a matured condition. The higher the temperature, the sooner the ova developed to a mature form, taking 4 at 37 °C, 5 at 35 °C, 20 at 30 °C, 22 at 24 °C, and 450 (about 18 days) at 18 °C. After being kept at 4.2 °C for 10 days 31% of the ova were still alive. Larva in egg shell kept their motility for 48 h in iced water. In tap water at room temperature, the ova generally reached matured state in 3 days and were gradually destroyed soon after the automatic hatch. The ova survived for 24 h in various concentrations of cleaning soap, 30% in 1 N solution of sodium hydroxide, and more than 50% in 5% – 10% solution of sodium hypochloride. The ova were destroyed within 48 h in 30% solution of sodium chloride, and within 24 h in soy sauce, while 15% of them survived for 24 h in vinegar. Sauces do not have any ovicidal action in the same hour, and 10 gm/dl of onion, garlic, red pepper, and mustard each showed no appreciable ovicidal effect during a period of 4 days.

12. Trichostrongylus orientalis
(ICD 127.6)

Of the genus *Trichostrongylus*, *T. orientalis* has been the only species identified as infecting humans in Korea. A recent report on the incidence shows 10.9% in urban and 5.9% in rural areas [128]. It is not certain why the incidence in urban areas is much higher than that of rural. Soh et al. [172] have suggested that the mass application of pesticides might destroy the larval stages of the parasite in the soil of rural areas.

In spite of its high prevalence in Korea, it has been almost ignored by medical people because of its obscure clinical manifestation. However, Lee et al. [175], in a clinical investigation with 66 *T. orientalis*-infected cases, found the most common symptoms to be neurosomatic (headache, vertigo, insomnia, nervous irritability, etc.) and gastrointestinal disorder such as nausea and indigestion. Eosinophilia (7.0% and over) was observed in 41% and lymphocytosis (30% and over), in 93.4% of the 46 examined.

13. Trichomonas vaginalis
(ICD 131)

Shin et al. [154] reported 35.8% – 42.0% positives from a total of 1114 leukorrhea cases of patients ranging in age from 14 – 60 years. Incidence seems most common among prostitutes and entertainers. Chung et al. [155] made an epidemiologic survey on *T. vaginalis* and *Candida* in 241 vaginal swab-specimens from prostitutes in the ports of Yeosu and Gunsan from July to August 1969. The vaginal flagellate was detected by direct-mount examination method, and all the vaginal specimens were cultured on Sabouraud's slant medium for detection of the *Candida sp.* Of the 241 prostitutes percentages of *T. vaginalis* infection were 38.2% at Yeosu and 27.3% at Gunsan. Percentage of *T. vaginalis* infection revealed 32.7% from the unlicensed group, 24.0% from the licensed group, and 40.0% among employees in restau-

rants. The highest prevalence rate was observed in the 20 – 30 age group (32.61% – 33.33%), and the age range of the prostitutes surveyed was 20 – 47 years.

Eighty one positive cases of *Candida sp.* were detected among 241 prostitutes (33.19%), and *C. albicans* was identified in 21 of the 81 positive cases.

It is important to note that employees in restaurants show a higher incidence than the licensed group of prostitutes obliged to have regular examinations. The male group may act as a transmitter in some sense. Chyu et al. [220] reported that 0.5% (47 of 9617 urine specimens examined) of male patients in Kyung-Hee University Hospital had *T. vaginalis* in their urine.

IX. Infections Due to Other Human Parasites Reported in Korea with Rare Incidence

1. Fasciola hepatica
(ICD 121.3)

Fasciola hepatica is not uncommon among cattle in Korea; no scientific report has been published on *Fasciolopsis buski*. Brooke et al. [135] reported six positives among 1726 examined at the prisoner of war camp on Kojedo island; however, the egg could not be differentiated as either *Fasciola* or *Fasciolopsis*. Extensive study was not done as to whether the infections were real or spuriously caused by ingestion of animal liver containing the egg. However, recently two human cases of *Fasciola hepatica* infection were reported by Dr. Cho et al. [278] and Dr. K. B. Hur in Soon-Chon Hyang in Seoul (not yet reported in written form).

2. Heterophyes heterophyes
(ICD 121.6)

Muda [205] reported several cases of *Heterophyes heterophyes* infection, but Kobayashi [205] did not agree with the report. Due to its minor importance and morphologic similarity to *Metagonimus sp.* or *Clonorchis sp.*, its eggs might be mistaken for other species. The metacercaria of *H. continuus* was observed in brackish-water fish, *Lateolaborax japonicus*, *Mugil cephalus*, and *Acenthogobius flavimanus*. The metacercaria is characterized by a relatively large ventral sucker. The metacercariae of *Pygidiopsis summus* were detected in *Mugil cephalus* and were very similar to those of *Metagonimus* [206]. Small intestinal flukes, such as *Echinostom* and *Echinochasmus* are also occasionally found in Korea, but are without medical importance.

3. Dicrocoelium
(ICD 121.8)

Dicrocoeliidae includes *Dicrocoelium dendriticum* and *Eurytrema pancreaticum*. Less than 100 sporadic cases of human infection have been reported elsewhere in the world. Infections of sheep and cattle with *Dicrocoeliidae* are commonly found among animals at abattoirs in Korea. *Eurytrema pancreaticum* infection in humans has never been reported. The eggs of *Dicrocoeliidae* were found in the stool specimens of one soldier out of 1990 soldiers examined [203]. This was considered a spurious infection from eating livers or other animal organs containing the eggs. The eggs of *Dicrocoelium dendriticum* are indistinguishable from those of *Eurytrema pancreaticum*. The 25-year-old soldier had no history of ingestion of ants, the intermediate host for development of the meta-cercarial cyst of *Dicrocoelium dendriticum* nor of ingestion of the land snail or grasshopper, intermediate host of *Eurytrema pancreaticum,* nor did he have a history of dysentery. The prominent symptoms observed were gastrointestinal disturbances in the form of flatulence with nausea, loss of appetite, and dizziness.

4. Diphyllobothrium latum
(ICD 123.4)

Since the scientific record compiled by Hara and Himeno in 1924 [208], over 20 cases of this infection have been reported in Korea. Geographically, most of the cases were found along the southeast coast of the country, thus paralleling distribution of the second intermediate hosts; *Onchorhynchus masou* and *O. keta.* The fish come down from Aryusian islands in the spring season and migrate upstream to lay eggs along the coast. The clinical manifestations of the reported cases were rather mild, with no severe signs of anemia.

5. Sparganum
(ICD 123.5)

Sparganum denotes the plerocercoid stage of an unknown species of *Spirometra* which causes sparganosis in humans. In Korea the majority of the plerocercoids producing sparganosis are considered plerocercoid of *Dyphillobothrium mansoni, Ligula mansoni,* or *Sparganum mansoni.* But the identity of the larvae of the species is not known. Although it is not uncommon in Korea, written cases are relatively few. Roughly 30 – 50 human sparganosis cases have been reported so far in the literature [209]. Human sparganosis can probably be acquired in several ways: by drinking water in which procercoid-positive cyclops exist, by application of frog poultices containing plerocercoids for wound therapy, and by ingestion of plerocercoid-positive frogs or snakes [210]. The habit of eating snakes and frogs either raw or in an insufficiently cooked state is often practiced among phthisic patients. Some reports on the incidence of sparganum in frogs and snakes endorse this explanation for the larval infection being to common in Korea. Kim et al. [209] found 61 (17.5%) plerocercoids in 348 frogs examined and confirmed that they belonged to subgenus *Spirometra* by experimentally infecting cats and dogs with them. Chang and Sung [212] found them in 7 (29.2%) of 24 snakes (*Natrix sp.*) on Ganghwa island. Cho et al.

[211] found 171 *Sparganum mansoni* in 75 snakes of 7 species: *Elaphe schrenki, E. rufodorsata, E. dione, Dinedon rufozonatum, Natri tigrina lateralis, Zamenis spinalis,* and *Agkistrodon halys.* Kim et al. [209] examined sparganum in the frog (*Rana nigromaculata*) in several districts of Korea, Kupo, Ham-an, Cheongpyong, Nunggock, and Naju. Of 348 frogs 61 were infected with sparganum, yielding a natural infection rate of 17.5%. The highest incidence was found in Cheongpyong, a mountainous district; lower incidences were found in suburban or plain areas. Of those infected frogs, 70% were infected by only one or two worms, although one frog was infected with 17 spargana. The majority of the spargana were found in the right and left inguinal regions; 74% were below 10 cm in length. Two cats and a dog were experimentally infected with the plerocercoids, and the resultant growth was characteristic of the subgenus *Spirometra.* All clinical reports indicate that the symptoms characterizing the infection are pain and swelling in the vicinity of the worm, a sensation of an object "crawling" within the inflamed mass, and itching of the skin over the involved area.

6. Hymenolepis (diminuta) (nana)
(ICD 123.6)

Even though these tapeworms are common among rodents in Korea, relatively few cases of human infections have been reported. The first report of a human infection by *H. diminuta* was reported by Dr. Il Chyu. In addition to those three cases, he also reported on one case of a 10-year-old who complained of slight psychotic distress and abdominal pain (unpublished report, 1966 [218, 129]).

In contrast to the *H. diminuta, H. nana* infection is rather common in humans. Soh et al. [133] identified 32 cases in 14 682 stool specimens in Severance Hospital Laboratory and found that 88% of the cases were below 10 years of age. Although an epidemiologic study has not been performed yet, an unsanitary environment is regarded as the primary factor.

7. Mesocestoides sp.
(ICD 123.8)

According to the literature, about eight cases of human infection have been reported elsewhere in the world so far. In addition to these cases, Choi et al. [216] found the first case of *Mesocestoides* infection in Korea. A 45-year-old man visited the outpatient clinic of St. Mary's Hospital with complaints of intermittent indigestion and abdominal distension lasting for nearly 1 year. The source of the infection was not clear, but the patient had ingested 15 raw snakes 1 year previously.

8. Strongyloides stercoralis
(ICD 127.2)

The first report appeared in 1914 with three cases of the infection in the northwest part of Korea [173].

Brooke et al. [135] found 0.1% – 0.5% of positive cases among the 2642 prisoners of war on Geoje island, Gyeongsang Nam Do. Soh [174] reported one ascites case due to strongyloidosis. According to the literature, some 70 cases of the infection were recorded sporadically during previous decades, but no case has been reported since 1960, nor has any epidemiologic study been performed in Korea.

9. Thelazia callipaeda
(ICD 128.8)

Human thelaziasis has been reported from various countries to involve two different species, *Thelazia callipaeda* and *T. californiensis*.

In Korea, *T. callipaeda* was first described by Nakada as oriental eye worm in 1934; this was followed by several case reports [274]. Recently Im et al. [183] isolated two female *Thelazia sp.* worms in a 20-year-old boiler man, who had lived in a rural area where he was presumably infected 7 months previously. Neither he could have had intimate contact with dogs or other animals nor have received any injury from insects. At the time of the first examination, he complained of a foreign-body sensation in right eye, lacrimation for 3 months, and blurred vision for 4 months, although no conjunctivitis was experienced during the period. The symptoms cleared up shortly after removal of the worms. On physical examination, pupillary reaction was normal. Fundus revealed no change, and the nasolacrimal duct was patent. On turning back the upper lid, a round worm was observed in the right upper cul-de-sac, the worm seemed to hide beneath the eye lid. It was picked up from the upper conjunctival bulbar surface with an applicator. The worm, whitish in color and threadlike, was gliding actively with a serpentine movement across the bulbar conjunctival surface. On the following day another worm was removed directly by the patient.

10. Pneumocystis carinii
(ICD 136.3)

A description by Lim [149] of 27 cases was the first report of this infection in Korea. The patients were 2 – 5 month old with complaints of diarrhea and atypical pneumonia. Fatality rates registered 17% – 35% and no difference according to sex was determined. No case was observed earlier than 1 month after birth or later than 6 months. The infection is believed to be transmitted by air. Pneumocystic carinii infections have also been reported by pathologists at post-mortem examination [150, 151].

X. Other Protozoal Intestinal Diseases
(ICD 007.8)

Entamoeba coli is frequently found in routine examination in the laboratory. Various reports indicate 20.5% to 27.1% incidence in the general population [139], but

Rim et al. [140] found 40.5% in an institution for mental retardation, *Seoul Gak-sim Won*. A high rate is observed in the upper age group 41 – 50, but no sex dependency is evident.

Endolimax nana has been indicated in various reports to infect about 5% – 10% of the population. Again, the 41 – 50 year age group shows the highest prevalence [139], no sex dependency was noted.

Iodamoeba buetschlii shows an incidence ranging from 0.5% to 16.4% in different reports [139]. It is relatively high in the infant group less than 5 years of age, but no sex dependency is evident.

Dientamoeba fragilis is rarely found. Choi [132] found only one case in 2000 specimens in Severance Hospital. Soh et al. [133] found no positive case in 10 320 fecal specimens in Severance Hospital Laboratories.

Enteromonas hominis (0.5%), *Embadomonas hominis* (0.5%), *Chilomastix mesnili* (0.4% – 4.4%), and *Trichomonas hominis* (3.6%) have been reported occasionally, but no pathogenicity has been proven [153]. Kim et al. [134] reported *Trichomonas hominis* (1.1%) and *Enteromonas hominis* (0.7%) in several localities.

XI. Anthropozoonoses

1. Anthrax
(ICD 022)

Anthrax in domestic animals has appeared in Jeju Do and in other parts annually for a long time. During 1950 to 1961, the yearly reported cases ranged from the lowest, 1, in 1956 to the highest, 105, in 1952. Fatalities were almost 100%.

The occurrence of this disease is unknown before 1950. Anthrax has been a notifiable disease for domestic animals and an endemic disease in Korea for many years. Particularly on Jeju island this disease results in many deaths of cows and horses annually.

2. Brucellosis
(ICD 023)

No report of any occurrence before 1956 could be found in Korea. This disease showed an annual occurrence among cattle since the start of importing animals from the United States. Mass attacks have been reported among imported animals, the sites of occurrence being the National Jeju pasture, Anyang Animal Farm in Gyeong-gi Do, and Gyeongsung orphanage in Anyang [3]. In August 1973, 6 cows died in Jeju Do due to brucellosis, and it was the last occurrence in Korea, up to present time.

3. Rabies (Hydrophobia)
(ICD 071)

After World War II and particularly immediately after the Korean War, this disease seemed to be disappearing. Recently, however, as the number of homebred dogs have increased, the disease has again appeared, and the number of victims is apparently considerable.

One of the difficulties in preventing this disease is that people tend to avoid being inoculated because of the side effects of the vaccine, even if they are bitten by dogs suspected of having hydrophobia. Although 99% of hydrophobia in Korea is transmitted by dogs, dog breeders often neglect to inoculate their dogs. The Ministry of Agriculture itself does not seem to pay serious attention to this disease.

However, its fatality rate is 99%; the percentage of cases due to negligence to receive complete inoculation after being bitten by infected dogs totals 11% in children and 1% in adults.

Certain epidemiologic peculiarities are found in Korea. Since the Korean peninsula lies in contact with the Chinese continent, Korean wild animals are closely related to Chinese wild animals, and many cases of hydrophobia occurring in both countries are transmitted by the same wild or domestic animals. In 1957 a rabid wolf came down into a village of Sancheong Gun, at the foot of Jiri mountain. Five deaths from rabies occurred among the 18 bitten villagers [125].

Even though it is old, the data furnished by the Laboratory of the Korean Governor General's Office in Seoul during 1928 – 1938 by testing dogs suspected of being infected by this disease stated that 56% of the total 2974 cases were proven to have Negri bodies, and 11% of the Negri-negative cases were shown to contain virus that could be isolated. The number of occurrences of rabies-infected dogs showed an annual average of 500 to 600 before 1945, which dropped to 9 – 81 during 1950 to 1960.

According to the data of the Korean Governor General's Office on the occurrence of rabies before 1945, the monthly distribution showed a rise in February, March, April and May. Regionally, the occurrence was high in Jeonra Nam Do and Gangweon Do along the Taebaeg mountain ranges.

The Statistics of 1926 – 1940 the annual occurrence of hydrophobia among persons bitten by dogs climbed from 749 in 1926 to 2244 persons in 1940, with the number of occurrences of hydrophobia from 15 (1926) to 17 in 1940. From 1971 to 1974 there were only four cases of hydrophobia reported.

The exact number of domestic dogs in Korea is not known, but it is roughly estimated to exceed one million. The number of people bitten by dogs can be estimated of 10 000 annually in recent years.

Vaccines are produced in such small quantity that they would not be sufficient to inoculate one-half the estimated homebred dogs in this country. The waste of the vaccines might be considerable because, if a person is bitten, injections are given blindly without any scientific observation of the dog.

4. Toxoplasmosis
(ICD 130)

Of 373 cases examined by intradermal reaction Soh et al. [142] reported 21 (5.8%) positive. The rate was seven times higher in males than in females, and the incidence showed occupational differences: butchers, fishermen, and students had rates of 7.8%, 9.5%, and 1.3%, respectively. Although some research from a related epidemiologic viewpoint have been carried out in Korea, no clinical case has yet been reported.

However, a seroepidemiologic study on human toxoplasmosis among neurologically and physically deficient groups in the Seoul area suggests that *Toxoplasma* infection may be one of the important causal agents of mental and physical retardation [143]. Throughout the study by Soh et al. in 1975 serum titer 1 : 16 and above were regarded as positive for indirect fluorescent antibody test (IFAT). Of a total of 166 mental patients, 16 cases (9.6%) were positive, and these were found only among schizophrenic cases; the other cases were all within normal range.

Three (4.7%) of 54 cerebral palsy cases showed positive titers, and they were spastic quadriplegia and progressive muscular dystrophy cases. Of 74 cleft-lip cases, 21 (28.4%) were positive. A much higher positive rate (38.3%) was seen in males than in females (11.1%). Of 104 maternal serum samples of prenatal care cases, 3 (2.8%) showed positive titers, but only 1 of 76 cord blood specimens (1.3%) from normal delivery cases showed IFAT-positive. Four (3.5%) of 114 control sera cases revealed positive reactions. Of the various groups examined, higher positive rates were seen in schizophrenia and cleft-lip cases, whereas the other groups showed more or less the same rate as the control group.

XII. Zoonotic and Other Infectious Diseases not yet Confirmed in Man in Korea

Zoonotic contagious diseases are classified in five groups according to their biologic characteristics [217]. In the first group, the existence of the following diseases were hypothesized from a global epidemiologic viewpoint, but their occurrence has not yet been reported in Korea: glanders, melioidosis, plague, dengue, and foot and mouth disease. The second group of zoonotic diseases, botulism, dermatophilosis, geotrichosis, encephalomyocarditis, psittacosis (bedsoniasis), and cat scratch fever might exist, but have yet to be confirmed either directly or indirectly. The third group of zoonotic diseases, leptospirosis, rat bite fever, tularemia, Q fever, rickettsial pox, scrub typhus, lymphocytic choriomeningitis, is recognized serologically and pathogenically among animals, but no human case has yet been confirmed. The fourth group of zoonotic diseases, extant in the past but no longer found today, includes louse-borne relapsing fever and cow pox.

Diseases occurring in Korea sporadically, endemically, or epidemically are included in the fifth and last group. Altogether 24 kinds of domestic-animal-diseases are listed at the Livestock-Policy Bureau, the Ministry of Agriculture and Forestry in Korea, as notifiable infectious diseases of livestock which threaten the livelihood of domestic animals. Of these, nine diseases, anthrax, bovine tuberculosis, brucellosis, foot and mouth disease, swine erysipelas, glanders, rabies, pullorum, and Newcastle disease, are zoonotic diseases affecting man, but foot and mouth diseases and glanders do not exist in Korea.

In accordance with the promotion of livestock raising in recent years, zoonotic contagious diseases have attracted increasing attention in this country, not only in the medical field but also in the public.

Some notifiable infectious diseases of domestic animals and those zoonotic contagious diseases not yet confirmed as infecting humans are: leptospirosis, rat bite fever, botulism, tularemia, Q fever, Russian spring summer encephalitis (in South Korea), psittacosis, ornithosis, cat scratch fever, and lymphocytic choriomeningitis.

Rat-bite fever has long been known to exist in various parts of the country, but there is no way of finding out the actual number of occurrences or of setting up a general concept of the conditions, since the disease is not notifiable.

The cause of rat-bite fever existing in this country is suspected to be *Spirillum minus* (the cause of the Japanese *rat-poisoning*) on several clinical grounds.

Only one case of dracunculosis was reported [272] in 1926, but no epidemiologic findings were described.

Ova of *Schistosoma japonicum* were detected in an inguinal hernia sac of a 70-year-old Korean male, but unfortunately the history of the case could not be traced [207]. However, the patient was presumed to have been infected in the endemic area in Japan. It is generally agreed that there is no autochthonous infection in Korea, since the intermediate host, the *Onchomelania sp.* is not distributed in the land.

Trichinella spiralis has never been observed in Korea, even though the animal hosts exist, such as hogs and rats.

Echinococcus cysts are readily found in cattle on Jeju Do island, but no form of hydatid disease among Korean people has been officially reported. *Toxocara canis* and *T. cati* have recently been recognized as agents of visceral larva migrans. These cosmopolitan species are also found commonly in Korean domestic dogs and cats, but a human case of visceral larva migrans has not yet been reported in Korea.

XIII. Noncommunicable Diseases

When communicable diseases, which are prevalent in countries with low standards of sanitation, are effectively controlled, then interest and efforts should be directed toward the control of degenerative chronic diseases, such as senile diseases, malignant plasms, industrial diseases, metabolic diseases, and inherited diseases. In Korea trends of morbidity and mortality in recent years show some signs of converting from the former to the latter.

1. Malignant Neoplasms
(ICD 140 – 239)

The Korean Cancer Association collected data on malignant diseases from hospitals in Seoul and the provinces in 1976. A total number of 4976 cases was analyzed in various ways (Table 25). Occurrence of the diseases were high among the age groups 55 – 59 years in males and 45 – 49 years in females. The registered cases amounted to 67.1% from urban and 37.9% from rural areas. The urban cases showed malignant diseases related to lung, breast, ovary, and thyroid, whereas rural patients presented diseases affecting esophagus, skin, stomach, liver, and lymph glands. Of the registered cases 47.9% were male, and 52.1% were female; thus, females show a slightly higher morbidity than males. In general, gastic cancer is the most common malignant disease, comprising 23.5% of malignant diseases in males. The other nine most common malignant diseases among the male population in decreasing order of incidence are hepatoma, lung cancer, rectal cancer, malignant lymphoma, laryngeal tumor, esophageal cancer, osteomyelitic leukemia, renal cancer, and pancreatic cancer. This provides a striking contrast data from the United States, where lung cancer is the most frequent of malignant diseases. In the female population, cancer of the uterus comprises almost half, 40.6%, of the total malignant diseases registered in hospital clinics. The other nine diseases in decreasing order of incidence are breast cancer, stomach cancer, hepatoma, ovarian cancer, thyroid tumor, lung cancer, rectal cancer, malignant lymphoma, and chorioepithelioma. The high incidence of stomach cancer especially among the male group has been ascribed to the consumption of bean mash contaminated with *Aspergillus flava;* however, this has not yet been proven scientifically.

Table 25. *Analysis of Neoplasmosis in Korea (Korean Cancer Association, 1976*

Order of incidence	Sex			
	Male (%)		Female (%)	
1	Stomach cancer	23.5	Uterine cancer	40.6
2	Hepatoma	16.4	Breast cancer	12.3
3	Lung cancer	11.2	Gastric cancer	12.0
4	Rectal cancer	3.9	Hepatoma	3.2
5	Malignant lymphoma	3.9	Ovarial cancer	2.8
6	Laryngeal cancer	3.3	Thyroid cancer	2.7
7	Esophageal cancer	3.1	Lung cancer	2.7
8	Leukemia	2.7	Rectal cancer	2.4
9	Renal tumor	2.7	Malignant lymphoma	2.1
10	Pancreatic cancer	2.3	Chorioepithelioma	1.7
11	Others	27.0	Others	17.5

This analysis was based on 4976 cases reported to the Korean Cancer Association in 1975 (referred from Dong-A Newspaper of September 2, 1976). The reason for a lower incidence of lung cancer in females (7th in order) compared with males (3rd in order) is not known, but more frequent cigarette smoking among the male group might partially contribute to the higher incidence.

2. Endocrine and Metabolic Diseases

a) Disorders of thyroid gland (ICD 240 – 246): This disease is of interest only in the clinics, and no appreciable epidemiologic study has been performed [237, 238]. During the period 1962 – 1972 a total of 3704 cases of hyperthyroidism were treated with radioactive iodine [238]. Professor Min in Catholic Medical College, Seoul, collected 292 surgical specimens of thyroid from eight hospitals in 1972 and proved that 14 cases (4.8%) had had Grave's disease. On the basis of various materials, he estimated that annually about 6000 cases with thyroid trouble visit clinics, and 1500 of them may possibly have Grave's disease. Lee et al. [237] classified 304 cases of thyroid disease at the radioisotope clinic of Seoul National University Hospital from May 1960 to August 1961: 127 cases were diffuse toxic goiter (41.8%), 82 cases, nodular nontoxic (27%), 54 cases, diffuse nontoxic (17.8%), 28 cases, nodular toxic (3.2%), 7 cases, thyroiditis (2.3%), 5 cases, thyroid tumor (1.6%), and 1 case, thyroid cyst (0.3%). Toxic goiter comprised 53% and nontoxic goiter, 47%. Most of the goiter patients older than 30 years of age were found to have nodular goiter. The ratio between males and females was 1 (20%) : 4 (80%). No data is available on endemicity of the disease. Geographic analysis and examination of water from the suspected area should be performed to clarify the causal factor of the disease.

b) Diabetes mellitus (ICD 250): The exact data on the incidence of this disease in Korea has not yet been calculated. However, several reports, although fragmentary, may provide some understanding of the present situation. Several surveys of both rural and urban inhabitants indicate that the morbidity according to groups are 0.2% for 20-year-olds, 4.3% for 30 years, 11.1% for 40 years, and 7.8% for 50-year-olds. The general trend shows the highest incidence in highest age groups, i.e., 40 – 50 [223, 224, 225]. The incidence is far higher among urban inhabitants than rural [226]. Glycosuria-positive rates were rural, 1.6% and urban 5.0% at 20 years of age; 3.8% and 18.5% at 30 years of age; 4.5% and 25.7% at 40 years of age; and 1.8% and 25.0% at 50 years of age, respectively. Clinically diagnosed morbidity of diabetes mellitus according to age and area were 0.37% and 7.8% at 50 in rural and urban areas, respectively. A repeated diabetic detection survey was made in four consecutive years, from 1969 – 1972, among bank employees in Seoul [27]. The number of positive glycosuria was 86 (7.3%) in 1969, 70 (5.2%) in 1970, 84 (5.9%) in 1971, and 86 (5.8%) in 1972; 29.3% of the positives were diagnosed as diabetes. In the survey, 18 of 64 cases were previously known patients, but 43 were new cases. Reasons for the difference between rural and urban population and the high incidence among those 40 – 50 years of age are not yet clarified, although food habits and some physical factors might be involved.

c) Inherited metabolic diseases: Very few physicians in Korea are interested in inherited metabolic diseases [228]. In the literature since 1963 the following have been reported among Koreans:

> Ehlers-Danlos syndrome
> Wilson's disease
> Morquio's disease
> Acute intermittent porphyria
> Hurler's syndrome
> Gout
> Marfan's syndrome
> Fanconi's syndrome
> Pierre Robin syndrome
> Pseudoxanthoma elasticum
> Albinism
> Methemoglobinemia
> Gierke's disease
> Laurence-Moon-Biedl syndrome

Not specific to Koreans, the above diseases are reported to occur elsewhere in the world. The number will increase when physicians pay more attention to analyzing unknown chronic diseases cytologically and biochemically and keep carrier detection of the family in mind [228].

3. Nutritional Deficiencies
(ICD 260 – 269)

Conditions arising from nutritional deficiency were analyzed by Professor K. Y. Lee and her associates (Yonsei University 1967 – 1968) [16].

In the mountainous area (Anheung, Weonju Gun, Gangweon Do), they determined manifestation of malnutrition for both sexes and their prevalence rates: angular stomatitis (25% – 77%), cheilosis (15.4% to 55.5%), hyperkeratosis (7.7% – 31.9%), angular scar (3.6% to 20.6%), and conjunctivitis (3.0% – 14.9%). Other minor conditions showed no relationship to sex and age.

The type of condition and prevalence rate observed among both male and female children in the coastal area (Guryongpo, Yeong-il Gun, Gyeongsang Bug Do) were angular stomatitis (27.6% – 54.5%), angular scar (9.1% to 48.2%), hyperkeratosis (24.8% – 45.5%), and cheilosis (17.3% – 45.5%). Some minor conditions occurred in the following order of incidence papillary atrophy of the tongue, impetigo, and eczema; all seemed to be unrelated to age and sex factors.

Nutritional deficiency diseases found among children in the plains or farming area (Gaecheong, Oggu Gun, Jeonra Bug Do) were angular scar (20.5% – 57%), angular stomatitis (12.3% – 83.8%), hyperkeratosis (18.2% to 88.9%), cheilosis (6.7% – 33.3%), and other minor conditions showing no correlation with age or sex.

Nutritional deficiency diseases noticed among children in the urban city area (Seoul) were angular scar (10.5% – 33.0%), cheilosis (7.7% – 26.8%), angular stomatitis (13.3% – 24.4%), hyperkeratosis (7.7% to 45.4%), conjunctivitis (3.2% – 14.3%), and some minor conditions like papillary atrophy of the tongue. Age and sex did not seem to affect the prevalence rate of the diseases.

4. Diseases of Blood and Blood-forming Organs (Hemophilia)
(ICD 280 – 289)

The Korean Society of Hematology collected data on 200 cases of hemophilia from several general hospitals in Seoul and analyzed them. All the cases, except one, involved males, and 47.4% of them had family histories (Eui-Sa Shin Mun No. 1456, Dec. 6, 1976). Children comprised 164 cases, and 88 of them were below 5 years of age, even though hemophilia is characterized by a lifelong tendency to bleed. It seems more likely that most older persons with hemophilia have always had mild disease or are subjected less trauma. The types determined were hemophilia A 101 (80.8%), hemophilia B 19 (15.2%), and hemophilia C 5 (4.0%). The main symptoms were petechiae, epistaxis, long-lasting bleeding after injury, gingival bleeding, and arthritis. Epidural hemorrhage was recorded in 8% of the cases. Arthritis was more common in earlier periods.

5. Mental Disorders
(ICD 290 – 318)

Mental health has become an important subject in Korea along with the rising standard of living. However, it is still in its infancy owing to the prejudice prevalent among people that mental illness is incurable. A lack of understanding and cooperation characterizes even educated persons.

Scientific thought in the field of treating mental patients in Korea was introduced only several years ago. The nature of psychiatry in those days was largely descriptive, and emphasis was laid mainly on classification, isolation, and various somatic treatments such as electroshock and insulin. Psychological aspects and environmental factors in treating mental illness were paid little attention until the time of the Korean War when the practice of dynamic psychiatry was introduced through military psychiatrists among the U.N. forces. During the Korean War, many psychiatrists had an opportunity to study abroad and thus strengthen the mental health movement.

It has been estimated that there are approximately 340 000 persons with major psychoses and two million with psychoneuroses and personality disorders in South Korea. About 40 000 persons with psychoses are in need of immediate institutionalization.

A study of the prevalence of the major psychoses was carried out by means of an intensive census taken in a Korean rural village (Wacheon Myeon) located 24 km from Daegu City, Gyeongsang Bug Do [229]. The total number of residents in the village was 6846, and the majority of them (87%) had less than middle school education. The study, performed in September 1972, yielded the following results: 23 cases of schizophrenia, 2 cases of manic depressive psychoses, 11 cases of epilepsy, but none of general paresis were observed in the total population. Thus, the percentages were 0.33% for schizophrenia, 0.03% for manic depressive psychoses, and 0.16% for epilepsy. The corrected prevalence rate calculated by Weinberg's method was 0.87% (age 16 – 40) for schizophrenia, 0.1% (age 21 – 50) for manic depressive psychoses, and 0.26% (age 5 – 30) for epilepsy.

6. Cardiovascular Diseases
(ICD 401 – 429)

Hypertension: Blood pressure varies somewhat depending on age, race, environmental conditions, diet, and emotional factors. Although no reliable study on the epidemiology of hypertension has been conducted in Korea, various reports on the subject suffice for an outline of the situation [221, 222].

From an etiologic viewpoint, hypertension among Koreans has been classified by Sohn [221] on the basis of 112 inpatients: essential hypertension, 91 (81.2%); parenchymal renal disease, 21 (18.8%); chronic glomerulonephritis, 7 (33.3%); pyelonephritis, 12 (57.1%); diabetic renopathy, 1 (4.8%); and renal tuberculosis, 1 (4.8%).

Mean blood pressure: Systolic and diastolic pressures increase with age. Normal systolic pressure values in the male population by age group are 125 in 20 – 29 year olds, 129 in 30 – 39, 138 in 40 – 49 year olds, and 145 in those above 50 years of age. In the female population 117 in 20 – 29 year olds, 123 in 30 – 39 year olds, and 131 in 40 – 49 years of age. The pressures rapidly increase after 40 years of age in both sexes. Higher values are seen in Korea than in other nations: Korean, male, 138 and female, 131; American male, 128 and female, 129; Japanese, male, 129 and female, 127.

Prevalence: WHO criteria define hypertension as blood pressure exceeding 160 mm Hg systolic or 95 mm Hg diastolic pressure or both. The pressures increase sharply after 30 years of age in both sexes: in males the prevalences of an excess are 7.8% at ages 20 – 29, but 18.2% at ages 30 – 39, and 37.3% at 40 – 49 years of age; in females it is 6.9% in ages 20 – 29, 17.6% in ages 30 – 39, and 41.6% in 40 – 49 years of age.

Diastolic pressure exceeding 100 mm Hg is more common among females of all ages than among males. Of cases in Kyung-Hee University Hospital admitted due to heart disease, 58.5% – 67.7% were hypertensive heart diseases. In the Clinic of Internal Medicine hypertensive heart cases account for 4.4% – 9.6% of the total patients [221]. Thus, hypertension is becoming one of the serious problems in present-day Korea.

Characteristics: A familial factor (genetic factor) is suspected of being involved in hypertension; however, no reliable data is available on this matter. Distribution of hypertensive patients according to retinopathy by Keith-Wagner Classification showed grades I and II were 35.7% and 56.3%, respectively, but serious cases above grade III were 1.2% (8 cases) of 699 in a mass survey [221]. Scheie S. III – IV showed 2.8% in the 40 – 49 year age group, and 4.4% in the 50 – 59. This is much higher than the same age level of Japanese, 0.2% and 1.5%, in the same age groups, respectively. All figures indicate that hypertension cases among Koreans show a high rate of sclerosis signs in arteries of the orbital fundus. Abnormal pulse wave velocities of 9 m/s and above have been reported in Koreans by Sohn [221]. In nonhypertensive Koreans, prevalences by age group are 10.3 in ages 40 – 49, 15.8% in ages 50 – 59, 43.5% in ages 60 – 69. These are also far higher than the same age groups of Japanese, 6.7%, 15.4%, and 17.0%, respectively. Especially in Korean hypertension groups, the rates are 16.0% in ages 40 – 49, 33.3% in ages 50 – 59, and 50.0% in ages 60 – 69 years of age.

The overall reports suggest that the grade of sclerosis may be increasing in Korea. Abnormal ECG distribution in the various age groups of the normal population are 13.5% in ages 40 – 49, 25.4% in ages 50 – 59, and 33.0% in ages 60 – 69 year [221].

Hypertension is ultimately related to disturbances of cerebral vessels, the heart, and the kidney. The relatively higher incidences of apoplexy and heart-insufficiency in Korea may be due to the poor management of hypertension. Sohn [221] found that 35 (19%) of 156 hypertension cases did not receive treatment previous to that in the clinic. Cholesterol levels in Korea are rather low compared with other nations, perhaps due to the low intake of animal fat sources. Serum cholesterol levels below 225 reach 91.5% in the general population and 79.3% in the patients admitted to the hospital.

It is a conundrum that so many hypertension cases have been encountered recently in spite of low cholesterol levels. Diet may play some part; however, no scientific evidence is available to support this supposition. In general, a Korean meal consists of a large amount of rice, the staple food, and small amounts of accessory foods, but a large amount of salt. The daily intake of salt ingested with food averages 20 g in Korea, far exceeding the amount in other countries: American, 10 g; Japanese, 15 g; and the WHO recommends 10 g per day. Excessive salt intake may damage the kidney function and promote continuous tension on the blood vessels (Dr. Sohn Wi-Sock, Josun Il Bo, January 4, 1977).

Another role may be played by the toilets in a Korean house; these differ from the western style and are unsuitable for hypertension cases. Since they are not seat-style, sudden congestion of blood in the brain may causes apoplexia when the person rises from the squatting position. Although no data is available on incidence, many apoplexias in clinics have confessed that the onset occurred while defecating in the toilet.

7. Industrial Accidents, Occupational Diseases and Poisoning

a) Industrial accidents: The 1972 statistics indicate that 45 673 (4.6%) of 987 856 employees in 9375 places (each with 30 or more employees) required medical treatment for 8 days or more due to occupationally acquired physical disorders [230]. The percentages of work-connected injuries were categorized as 9.8% of accidents involved insured workers, and 47.4% of accidents involved uninsured. Accidents categorized by injured parts of the body, in order of their frequency, in the manufacturing and coal mining industries were finger, foot, hand, face, abdomen, leg, arm, eye, head, chest, toe, back, and others. Poisonous factors according to size of the industry were benzene, toluene, xylene, sulfur dioxide, carbon monoxide, ammonia, chlorine, formalin, lighting, noise, and dust. Prevalence of major occupational diseases organized by industry were textile, paper, chemical glass, chinaware, metal and machines, electrical appliances, transportation, lumber, and lead. In 1968 a total of 2630 working in 60 different factories were sampled; 480 diseases were detected with a prevalence rate of 18.2%. Such a high rate of hazards is considered to be due to lack of safety control measures and poor facilities [230].

b) Occupational diseases: Pneumoconiosis is one of the common occupational diseases of employees who work in conditions of poor ventilation. Of the 6588 cases of occupational diseases reported to the Ministry of Health and Social Affairs in 1960, the highest rate was seen in manufacturing; the second highest was in mining. Diseases with a high incidence were silicosis and dermatitis due to mechanical irritation or high temperature; eye trouble due to light and radiation; radiation sickness; arthritis and neuralgia due to excessive work; healing disturbance due to noise. Less significant diseases were poisoning caused by nonmetallic or metallic substances and dermatitis and poisoning caused by organic substances. The incidence of these diseases will increase if working environments are not improved. According to one report, 84% of all establishments had unsatisfactory or very poor working environments. Compensation for occupational diseases should be made in accordance with regulations, but the regulations covering maximum allowable concentrations of industrial contamination have not yet been established.

Establishments with dining halls, washing facilities, and clothing changing rooms occur in 91% of the mining companies and 48% of the electricity and gas industries. The law requires pre-employment physical examination and annual checkups, but a thorough examination is not being carried out. Aptitude tests and manpower surveys are not at all practiced at present.

c) Industrial poisoning: Problems of industrial poisoning commonly accelerate as industry develops. This is because the employees in factories usually have contact with various chemicals in the form of gas, vapor, mist, fog, dust, fumes, or aerosol. There are roughly 400 kinds of noxious chemicals in the factory atmosphere which may cause various acute or chronic occupational diseases in the employees. Although the hazards may depend upon the concentration, exposure time, and working conditions, it is generally accepted that many of the factories in Korea exceed the threshold limit value and maximum allowable concentration.

Outbreaks of lead poisoning have frequently occurred in lead mines, lead-refineries, battery-maker, printing offices, and porceline factories. Chung [231] examined 217 employees working in these factories and found about 26.7% distressed from the atmospheric concentrations exceeding 0.2 mg/m³. Poisoning from organic solvents is also common these days. Aplastic anemia from chronic exposure to benzene is one of the serious examples [233], and there is a strong possibility that the incidence may increase, because there are many factories, including home industries, where benzene is used as a solvent. Metal poisoning from mercury, chromium, cadmium, beryllium, zinc, magnesium, and manganese also show an increasing trend. Especially the latter three cause "metal fume fever" leading to leukocytosis [232].

d) Carbon monoxide poisoning: The majority of people utilize coal briquettes (local name, *Yon-Tan*) as their main fuel source. But the carbon monoxide produced during combustion of the *Yon-Tan* raises a serious problem, noxious gas poisoning. Accidents occur mainly at night while sleeping on *Ondol* (earthen hot-floor). Sometimes, crevices are formed between the *Ondol* and Yon-Tan site through which the gas is introduced into the room. In this way many accidents have occurred, especially in the winter and spring seasons, although the statistics are obscure. A newspaper, *Dong-A Il Bo* (1 February 1977), reports that the number of deaths due to poisoning was 580 in 1973, 468 in 1974, 517 in 1975 and 1013 in 1976. Also closely related to this problem is low atmospheric pressure. Even when not acutely poisonous, the gaseous environment in the house may cause other noxious sequences such as headache, neurologic disorders, and chronic systemic effects.

The health of housewives exposed to carbon monoxide polluted air when attending to the coal briquettes is posing a serious problem these days. An investigation was made in Seoul from 2 – 30 October 1972 to find out the amount of carbon monoxide in kitchens of 184 houses when housewives cook [234]. The major findings obtained were as follows: the mean concentration of carbon monoxide was 59 ppm, whereas the carbon monoxide concentrations of 15 households (8.1%) exceeded 100 ppm, known to be the maximum permissible concentration. The mean concentration of carbon monoxide was 53 ppm in Korean-style houses, 68 ppm in western-style houses, 46 ppm in apartments, and 71 ppm in mixed Korean-western style houses. The concentration decreased proportionally with the decrease of kitchen space. The mean carbon monoxide concentration remained at 93 ppm for 2 h after the coal briquette was changed. A higher concentration of carbon monoxide was observed in the higher temperatures and lower atmospheric pressure.

e) Shellfish poisoning: Shellfish is one of the favorite foods of Koreans. It is a reasonable conjecture that such a food habit may occasionally result in shellfish poisoning. In the past, several suspicious outbreaks have been reported, although scientific proof was lacking. However, recent outbreaks of shellfish poisoning caused by clam poison, namely *venerupin,* in the coastal area of Okpo Bay, Koje island, attracted attention, and the epidemiologic features were analyzed [235, 236]. The incidence in the area between 1968 and 1971 revealed 319 cases with 21 deaths. The outbreaks were frequent during March and April of 1968 and 1969. Clinical signs were headache, nausea, vomiting, malaise, abdominal pain, fever, epistaxis, gingival bleeding, coma, and other symptoms. Death occurred within 5 days after ingestion of shellfish, but the survivors recovered within an average of 6.2 days. The clam poison has organotropic rather than neurotoxic. As the causal agent, an organic phosphate substance was suspected [236]. Furthermore, it has been proven that shellfish poison is derived from the ingestion of plankton. Mollusks collected in the area of the Okpo Bay were *Meneinella clavigera, Tegula lischkei, Mytilus edulis,* and *Tapes philippinarum.* Since 1971 there has been no incidence in the bay area due to prohibition of clam collection. A toxicologic survey was conducted by Hong et al. [236] on shellfish obtained in coastal towns of South Korea. But poisonous *venerupin* shellfish were not found except in the area of Okpo Bay. Nevertheless, it is still early to conclude that the problem of shellfish poisoning has been eradicated from the land until the epidemiologic and ecologic characters of shellfish and plankton are clarified.

D. Factors Affecting Health Problems

I. Cultural Factors

1. Cultural System

Two significant cultural systems in Korea affect the traditional concept of disease: shamanism and oriental medicine.

Shamanism is an aboriginal religion in Korea, existing even before recorded history. Buddhism, introduced into Korea in 372 AD, Confucianism, also appearing in the fourth century, and Christianity in 1784 AD, have spread in this country to become the three major religions. But even today in the midst of vigorous modernization the religious attitudes and perceptions of the masses are mostly under the influence of shamanism. Consequently, personality formation of the masses and folk culture, aside from the elite culture, have largely been under its influence.

Oriental medicine, well known as *Hanbang*, is a kind of folk medicine, an amalgamation of Korean traditional folk knowledge, aboriginal ancient medicine, Chinese medicine introduced into this country in 561 AD, and Indian medicine brought in with Indian Buddhism in 372 AD. No less than that of shamanism itself, oriental medicine has played a great role in the formation of the concept of disease in Korea [242, 243, 244, 245, 246].

2. Native Festivals

As in other nations the immortality of the soul has been a basic belief in this country. Indeed, this belief has induced the people to revere a corpse even after its burial. The inside of the grave, in the Koguryo era, was provided with a beautifully painted stone wall. Moreover, the selection of the grave site was one of the big events in the life of a family. The grave site should be in the middle of hills on its left and right. This prescription based on the oriental belief in the wind and water theory (*Poong-soo swol*), according to which a suitable resting place for the soul should be on a hill facing south and at the midpoint between the blue dragon on the left and the white tiger on the right (blue dragon and white tiger are symbolic expressions meaning hill). If the body with its soul is buried correctly, the soul will be pleased and will take care of all fortunes of its progeny. With the exception of religious sects, the majority of the people perform ceremonies for 3 years following the death of their parents. On the 1st and 15th day of each month, the children prepare special food for the deceased parents and offer it on a specially prepared meal table. In addition, they perform a worship ceremony once every year on the anniversary of the death. On the day of the full moon in August by the lunar calendar, the descendents visit the grave to worship. In these ways, the generation of elders and ancestors has become a respected custom in Korea. Since Korea has traditionally been a land of agriculture, several ceremonial days are related to agriculture: *Tan-0* (May 5th on the lunar calendar), *Chilseock* (July 7th on the lunar calendar), and *Choo-Seock* (August 15th on the lunar calendar). *Tan-0* is the time for planting. All activities in the paddy fields and gardens, even weeding, are finished by *Chilseock*. *Choo-Seock* is the beginning of the harvest, when the people put on clean clothes and worship at their ancestors' graves. The festival bears some resemblance to Thanksgiving day in Christian society. In addition, there is a festival day in almost every month of the lunar calendar: Jarnuary 15, February 1, March 3, April 4, May 5 (*Tan-0*), June 6 (*Yudu*), July 15 (*Beak-choong*), September 9 (*Joongyang Jeol*), October 1–10 (*Sangdal*), and November (*Dongji*-winter solstice). Farmers have had the custom of observing these festivals annually, commemorating each festival day from seed time to harvest.

3. Religions

Statistical information on religions in Korea can be categorized into eight main religious groups: Buddhist, Protestant, Catholic, Confucian, Chun-Doh-Gyo, Won Buddhism, Tae-Jong-Gyo, and other. In 1972 the Ministry of Culture and Information released data on religious beliefs in Korea (see Table 26). Of a total population of

Table 26. Statistical Information on the Various Beliefs [247]

Religion	Practicing clergy	Believers
1. Buddhist	18 629	7 990 000
2. Protestant	17 562	3 460 000
3. Catholic	3 478	790 000
4. Confucian	11 831	4 420 000
5. Chun-Doh-Gyo	1 526	720 000
6. Won Buddhism	850	680 000
7. Tae-Jong-Gyo	55	150 000
8. Other	4 102	1 410 000
	57 997	19 610 000

35 million, the buddhist and Won Buddhist sects comprise a total of 8 670 000 followers; Protestants and Catholics, a total of 4 250 000; Chun-Doh-Gyo and Tae-Jong-Gyo, religions founded in Korea, a total of 870 000; and various other types of religions, e.g., newly-formed beliefs, a total of 1 410 000. Confucianism, which was propagated among the people after the establishment of the Yi Dynasty, was in its essence a political ideology more than a religion. Since Confucian "Clergy", estimated at around 11 831, are not discernable as clergymen by any manner of dress or form, it is impossible to know who they are.

Data relating to folk religions is entirely absent from this description because they are considered superstitions and not religions in the regular sense. However, in 1973 members of an organization composed of *Mudang* (Shamans) and fortune tellers, known as *Kyung-Shin Hoe* and having a total following of 48 980, claimed a membership of an estimated 208 424 practitioners (*Mudang* and fortune tellers) around the country; however, a number of the management staff asserts that because 80% of the members are uneducated, guidance and organizational structures are difficult to maintain. Nevertheless, shaman culture seems to be gradually disappearing. Indeed, the reality of Korean folk religion today is a lamentable one as western civilization infiltrates into this feudal society [247].

4. Ceremonials

A Confucian ceremony is held by blood relatives of the upper-class, male-dominated paternal family line. On traditional holidays, all male descendants, including the first, second, and third cousins, gather to hold a brief ceremony for the four generations of the departed grandparental ancestors; a memorial ceremony is also held for them on the day of their death. For the fifth generation and over, a combined grave-site ceremony is held at each grave during the 10th month of the lunar year. The eldest grandsons participate in these grave-site ceremonies, but all descendants are expected to participate in brief religious services and memorial ceremonies. The women of the family may prepare the things to be used in the ceremony, such as the food and wine offerings, and serve them, but ordinarily their actual participation in the ceremony is prohibited.

The Korean belief in maintaining respect for one's ancestors and the aged and the separation of men and women is a custom that is rare to find among the peoples of the world today. The Confucian practices as a whole have their merits as well as defects. In former times, the Confucian literati themselves were responsible for the flunkeyism (manifested in Korea's vassal state posture toward China), factionalism (among different clans), ill-effects of the family-centered system, rank-consciousness, effeminacy, and contempt of any industrial capacity that existed (merchants and laborers were looked down upon by the upper-class literati, who regarded manual work of any sort as mean). Their critical attitude to everything, along with the aforementioned, contributed to the defects that still crop up in the many-faceted society of today.

Because the Confucian ceremony of paying respects to ancestors was held exclusively by the blood relatives of the male-dominated paternal family line, the alternative religions adopted a joint ceremony held by small neighboring villages which maintained an existence apart from the Confucian-style ceremony. However, because of the strong Confucian influence, religion for females was completely disregarded by the Korean populace and misunderstood as being composed of nothing but superstition. Chun-Doh-Gyo, the religion of the "Heavenly Way", is found in Korea since fairly recent times. Won Buddhism is a newly created sect of Buddhism which does not believe in the use of icons; further, the clergy do not don the traditional garb of Buddhist monks, and the chants and sacred songs are sung in the western-tempered scale, much like that of Christian hymns. Tae-Jong-Gyo is a religious sect based on the teachings of Tangun, the mythical founder of Korea.

5. Superstitious Religions

a) New religions: Since the nineteenth century a large number of new religions have been appearing in Korea. Their number today is estimated at around 240 with the number of adherents ranging from a handful to 600 000. These new religions have generally appeared in times of social turmoil, e.g., in 1884 during the Tonghak Revolution, which tried to eliminate foreign influences from the country, after the Japanese occupation in 1910, following Independence in 1945, after the Korean War in 1950, and the Student Revolution in 1960. Many of these new religions share the following characteristics: (1) *Syncretism* — Most claim to represent a union of the established

religions and contain elements of shamanism, Confucianism, Taoism, Buddhism, and Christianity. (2) *The beginning of a new world* — All contain the belief that the end of the world is at hand, and a new one is about to begin but only the believer will inhabit this new world without disease, poverty, or social class. (3) *The chosen people and Korean nationalism* — There is confidence that in the new era, Korea will be the political and religious center of the world. (4) *Sexual activities* — Three types of sexual activity are associated with these new religions. First, sexual intercourse is understood as an esoteric rite between a charismatic leader and some of his fanatic female believers. For example, the leader of one sect believes that he must have intercourse with 1000 female believers to hasten the end of the world. Second, "chaotic" sexual activities between religious leaders and believers are engaged in. These various sexual practices have been censured by public opinion. (5) *Concern with "this world"* — Most new religions see the new era as materializing in this world and not in some other heavenly realm. (6) *Union of politics and religion* — Most of the new religions expect to govern Korea themselves or even the world. (7) *Faith healing* — Often these religions have a prominent faith healing element based on primitive, magic beliefs, e.g., drinking cold water, drinking lessed water, laying on of hands, spells, and starvation.

These new religions are interpreted as wish fulfillment of deprived people: 80% – 99% of their members come from the lower socioeconomic classes, and about 80% are semi-illiterate. In some groups, membership is almost entirely female, and most are at least covertly anti-government. Some founded by female leaders envisage the new world as one in which men are dominated by women. Often members are the sexually deprived such as widows, separated women, or women with impotent spouses. Korean culture has traditionally inculcated shame about sexual needs.

A number of explanations have been offered for the burgeoning of these new religions. One reason is the failure of the older religions in a variety of ways. Shamanism is too primitive to appeal to civilized people, and it is also overly concerned with money and power. On the other hand, Confucianism is too formal and impersonal to help people in a rapidly changing society. Another important feature is the effect of rapid acculturation. Korean society is rapidly moving from inhibited sexual attitudes to greater freedom, from the extended to the nuclear family, from interdependence and mutual cooperation to vigorous individual competition, from an other-directed society to an inner-directed one. These rapid changes have given rise to profound confusion and anomie. The new religions are an attempt to find new values and gratifications.

b) Shin (gods): Beliefs as practiced by females in the household and shamanism are rather diverse. The womenfolk of Korea have always lived to serve their husbands and sons with the utmost sincerity, attempting to fulfill their every wish and desire. This is the object of their supplications and prayers before the various icons that adorn the household. Although some differences may be found by localities, in the kitchen revered the fire god whose name is *Cho-Wang,* and in the household proper, other gods are supplicated to to prevent misfortune. In the inner chamber, *San-Shin,* the god of birth, or *Cho-Yong,* must be attended upon if the children are to grow well and the crops to be abundant. In the wooden floor room (Korean-style houses are composed of two types of rooms: that of hot-floor rooms heated from beneath, and wooden-floor rooms, which are unheated), *Sung-Ju,* the god of the household, must be served if family and household peace are to reign. The back of the house is the place where the talisman is set and supplicated to for good fortune (in Korea, such creatures as weasels, toads, and large snakes are regarded as such). Before the food storage area, *T'uh-Ju,* the god of the ground, is prayed to, and *T'aek-Ju Shin's* (the good of the house ground) appeasement is sought.

The religious practice of women in revering these various gods adheres very closely to shamanist ritual, specifically that of the more important procedural steps of the aforementioned *Cho-Wang, San-Shin,* and *Sung-Ju* rituals as well as that of the *Tae-Gahm Nori,* the Play of the General (a shaman ritual of soul held usually for good fortune). Here are also found rituals for keeping away the evil spirits bringing measles and smallpox, and supplicating to the god who wards off various demons in all the five directions (north, south, east, west, and center); the ritual for the ancestor spirit, *Tangun* (the mythical founder of Korea) and the god who wards off evil throughout the 12 months of the year; the rituals for the god of good fortune, the mountain spirit, and the god of long life; the rituals for *Ch-il-Sung* (The Big Dipper god, the god of life), for the god of the sacrificed food for the dead soul for *Bari-Kongju* (the god of the dead who consoles the departed soul and is also believed by the *Mudang* to be the founder of Mu-Sok, for the god of birth, for the Ten Kings (the Ten Kings of Hell) who pass judgement on all departed souls, and rituals of all different sorts depending on the purpose for which they are held. Also the Korean shamanist ritual, called *Koot,* is structured with a ritual beginning with the devotee entreating the various gods to reveal the cause of the present unrest that has come about and ending with her supplication of appeasement.

c) Mu-Sok: Of all the religious phenomena existing or coexisting in Korea, including Buddhism, Taoism, the Confucian ceremony, etc., that which boasts the longest history and is in fact the only native religion is Mu-Sok. Although Mu-Sok has long suffered under the stigma "superstition", and the ensuing problems, the roots of this folk religion are still deep today and are embodied in the entire religious thought of the Korean people.

The very beginnings of Korean Mu-Sok are obscure; the first piece of clear evidence we have is the gold crown of Shilla. In the past, Mu-Sok functioned as a household religion, which unified the people and give rise to various forms of art. Although Mu-Sok of the Koryo period harmonized well with the state religion of

Buddhism, the Confucian Scholars of the time had already begun to voice opposition to it. Whereas some *Mudangs* outside the court were banished for having deceived the people, there were some upper-class female shamans who received the official title of *Sun-Kwan* (fairly officer).

During the establishment of the Yi Dynasty, the policy of "Respect Confucianism — Oppose Buddhism" was in effect. It was prescribed in the first book of rules entitled *Kyung-Kook Tae-Jun* (A Compilation of the Laws of the Land) that the wife of a nobleman shall not visit a shrine or tomb, nor participate in any form of supplication in the mountains or other outdoor areas, penalties being imposed for violators. Further, it ordained that likewise the "Eight Outcasts" of society should be prohibited from these activities: Buddhist monks, *Mudangs, Kwangdae* (male entertainers, e.g., actors, acrobats, clowns), *Kisaeng* (female entertainers), artisans, butchers, funeral pallbearers (a special profession), and servants. Although Confucianism had permeated to the level of the common people, the so-called brief religious service, the memorial service, and the tomb ceremony, although well organized and well ordered, suffered from overformalization, and as a result, the common peoples' desire for a true religious feeling could not be fulfilled. Also the disregard for women in this strongly paternal-line blood-relative oriented religion served to suppress natural religious feeling.

Therefore, Mu-Sok, sinking deep into the consciousness of the common folk, became particularly an inseparable part of the religion of the womenfolk, and it has survived oppression even until this very day. Under the policies of the Yi Dynasty's Confucianism, women's education was also neglected. As a result during the 500-year reign of the Yi Dynasty, Mu-Sok was mainly practiced by uneducated women and thereby underwent change. The women believers themselves do not think of their religion as being superstition (although it really is to some extent), but as only a creed that in the execution of its various rituals is followed by natural human instinct and desire. This is the element of realism making up the basic structure of Korean Mu-Sok.

Korean Mu-Sok is devoid of any form of religious theory, scripture, organizational or ethical code. The form of religious thinking has been directed only toward the invocation of blessing, longevity, wealth, the exorcism of evil, and the things of everyday life that are wanting. In other words, it is only these desire phenomena that exist foremost in the devotees. In addition, numerous gods are supplicated to by the devotees with modesty and sincerity.

d) The function of Mu-Sok: In Korea there are many small islands and mountain villages where hospitals, doctors, pharmacies, and even *Mudangs* are not to be found. In such places live people who function much the same as mentioned above. On small islands off the western sea coast live women between the ages of 50 and 60 called *Sun-Kuh-ri* (*sun* is an amateur; *kuh-ri* means ritual, or in this particular case, *Mudang*), and in the mountain villages of Gangweon Do are found spirit-possessed men between the ages of 50 and 60 who are called *Bok-ja* (fortuneteller). There main occupation is farming but, when called upon by the villages to do so, they act in the capacity of shaman and fortuneteller. These individuals are not only empowered to tell fortunes, but when a sick person is to be treated, they also conduct small-scale rituals, intone prayers of supplication, and indulge in acupuncture as well. In addition, they are called upon to conduct ritual supplications at the lunar new year for the peace and prosperity of the homes. These prayers of supplication for the village and the practice of acupuncture are done without charge; fortunetelling and small-scale rituals are done for remuneration less than 100 won (about U.S. 20 cents) plus a cheap pack of cigarettes as a gratuity; and all-night rituals for the peace and prosperity of the home (called *ahn-t'aek koot*) are performed for somewhere between 500 to 1000 won (U.S. $ 1 to 2) [244].

In Korea the functions of artistic creativity, dance, and musical entertainment other than in the service of Mu-Sok, the village ceremony, and the long-enduring folk religion abound. The many Buddhist temples, pagodas, sculptures, and paintings are religious works of art also created by the hand of the Korean people. In the case of Confucianism, which is difficult to regard as a religion in the strict sense of the word, it is not an exaggeration to say that almost nothing new in the way of religious art was created for it by the Korean people. However, we have found that Korean Mu-Sok, the shaman, and the fortuneteller have served to unify the common people so as to create works of art and produce dance and musical entertainment.

6. Shamanism and Hanbang (Korean Herbalist)

a) Projective systems: Shamanism can be described as a cultural system projecting the pain of the mind onto supernatural beings. *Hanbang,* on the other hand, is a cultural system projecting the pain of the mind onto the body organs of the other. These two are entirely different systems but are similar in that they are both projective [240, 241, 245].

A neurosis is declared to have occurred "because of a dead ancestor" in shamanistic society, whereas it is announced to have manifested itself "because of the heart, the kidneys, and liver, etc." in *Hanbang.* Many clinicians in Korea know by experience that the projective tendency is strongly developed among Korean mental patients. Where in the past spirit possession syndromes were numerous, recently a precipitous decrease in their incidence is taking place due to the acculturation process. The possession syndrome is an illness arising from the projection of one's psychic ailment onto the ("because of") gods.

However, the increasing tendency of projecting one's psychic conflicts onto other human beings has recently begun to predominate, i.e., the object of projection is

being transferred from the gods to human beings. On the other hand, mentally ill patients in Korea show a very strong tendency to complain of their psychic pains in the form of various somatic symptoms instead of accepting and overcoming psychic ailments. In other words, mental illness produces more definite somatic symptoms than mental symptoms. Somatic complaints such as "powerlessness of the abdomen", "vacant feeling of the stomach", "indigestion", "weak heart", "sick liver", or "chest distress" are the outcome of the somatization mechanism, projecting and converting the psychic pain into somatic pain. Here again, the sense of individual responsibility and insight is diminished.

Shamanism and *Hanbang* have played positive, contributing roles in Korean society, but they have also negatively influenced the Koreas personality in that both of them are projective systems. The disease concept of shamanism is a projective system in which all the emotional and physical distresses are attributed to supernatural causes. The concept of Hanbang is also a projective system in that all emotional and physical distresses are attributed to supernatural causes. The concept of Hanbang is also a projective system in that all emotional problems are attributed to somatic dysfunctions. In this way, the predominant tendency of projection, especially somatization of the Korean mental patients, has been influenced by the disease concepts of both shamanism and Hanbang.

b) Healing ceremonies: There are four kinds of shamanist healing ceremonies in Korea:

Sonbibim (simple praying). Sonbibim, the most inexpensive and simplest form of the four healing ceremonies, is recommended to the client who has a mild illness or poor economic status. With simple offerings such as boiled rice or cold water in a bowl, the shaman prays in regular sequence of the twelve gods, who are said to be present at the site, to drive away evil spirits from the patient and bring happiness. Occasionally a shaman casts magic spells to expel the evil spirit from the patient's body. Shamans give some suggestion of when the patient will be healthy. The seance can be held anywhere, at the shaman's house, at the patient's house, or even in the mountains. The patient or his family pays the equivalent of U.S. $ 12 – 40 for one seance.

P'udakkori: This healing ceremony is the commonest form, usually recommended to a patient who is not seriously ill or does not have enough money for a grand healing ceremony, *Koot.* After giving simple offerings and prayers as well as following the procedures of *Sonbibim,* the shaman sacrifices a cock or a hen with some magic gestures, or sometimes simply pantomimes the sacrifice. A shaman may give an oracle to the patient and his family. The term *P'udakkori* in Korean means to resolve problems with the sacrifice of a hen or cock. This ceremony is commonly held at the shaman's own house, and the client pays the equivalent of U.S. $ 25 – 50 for one seance.

Salp'uri: A patient suffering from acute illness is recommended this ceremony. The Korean term *Sal* means some evil spirit or power. Using more offerings, the shaman prays to evict the evil spirit or power from the patient's body while dancing vigorously with swords and spears in her hands. While dancing around the patient she mimes a stabbing of the evil spirit with swords and spears. Several other magic gestures usually accompany the dance depending on the nature of the case. This procedure suggests recovery. The *Salp'uri* is held at a handy place, either the shaman's or the patient's house. The shaman charges a fee ranging from the equivalent of U.S. $ 12 – 40 for one seance.

Koot (grand healing ceremony): A *Koot,* the grandest healing ceremony of the shamanistic approaches, is recommended when the cause and symptoms of illness are divined to be grave. A performance of all the 12 processes requires about 16 – 24 h. Each session has a special god to be worshiped, as every shaman believes in 12 gods. The particular god having the most intimate relation with the shaman differs from shaman to shaman.

Every session of this ceremony consists of three parts: dancing to accompaniment of drumming, an oracle intermingled with singing, and dancing with drumming again. Initially, dancing begins at a slow tempo along with the rhythm of the drum, but becomes progressively more rapid, and finally enters into simple and repetitive jumping. The shamanist meaning behind this procedure is to call and entreat the god to visit the seance. After incessant vigorous jumping, the shaman enters into a depersonalized state or into sham behavior of being depersonalized and gives an oracle in a dignified manner. From all appearances the shaman seems to be possessed. All participants pray to the shaman whose talk and gestures are believed to be those of the gods. The shaman's compassion and the close emotional communication between shaman and participants are clearly manifest. Finally, the shaman dances again to the tempo of the drumming; this corresponds to a farewell party for a given god. Such a process is repeated in every session with some variations of mood and technique in accordance with the character of the god who is the object of worship in each session.

During the session, comic dramas and plays are often performed by shamans for the purpose of entreating the god to visit, of amusing the god, or of diverting the participants' moods. Divination with various magic gestures are also frequently carried out at the proper time. At the peak of the ceremonial mood, the participants have the opportunity to dance with each other individually or collectively. This dance, called *Mugam,* begins also at a slow tempo, becoming faster and faster until they enter into a trance state. Emotional discharge is apparent.

The above description of a ceremony is applicable to the ordinary shamanistic ceremony or *Koot.* On the occasion of a healing ceremony, shamans perform the above ordinary process with some variations. *P'udakkori* and *Salp'uri* are frequently performed at an appropriate time during such sessions. The following procedures are also occasionally carried out according to the nature of the cause:

c) Other procedures: Che-ung: This procedure is a special magic for driving away the evil spirits who make trouble for people. In this procedure a straw puppet is used as a scapegoat. After rubbing the straw puppet over the patient' body, the shaman dances with a pair of swords and burns the puppet. The therapeutic significance seems to be that of suggestion.

Hwa-Jon: This fire dance is used for driving away the evil spirit who causes grave disease, especially psychoses. The fire dance is always held at night. Initially the shaman treats the patient as a dead body, pretending that he is really dead. After covering him with grave clothes and cadaver sheets and laying him in a supine position on the ground, the shaman, carrying fire in her hands, dances vigorously around the dead body for about 20 min. The emotional circumstance is so frightening that the patient and other participants may experience a feeling of terror. Although shamans interpret this procedure as driving away the evil spirits which are afraid of fire, three therapeutic significances are present: the suggestion of health, the effect of threats or punishment, and the symbolic realization of "rebirth."

Kilgallajugi: A sort of requiem, this procedure is performed just after a session for the ancestral spirit, when divinatory signs indicate to the patient that the ancestral spirit is the cause of illness. Shaman or patient runs rapidly forward and splits a long sheet of yellow hemp cloth into two pieces with her own body. The shamanist meaning of this procedure is to separate the dead ancestor from the living person and thereby avoid trouble. The emotional mood is so intensely sorrowful that all participants usually weep and relieve their guilty feelings and hostility toward their parents through catharsis and abreaction.

The above-mentioned procedures are performed in combination or singly during the healing ceremony *Koot* depending on the decision of the shaman. The shaman is accompanied by several other shamans and attendants at the *Koot,* which is usually held at the patient's house, but sometimes at the shaman's house. The regular charge for one series of *Koot* ranges from the equivalent of U.S. $ 120 – 400.

7. Future Scope

a) Statement: In summary, under the Confucian precepts of rationalism, formalism, and male domination, that thought to contain superstition of any sort or form was unconditionally rejected, thereby esablishing a prejudice in the minds of the Korean people during the 500-year reign of the Yi Dynasty. The succeeding Japanese annexation, the advent of Christianity and Communism, the desire of the younger generation of today to break with the past, and modernize, and the present streamlined government administration – all these have contributed to the gradual destruction of Mu-Sok. Yet, despite all, a tradition with such a long history and roots that penetrate as deeply as those of this village folk religion, continues to endure, despite the severe ill-treatment that has been accorded it, simply because it is so inseparably tied to the folk culture itself. Also it naturally has been forced to undergo change because of the painful oppression endured. Now a certain portion of the younger generation along with scholars, researchers, and the Ministry of Culture and Information of the Korean Government have recognized that folk religion is a worthy cultural treasure and are making every effort to preserve its traditions. However, in spite of all the merits, if any, these strong faith-healing elements, based on the undercurrent of so-called native religions, have hindered the honest acceptance of modern medicine in the scientific sense.

b) Practitioners of superstition: A Korean saying states that a "sick person should be treated as a baby whatever his or her age." The sick person is usually very feeble psychologically and is not able to reason. In such a state, some superstitious professionals attract the attention of the sick, causing him to delay medical treatment for some time. The prevalence of professionals dealing with the supernatural is higher in less-literate societies. Even though the educational level of Koreans is relatively high, most people are still not far removed from superstitions. Social factors are also related to disease prevalence. Statistics in 1961 indicate that the total number of superstition-related professionals was 19 295, including male witches, 1083, and female witches, 4246; diviners, male, 4246, female, 6842; physiognomists, male, 409 and female, 93; palmists, male, 159 and female, 107; phrenologists, male, 82 and female, 12; grave-site selectors male, 1746 and female, 18; fortunetellers, male, 834 and female, 194; and sutra-chanters, male, 1839 and female, 569. Of the above, it is noteworthy that witches and diviners, who are more closely related with the mentally and physically ill, have more female practitioners than males.

One way to cure this kind of social pathology is to raise the educational standard of the people and resolve the problems of social insecurity.

II. Unequal Distribution of Health Personnel

1. Changes in Rural-Urban Distribution

Population growth in the rural, urban, and metropolitan sectors over a 15-year period is noticeable. In rural areas, the population increased by 9.5% during 1955 to 1960 and by 8% during 1960 – 1966, but it decreased by 7% during 1966 – 1970. Urban population grew by 35.2% during 1955 – 1960, by 32.5% during the next 6 years, and then increased by only 7.3% between 1966 and 1970. The metropolitan population, on the other hand, has grown at an accelerating rate: by 37.5% during 1955 – 1960, 44.6% during 1960 – 1966, and 62.9% during 1966 – 1970. Over the entire period the rural population grew by only 9%, whereas the urban population increased

by 92.3% and the metropolitan population by 223.9%. These statistics clearly show that most of the recent growth of the South Korean population has been absorbed by the two metropolitan areas, Seoul and Busan, and to a lesser extent, by smaller cities. Details of different growth and the development of urbanization are to be found in Sect. A. V. 9.

The migration pattern has caused different growth rates among the nonmetropolitan provinces and between these provinces and the metropolitan areas over a period of time. During 1955 – 1960 Seoul grew at the unusually high annual rate of 9.2%, but the populations of the nonmetropolitan provinces increase at similar rates, more or less comparable to the national growth rate of 2.9%. This rapid but abnormal urbanization has brought many problems in its wake: inadequate housing and transportation, increased crime rates, increase in public nuisances, etc. Incidences of carbon monoxide poisoning, traffic hazards, and other various so-called civilization diseases have accelerated due to this growth phenomena. To reduce the various untoward effects, the government is enforcing a depopulation policy by dispersing governmental organs and factories outside the metropolitan cities.

2. The Medical Professions

a) Status: The Yearbook of Public Health and Social Statistics of 1974 indicates that 5735 (50.8%) of the total 11 286 medical facilities are concentrated in Seoul (4543) and in Busan (1192). The rates of myeon (administrative districts), i.e., population without medical personnel (see annex II), neither dentists nor herb-doctors, showed the reserve trend: none in the cities of Seoul and Busan, but the highest rate in Jeonra Bug Do with 46.8%, then Jeonra Nam Do with 46.6%, and Chungcheong Bug Do. Chungcheong Nam Do, Gyeongsang Bug Do, Gyeongsang Nam Do, Gangweon Do, Jeju Do, and the lowest 10.9% in Gyeong-gi Do. Statistics on hospitals and hospital beds in 1976 still show a similar problem (Table 27 and Map 4) [248].

Licensed physicians in 1974 numbered 15 722. Statistics in 1976 (*Eui-Hak Sin-Bo,* a medical weekly press, No. 518 and 552, 1976) reported that 12 416 medical professionals were concentrated in Seoul, and of this number 5522 were licensed physicians (male, 4566; female, 956), thus indicating that almost 35% of the licensed physicians work in the capital city [9, 17, 45, 55]. In 1974 the total number of board specialists numbered 5262, and 2255 (43%) of them were concentrated in Seoul and 430, in Busan. Even though the absolute number of medical personnel is not seriously insufficient, such an abnormal distribution may work against the principle of equal opportunity to receive health care benefits. In 1974 there were 536 myeons without any medical personnel, neither dentists nor herb doctors, for a population of 4 964 489, about 15% of the total population of 32 905 000 in South Korea, as estimated from a continuous demographic survey in 1973.

Table 27. Hospitals and Beds in Republic of Korea in 1976

City and province	Hospitals			Beds		
	Total	Public national	Private	Total	Public national	Private
Seoul	27	8	19	6 136	1 635	4 501
Busan	10	3	7	1 607	633	974
Gyeong gi Do	12	6	6	1 916	532	1 384
Gangweon Do	10	6	4	993	322	671
Chungcheong Bug Do	3	2	1	398	185	213
Chungcheong Nam Do	7	5	2	1 108	286	822
Jeonra Bug Do	6	4	2	906	347	559
Jeonra Nam Do	10	4	6	2 096	520	1 576
Gyeongsang Bug Do	11	7	4	1 792	708	1 084
Gyeongsang Nam Do	7	2	5	869	200	669
Jeju Do	3	2	1	131	82	49
Total	106	49	57	17 952	5 450	12 502

Source: Ministry of Health and Social Affairs (1976. 9.20); Eui-Hak Shin Bo, No. 552

b) Government Plan: To correct such adverse phenomena the government set up a 5 year plan to cover establishment of a comprehensive medical care delivery system, strengthening of preventive health measures and disease control, improvement of living environment, gradual development of a medical insurance system, effective management of national welfare, a pension system effective compensation for industrial hazards etc. The ultimate plans include a first-stage medical care by health centers or dispensaries with 60% coverage; a second-stage medical care by health centers, hospitals, and private clinics with 25% coverage; and a third-stage medical care on the national level by general hospitals and specialized hospitals with 15% coverage.

To help meet these needs, a formal request for loan support to establish County Health Care Centers was made by the Korean delegation at the 1972 conference of the International Economic Commission for Korea (IECOK). On the basis of this request, the United States Agency for International Development (USAID) dispatched a feasibility study team to Korea in November 1973. The team recommended conducting a survey of national health care needs, setting up a low-cost health care system, and establishing a health planning unit in the government to be exclusively in charge of planning. Consequently, a Health Planning Project Agreement was concluded between the governments of the Republic of Korea and the United States (USAID) in June 1974. Then in November 1974 the Economic Planning Board (EPB) further requested on AID loan of U.S. $ 5 000 000 for a Korean Health Demonstration Project. A loan agreement was developed, which the parties concerned signed in September 1975.

In December 1975 the law of the Korean Health Development Institute (KHDI) was promulgated, and a Presidential Decree for enacting the KHDI Law was approved and promulgated in April 1976. The official opening ceremony of the KHDI followed the promulgation of the law and the decree on 19 April 1976.

The functions of KHDI are the following:

(1) To investigate, research, and evaluate previous and existing projects and systems related to public health.

(2) To design and develop effective and low-cost health care delivery sytems as demonstration/research projects in selected geographic areas.

(3) To implement and then evaluate the effectiveness and utility/cost of the foregoing demonstration projects (Projects should address the following aspects of health care: prevention of disease, diagnosis, medical treatment, rehabilitation, and medical insurance).

(4) To investigate and evaluate the long- and short-range needs for health services.

(5) To support the improvement and maintenance of the health status of local community residents.

(6) To educate and train persons engaging in health care demonstration projects.

(7) To exchange information with, conduct joint research with, and/or provide support to, research institutions in Korea and in foreign countries, on national health care delivery systems.

(8) To investigate, research, and evaluate particular health service activities as requested by the government.

The demonstration program was outlined as follows.

Beginning in 1977, the KHDI will implement a series of research/demonstration (training) projects leading to a comprehensive health program in selected *Guns* (counties), covering a total population of 500 000. The selected *Guns* are Oggu (Gun in Jeonra Bug Do), Gunwi (Gun in Gyeongsang Bug Do), and Hongcheon (Gun in Gang-weon Do). Although the ultimate purpose is to find the most effective way to practice integrated health and medical care in the country, the respective activity in the three Guns is different in content: in Hongcheon the project is concerned with nurse practitioner demonstration, in Oggu with health insurance, and in Gunwi the maternal health system.

c) Illegal medical practitioners: Violations detected in 1974 totaled 1697 cases, less than the 2734 cases reported in 1965. Improvement of social conditions may have influenced this decrease. These violations may be classified as practicing without a licence (232), illegal transfer of licence (4), posing as specialists (123), illegal advertisement (110), practice outside or registered area (440), nonregistration (50), and others (738). Of these violations the one of practicing without a licence is a real hazard to good medical practice. These charlatan physicians, dentists, or herb-doctors may deceive not only the patients, but also society. It is estimated that more violations occur than those exposed by the governmental inspectors, especially in doctorless areas. To treat such illegal medical practices, strict inspection, adequate disposition of medical personnel, establishment of a medical insurance system, establishment of at least medical facilities even in remote areas, and improvement of the transportation system are courses to be followed.

3. Expenditures for Medical Care

Budget allocations in 1976 for Health and Social Affairs in the Republic of Korea were 40.5 billion won (1.99%) for social affairs, 16.7 billion won (0.82%) for medical services, and 1.7 billion won (0.087%) for disease prevention from a total government budget of 2036.1 billion won. However, data are scarce in the estimates of gross national health expenditures. The total national health expenditure amounted to 191.2 billion won in 1974, which was three times the 64.7 billion won spent in 1970. The expenditure as a percentage of the GNP increased from 2.5% in 1970 to 2.8% in 1974 [249]. Health and medical care expenditures per capita increased from 2000 won in 1970 to 5724 won in 1974. The Social Security Subcommittee of the Health Sector Task Force use 5611 won as the baseline figure for average per capita medical care expenditure in 1973 (Table 28).

In 1974 the public sector consumed 11% or 21 billion won of the total national health expenditure, and the private sector, 89% or 170.2 billion won. The sources of health expenditures are increasing in the private sector. In 1970, 48.2% of public expenditures came from

the central government, and the remaining 51.8% came from the local governments. In 1974 the proportion of expenditures from the local governments increased to 54.7%. The public sources are increasingly to be found in local government. Local government's average per capita expenditure on health and medical care amounted to 386 won in 1973. Gangweon Do spent the most, 590 won per capita: and Jeonra Nam Do spent the least with 274 won per capita. The proportion of out-of-pocket expenditure on health and medical care increased from 2.3% of total private expenditure in 1970 to 3.5% in 1974. The proportion of national health expenditure to GNP is predicted to remain at 2.8%; the gross national health expenditures will increase from 250 billion won in 1975 to 662 billion won in 1981. If the rate of national health expenditures to GNP increases annually by 0.1% – 3.5% in 1981, the gross national health expenditures will increase from 259 billion won (W) in 1975 to 828 billion won in 1981.

Private consumption expenditure on health and medical care are predicted to increase from 213 billion won in 1975 to 583.3 billion won in 1981.

4. Utilization Data

According to figures issued by the Korean Pharmaceutical Economic Institute on the basis of information collected by the Korean Bank and the Agricultural Bank the monthly medical expenditures of a family in 1975 were W 1500 in urban areas and W 900 in the rural areas. However, figures vary from report to report (Hwan (Hw) was used before 1962. Won (W) is the currently official money unit.)

According to a study by Yang based on a sampling of 5159 persons in the Seoul area, the total cost of medical care for 1 year was Hw 2100 per family and Hw 400 per capita [250]. This cost of medical care was 7.2% of the average annual income of a family in Seoul city at that time. A breakdown of the cost of medical services showed that 73% was paid to physicians, 22.3% to herb-doctors, 3.6% to pharmacists, and 1.1% to others. Moreover, the study revealed that only one-fifth of the total illnesses were being cared for by medical personnel and related facilities. Of those whose illnesses were not cared for, 78% claimed economic difficulty as the underlying reason.

In a similar study in Jeju Do, where the sanitary conditions were much poorer than in Seoul, the total cost of medical care per year was Hw 2980 per family or Hw 640 per capita, i.e., 14% of the average annual income of one family on Jeju Do. This proportion of medical cost to income on Jeju Do is remarkably higher than the figures in Seoul. It indicates that low-income families spend a greater portion of their income for costs of sickness than do the well-to-do families.

The latest surveys conducted by Dr. Huh and his associates analyze the pending problems of medical care expenditures in rural and urban areas in present day Korea [251, 252].

Table 28. Comparison of Medical Care Expenditure (M.C.E.) {249}

	Korea (1974)	Japan (1973)	United States (1974)
Total national M.C.E.	191.6 billion won	3 949.6 billion yen	104.2 billion dollars
Total national M.C.E. per GNP	2.58%	3.56%	7.46%
M.C.E./capita	5 726 won (U.S. $ 14)	36 544 yen (U.S. $ 130)	492 U.S. $
GNP	6 779.1 billion won	111 033.9 billion yen	1 397.0 billion U.S. $
Ratio: population to 1 physician	2 500 : 1	850 : 1	615 : 1

a) Rural medical care: A service of medical care expenditures by rural areas in Korea was conducted in 1700 households comprising 16 655 persons during a 1-month period from 1 August to 31 August 1973 [251, 252]. Fourteen *Guns* were randomly sampled from the 140 Guns in the country. Of the total population age 5 and above (excluding children below the school age), 17.8% were illiterate, only 1.9% had received a college education. Of the total population age 15 and above, those engaged in farming and fishing constituted 18.5%. Jobless persons, including the unemployed, those engaged in housekeeping, and students, accounted for 60%.

Medical care expenditures were classified into direct medical costs and indirect medical costs. Direct medical costs constituted 70.6% of the total amount of medical care expenditure. Supply costs constituted 38.5% of the total amount of indirect costs, nursing services 26.2%, transportation, 35.3%. The average amount of medical care expenditure was W 1315 per treated case of W 151 per capita. The higher the educational level the greater the amount spent on medical care expenditure, but no significant difference was noted in medical care expenditures by type of medical care amounted to W 3178 for hospitalization, W 2345 for herb drug store, W 1683 for herb clinics, W 1685 for doctor's home visit, and W 1250 for superstitious practices.

During the survey period, 45.8% of the sick were "completey cured", 37.9% were "undergoing treatment", 8.6% showed "no change", and 6.5% showed "improvement". It is noteworthy, that as reasons for incurability, 51.5% cited financial straits.

b) Urban medical care: A study of medical expenditures by urban residents involving 1700 households comprising 7896 persons took place during a 1-month period from 1 August to 31 August 1973 in Seoul, Busan, and Daegu [252]. The average number of persons per household was 4.64. Of all the persons surveyed, illiterates accounted for 9.7%, those barely able to read the Korean alphabet, 10.2%, primary school graduates, 7.3%, middle school graduates, 20.8%, high school graduates, 29.6%, and college graduates, 22.4%. During the survey period prevalence rate per population of 1000 persons was 200.2 (161.1 for males and 242.5 for females). Patients wanting medical care, numbered 1182 of a total of 1581 reported sick cases, thereby showing a medical care demand ratio of 74.8%. The greater portion or 48.0% of the patients utilized pharmacies. Those visiting hospitals or clinics constituted the second greatest portion or 29.5%, and hospitalized patients, 10.1%. Those utilizing such non-western therapies as oriental clinics, oriental pharmacies, and folk therapies accounted for 9.9%. The total medical expenditure of all those surveyed was broken down into 72.9% for direct medical costs. The average medical expenditure per case was W 1983 (W 2155 for males and W 1860 for females). Medical expenditure per case for medical treatment amounted to an average of W 2637 (W 2935 for males and W 2433 for females). Monthly per capita medical expenditures reached W 397 (W 347 for males and W 451 for females), and the average

monthly medical expense per household was W 1842. The greatest medical expenditure was for the age group 0 – 9 years. The smallest expenditure was for the age group 10 – 19 years, and beyond this age the amount of medical expenditure tended to increase in proportion to age. The average medical expenditure by primary school graduates and those with lower educational standards remained below W 2000, a smaller amount than that spent by middle school graduates and those with higher educational standards. Thus, the higher the educational standards the greater the medical expenditure. An average of W 8333 was spent per medical case of diseases of the blood and hematopoietic organs, W 5949 for neoplasms, W 4662 for contagious and parasitic diseases, and W 4369 for pregnancy, child delivery, and obstetric complications. As reasons for failure to receive medical treatment for sickness, 47.4% or the largest proportion said, "I can stand it for the time being but will get medical treatment if it gets worse." The second largest proportion or 33.6% were awaiting natural healing because of the mild nature of the case. Those who did not receive medical treatment for financial reasons constituted 14.3%.

5. Economic Problems

Ignorance of environmental sanitation may widen the gap between rich and poor and urban areas and rural and may confer a higher morbidity rate on the poor and rural population. It is reasonable to generalize that unsanitary conditions prevail in low economic levels. Likewise, the prevalence rate of parasite infection will also correlate with living standards. The incidence differences among the people living in sanitary areas where flush systems are in use (A), unsanitary areas where rudimentary privies are used (C), and (B) the junction between (A) and (C) areas of Yonhee-Dong, Seoul, provide an example (Table 29) [128].

Most of the residents living in the C area are in the poor, tax-exempt group. The difference in prevalence and intensity of *Ascaris* and *Trichuris* infections between A and C presents quite a contrast. For *Ascaris* the incidences are 28.6% in A and 48.4% in C, and the intensities by mean EPGF (eggs per gram feces) bases are 3500 in A and 24 795 in C. *Trichuris* infection also shows a similar tendency. Thus, living standards and environmental conditions may give rise to important factors concerning morbidity rates. Mean values of EPGF among hospitalized cases in Severance Hospital who came from various parts of Korea are different according to the ward classes. *Ascaris* infected cases show a mean EPGF of 5936 in the low economic ward, 4551 in the middle ward, and 1846 in the private ward.

Hospital charges are too expensive for the majority of the population. If a person with an annual income of U.S. $ 500, which is a little above the per capita income of Koreans, is admitted to a hospital costing U.S. $ 20 per day for admission, drugs, and other medical cares, this means that he may consume all of his annual income in about 3 weeks. In view of the expense of private

Table 29. Soil-transmitted Parasites of the Inhabitants in Yon-Hee Dong, Seoul {128} [a]

Localities	Examined	Positive cases (%)					Total	EPGF		
		A. l.	T. t.	H. w.	T. o.	Taenia			A. l.	T. t.
A	56	16 (28.6)	18 (32.1)	0	0	0	25 (44.6)	Md	2 400	1 000
								Mean	3 500	1 900
B	67	18 (26.9)	26 (38.8)	2 (3.0)	1 (1.5)	0	35 (52.2)	Md	4 800	2 400
								Mean	5 250	3 739
C	122	59 (48.4)	71 (58.2)	6 (4.9)	0	2 (1.6)	101 (82.8)	Md	17 400	1 800
								Mean	24 795	3 410

[a] A, Flush system is facilitated, sanitary; B, Junction between A and C, relatively sanitary; C, Privies are used, unsanitary; ( ), Positive rate; A. l., Ascaris lumbricoides; T. t., Trichuris trichiura; H. w., Hookworm; T. o., Trichostrongylus orientalis E.P.G.F. Eggs per gram feces

hospitals, it is easy to understand why a large proportion of the population hesitates to visit western clinics and instead seeks other inexpensive ways to treat their physical problems such as visiting fortunetellers or herb drugists. However, it is fortunate that the government has established a new plan to assist the economically poor class by treating them without charge or adopting a loan system in the case of expensive surgical treatment. For this purpose, the government has allocated a budget of U.S. $ 22 million (11 178 383 000 won) for 1977. The people expected to be eligible number 14 063 in Seoul and 304 141 in other areas. The medical institution for the charity service will be designated by the government. Unfortunately, such assistance may not be the fundamental way to solve the pending problems, which lie deeper in society.

III. Environmental Sanitation

1. Administrative Organization

The Sanitation Section of the Bureau of Public Health, Ministry of Health and Social Affairs is authorized to control sanitation problems. Its administrative activities include sanitation education; enforcement of laws and regulations; inspection of the sanitary condition of water, sewage, foods, beverages, and commodity products; control of sanitary facilities, food manufacturing plants, swimming pools, public baths, restaurants, hotels; and sanitation guidance for model health towns.

2. Drinking Water and Water Pollution

Korea has water resources reaching about 114 billion tons annually; of the total rainfall, about 63 billion tons flow into the rivers.

In 1968 the water demand for household use throughout the country was estimated at 414 million tons a year; agricultural water, 7.27 billion tons; industrial water, 533 million tons; and other water applications, 1.04 billion tons. The total water demand was 9.25 billion tons, about 14% of the total river flow. In 1971, water supplies

for household use, agriculture, industry, etc., increased to 579 million tons, 10.16 billion tons, 863 million tons, and 1.04 billion tons, respectively, totaling 12.65 billion tons or approximately 19% of the total river flow of 63 billion tons.

In 1981 when the 4th Five-Year Economic Development Plan ends, four major river basins will have been developed, thus greatly increasing the water supply and supplementing the presently inadequate water resources for national needs. By then 17 714 billion tons of water will be available for use, comprising about 28% of the total river flow. Household water use will triple by 1981 to about 1.7 billion tons, and industrial water demands will increase about 3.5 times to 3.05 billion tons.

About 87% of the population of South Korea is dependent on roughly 238 000 private wells and 145 000 public wells, which have inadequate facilities and water quality. The Ministry of Health and Social Affairs has allocated a national subsidy to rectify these inadequacies and construct 28 000 new wells and has instructed local governments to follow a standard design.

Another 13% of the population is supplied with treated water through 85 municipal water sources. The source of municipal water supplies is surface water from rivers or reservoirs, which are not adequately protected against contamination. Operations at the treatment plants include coagulation, sedimentation, filtration (slow and rapid sand filtration), and in some cases chlorination. A few treatment plants are using simple processes such as filtration or chlorination, and a few other plants include aeration as well. For example, for the chlorination of water supplies, 970 000 lb liquid chlorine and 352 400 lb calcium hypochlorite powder were allocated by the AID program through the government in the period 1956 to 1959 (United States Aid program for International Development).

Nevertheless, water treatment still remains at a relatively theoretical stage due to the shortage of both facilities and specialized technicians to cover all the areas. Especially blatant in this respect is that no treatment plan has been put into effect in the rural areas. In the reclaimed zone in Honam Plain, which is one of the largest plains in the country, inhabitants still utilize water from ditches or brooks for drinking and laundry. Such a practice is also found in Jeju Do. In villages where a pipeline system is not yet provided, people use open well-water for both drinking and laundry. In such cases it is easy to conjecture why contageous diseases of the intestinal tract are highly prevalent in these areas. In Jeju Do drinking water in storage jars contained *Entamoeba histolytica* cysts in 8 (6.8%) of 117 samples examined [138].

The Marine Office of the Korean government reported in 1970 that the loss in marine products due to water pollution amounted to about 200 million won. Some fishing areas on the Ulsan and Yeosu coasts were seriously damaged by acid and oil pollution. In the downstream areas of rivers where sewage and effluents are discharged, fish have been exterminated, and these spots are no longer suitable for recreation purposes for city dwellers.

Diseases caused by water pollution are intestinal water-borne diseases such as typhoid, infectious hepatitis, and some parasitic diseases. Polluted water also consumes more chlorine for disinfection, thus raising the costs of water treatment.

Water pollution is caused by municipal sewage, agricultural wastes, and industrial effluents. Population increase and urbanization also exacerbate the effluent situation. Various kinds of waste effluents, toxic chemicals, and agricultural chemicals pollute the soil, which in turn pollutes underground water resources.

In the case of household sewage and industrial effluents in Korea, the absence of a water-treatment system adversely affects the quality of water kept in reservoirs for industrial use or for public supply. When these effluents flow into the sea, they also contaminate fish, algae, and other marine products. The increase of water resources in turn results in an increase in sewage, industrial effluent, and waste. Unless some adjustment is made to control water consumption and the amount of waste, then the amount of effluent discharged from industry will triple. Water pollution will then emerge as a grave problem indeed.

A case in point is the Han River, which has an annual rainfall of about 31.0 billion tons and an average flow of about 17.0 billion tons. In 1968 approximately 11% of these water resources were used for household use, industry, and agriculture; in 1976 the total water demand was 3.5 billion tons or about 21% of the river flow. By 1981 the total water demand will reach 5.0 billion tons, constituting roughly 30% of river flow. The effluent discharge, if left untreated, will greatly pollute the water in the river.

An estimation of the water pollution caused by municipal sewage and industrial effluent will serve as a yardstick. For example, as of October 1970 the population of Seoul numbered 5.536 million, the amount of garbage and refuse reached 2.30 million tons a year, and about 2.50 million tons of night soil were discharged. These wastes went back to the farming areas near the city.

In many western countries, industrial emission and wastes are the main causes of environmental hazards, whereas in Korea environmental pollution is still mainly a result of household wastes. However, as industrialization progresses, industrial effluents or emissions will gradually become the primary source of pollution. The pollution level of the Han-gang River in 1970 was BOD (Biochemical Oxygen Demand) 13.0 ppm in the Gwangjangri area; near the first Han-gang bridge it was 39.5 ppm, and around the second bridge it was about 36.7 ppm. This means high-quality water treatment facilities are needed to supply clean water for urban dwellers. As a drastic measure to solve such a significant problem, a collecting sewer system is being planned for both sides of the Han River. All the sewers from the city of Seoul will therefore drain into these collecting sewers through to a cesspool, which will be constricted at the lower part of the city, where the sewage will be treated. It is expected to be completed by 1980. At the same time, an extensive plan for treating human waste is being activated. When completed, the treatment plant will manage about 1600 kiloliters effluent, more than 90% of the daily output from the city. The initial BOD is estimated at 20 000 ppm, but this will decrease to 2000 ppm after the first treatment stage, and reach 19 ppm by the end of the final process. The treated water will then be discharged into the Han-gang River.

3. Pesticide Poisoning

Korean agriculture has greatly improved technologically in the past 20 years in such areas as plant breeding, dense planting, and heavy application of fertilizer. But the vermin damage has also increased, requiring heavy use of agricultural pesticides.

In Korea 62 602 tons of pesticides were used in 1974 for agricultural purposes at an estimated 28.0 kg per hectare, a 20% increase every year since 1970 [254, 255].

It is almost impossible to figure out exactly the incidence and mortality of acute pesticide poisoning. However, a survey in 1974 reported that 33.5% of males and 32.7% of females who sprayed certain pesticides complained of several symptoms seemingly due to acute pesticide poisoning. Moreover, in more than 60% of the above cases the main cause of poisoning was supposedly dusting without protective covering such as masks, boots, and gloves. In particular organic mercury compounds used in preventing rice blight are absorbed by the grain and accumulate in turn in the human body; they contaminate agricultural waste, which is thrown into the water where it is absorbed by fish, which are in turn eaten, creating a food cycle.

Recently the amount of mercury contained in rice in our country was discovered to approach the limit of daily permissible intake set by the joint committees of WHO and FAO. Also some pesticides such as DDT or dieldrin have been detected in fruits and vegetables. Thus, wide use of agricultural chemicals, if left uncontrolled, will cause much damage as was apparent in 1952 in Kyungsan county in an episode of parathion poisoning. Control of agricultural chemicals is very urgent.

4. Noise Pollution

Urban noise mainly stems from construction, vehicles, factories, advertising, and crowds of people. Noise pollution is not as widespread a public nuisance as air and water pollution, but it can cause buzzing in the ear, lack of sleep, disorders in gastric function, and decline of thinking ability and business efficiency. In the big cities of developing nations, construction and factory noise make urban life very unpleasant. Of the 4615 factories in Seoul, 3944 factories are noise-producing. In 1969, 3119 places were reported as producing offensive noise. Of this number, the neighbors reported 1112 cases, and the authorities detected 3575 cases. About 70% of these were classed as public nuisances [255]. In the residential areas of Seoul are located 729 food-processing factories and 476 metal factories, 229 textile and weaving factories, 180 printing shops, and 194 chemical plants. Thus, about 2500 are concentrated in Seoul, which can be considered an industrial public nuisance as well as a source of noise pollution.

5. Air Pollution

Since 1960 Korean industry has been achieving rapid growth, setting the foundation for the economic development of the 1970s. Simultaneously, many problems have arisen such as urbanization, housing, transportation, plants, and factories. Wastes from these industries pollute the air and water, and noise adversely affects the cattle, plants and living organisms.

Air pollution increases as the demand for fuel increases, producing soot and dust from the smokestacks of the factories, emission gas, or particles. Consumption of fuel in Korea has kept pace with the growth of industry and population. Since 1956, widespread use of liquid fuel has also aggravated air pollution [255].

The estimates on fuel consumption throughout the country were about 10 million tons coal, such as smokeless coal, in 1965, and a total of about 1.5 million kiloliters of gasoline, kerosene, diesel or bunker C oil. Six years later in 1971 consumption of coal rose to roughly 18 million tons, and petroleum, to 8.25 million kiloliters, creating a situation that inevitably increased air pollutants. The total weight of the air pollutants is about 800 thousand tons, i.e., roughly 5.2 times the total emission of pollutants in 1965 which totaled 140 thousand tons.

About 95% of air pollutants consist of carbon monoxide, sulfur oxides, nitrogen oxide, and hydrocarbons. About 29% are emitted from factories, 10% from thermo-power plants, 25% from automobile exhausts, and 36% from space-heating systems. The mean amount of pollutants per 1 km^2 throughout the country is approximately 15 tons per year, a rate higher than the amount of pollutants in the United States in 1965 recorded at their highest level of air pollution, 13.5 tons.

The worst air-polluted areas are near factories, housing areas, and traffic intersections. In Seoul, fuel consumption is about 28% of the national total, and yearly approximately 600 tons of air pollutants are emitted per square kilometer. The major cause of air pollution is the automobile, followed closely by the use of diesel and bunker C oil which emit soot, toxic carbon monoxide, sulfur dioxide, and hydrocabons.

Limited road capacity in Seoul, only 9.5% of the national total, causes heavy traffic jams; therefore, only slow speeds are permitted. Due to the great number of bus stops, extensive exhaust fumes are emitted. Moreover, about 40% of the cars are very obsolete, and these emit more than twice the normal amount of emission gas. Each car runs approximately 10 times the distance of cars in other countries each day, heightening the amount of exhaust.

Sulfur contained in petroleum is about 3% in the case of bunker C oil and about 1% in the case of diesel oil. As long as we use these fuels, the concentration of toxic pollutants in the air will keep increasing. An analysis of the national air pollution picture predicts that the urban population in Korea in 1981 will comprise about 60% of the total population, and the number of heavy chemical industries will increase to two or three times the present number. The demand for fuel and energy will also increase proportionately, heightening the level of air pollution [255].

The air pollution in Seoul was surveyed by governmental research institutes and the concentration of sulfur dioxide in the air was found 0.01 ppm in 1965 in the factory area; in 1970 it increased to 0.06 ppm. Also in the residental area in the same period it went up from 0.003 ppm to 0.05 ppm. Nitrogen dioxide increased in the same period in the factory area from 0.085 ppm to 0.45 ppm and in the residential area from 0.03 ppm to 0.15 ppm. This means that the air pollution problem in urban areas is continuously escalating. Another survey in 1976 verified the problem again although nitrogen dioxide showed different figures (Table 30).

The National Institute of Health reported the densities of air pollution in the industrial area (Seong-Su Dong, etc.), commercial area (Bang San market, etc.), metropolitan region (Kwang wha moon street), residential area (Seo-gang area), and in park area (Samcheong park) in Seoul and found that whereas the carbon monoxide content was lower than that of the international standard, the nitrogenous matter, dust, and noise exceed-

Table 30. Public Nuisances by Area in Seoul, 1976 {255}

	SO$_2$ (ppm)	CO (ppm)	NO$_2$ (ppm)	dust (ton/km^2/ month)	noise (decibel)
Industrial area	0.07	6.70	0.023	28.65	67.1
Commercial area	0.05	9.75	0.042	31.93	75.7
Center of Seoul (Kwangwha moon)	0.06	11.3	0.055	23.68	76.1
Residenial area	0.05	6.5	0.012	22.6	64.9
Park area	0.01	0.9	0.00021	18.67	49.9
Average	0.04	6.7	0.026	25.11	66.7
Standard in Japan	0.04	10.0	0.02	20.0	65.0

ed that of Japan. Sulfurous substances, which cause respiratory diseases, were almost equal to the standard of Japan, but in the industrial area and business center they showed higher levels, 0.05 – 0.07 ppm. (Jeosun Il Bo, Dec. 1 1976). The higher level of nitrogenous substances may accelerate the smoggy phenomenon followed by the combination of nitrogenous substances and carbon compounds in the air. Smog may induce nausea, vomiting, or audiovisual diseases. Compared with Japan, dusts were higher in the business center, 31.93 ton/km²/months, as were zinc, lead, iron, and manganese, which are harmful to the human body (Table 30).

There is no doubt that air pollution adversely affects human health and property, but the criteria of measurement differ from one country to another. Health may be affected in many ways, e.g., irritation to the eyes, nose, and upper respiratory tract causing respiratory disease, laryngitis, nasal inflammation, bronchitis, and asthma. In the area where sulfur dioxide exceeds 0.05 ppm, the incidence of bronchitis and asthma increases by about 5%; lung cancer and heart disease also increase in the same proportion as the increase in air pollution. Nitrogen oxides and PAN (peroxyacyl nitrate), which are photochemical oxidents, also irritate the eyes and adversely affect plants, rubber, metals, and livestock [255].

6. Sewage and Health Problems

Except for several sewage plants in Seoul and some other main cities, no internationally standardized facility is provided in almost all other areas. All the liquid discharges are carried away from individual houses to open canals, ultimately reaching lake or sea via the rivers. Sewage used to be contaminated with fecal excrement. As an indirect measure for keeping the sewage relatively sanitary, insecticide or pesticide spraying is practiced in the summer season, but not regularly. In such an environment, sewage itself may serve as a very nutritious medium for propagation of pathogenic agents and insects. Under such circumstances, inhabitants are exposed to the dangers of polluted soils and vegetables, which are grown in the area, and consequently a high prevalence of various contagious diseases.

Even in the same city, the level of infection varies according to sanitation conditions. Kim [256] examined fecal specimens of school children in the downtown area and a suburban area of Seoul (Table 31). The suburban area belongs to the administrative boundary of Seoul Metropolitan city, but the inhabitants more or less engage in farming. They utilize human excrement as fertilizer to grow vegetables. Except for *Trichostrongylus sp.,* the incidence of soil-transmitted helminths was far higher in the suburban area as seen in Table 31. Even in the same district the intensity of soil-transmitted helminths differs according to sanitation conditions. Soh et al. [128, 256] estimated the worm burden of soil-transmitted helminths in two areas of Seoul (Table 32). A "sanitary" district (Huam Dong) was well paved, and the sewage system was well organized and well covered. Human excrement was removed regularly by city employees. An "unsanitary" district (a part of A-Hyun Dong) was located on a hillside. The privies sometimes overflowed due to the poor collecting system, and roadside defecation by children was common. The Ratio of EPGF of *Ascaris*

Table 31. Parasite Incidences of School Children in Downtown and Suburban Area of Seoul {256}

	Examined	Positive [a]	A. l.	T. t.	Hw	T. o.
Downtown	63 924	34 967 (54.7)	16 601 (26.0)	25 228 (39.6)	2 190 (3.4)	3 418 (5.2)
Suburban area	11 475	9 355 (81.5)	6 577 (57.3)	5 766 (50.2)	1 062 (9.3)	593 (5.2)

[a] Soil-transmitted helminths plus others. Abbreviations see Table 29, page 105.

Table 32. Number of Eggs of Soil-transmitted Helminths per Gram in Stool Tanks in Several Areas {128}

Areas	Year	District	Samples	A. l.		T. t.		H. w.		T. o.	
				Total	Average	Total	Average	Total	Average	Total	Average
Sanitary	1959	Huam-Dong	6	30 000 (1 600 – 10 800)	5 000	6 000 (400 – 1 800)	1 000	6 000 (400 – 3 000)	1 000	0	0
Unsanitary	1959	Bukahyun-Dong	6	247 800 (0 – 156 000)	41 300	10 800 (400 – 5 400)	1 800	4 000 (0 – 1 800)	666	600 (0 – 600)	100
Unsanitary	1973	Bukahyun-Dong	10	10 800 (0 – 2 000)	1 080	6 000 (0 – 2 000)	600	400 (0 – 400)	40	400 (0 – 400)	40
Unsanitary	1973	Haengdang-Dong	5	25 800 (2 800 – 7 600)	5 160	5 200 (800 – 1 600)	1 040	2 000 (400)	400	2 400 (400 – 800)	480

[a] ( ) range.

eggs from the privies was eight times higher in the unsanitary area than in Huam Dong, although no clear-cut difference was recognized in other parasites. EPGF in Huam Dong was 5000 and 41 300 in A-Hyun Dong. In 1973, fecal materials from privies in a part of A-Hyun, the same area as 1959, were examined by the author. The EPGF were *Ascaris,* 1080; *Trichuris,* 600, Hookworm, 40; and *Trichostrongylus,* 40. The ratio difference of the EPGF also indicates a remarkable reduction during the years 1969 – 1973. In the meantime, the sanitary environment, living standard, and education level in this part of A-Hyun Dong have been improved. The above data also suggest that the people's knowledge of parasite infection has improved compared with former decades.

7. Garbage and Waste Disposal

a) Primitive practice: Most garbage and waste in the city, estimated at 500 g per day per capita, is eventually buried. In Korea a yearly total of 4 – 5 million tons of garbage and waste are removed from the cities and towns by truck, ox-cart, and hand-cart by workers. Much of the human excreta is still stored in large collective tanks, provided in each area by the city government, and waits for secondary use in the surrounding districts. Human excreta in urban areas are collected by scavengers and transported via truck to a large collecting tank, provided by the city government. About 70% of the total night soil from cities and towns in Korea is utilized as fertilizer. Daily amounts of human excreta (night soil and urine) from urban areas are estimated at 14 120 tons, and 20 800 tons from rural areas (Bo-Kun Shin-Bo-Weekly Press, November 15, 1971). In rural areas excreta are used directly by the family unit as fertilizer. In this way, except in certain urban areas where the flush-toilet system is used, almost all the population still use the conventional latrine. As mentioned above, raw excreta is transported ultimately to the farms without any previous sanitary treatment. Such a primitive practice is still seen even in the peripheral zones of large cities. For this reason, a law was passed at a Cabinet Meeting 28 January 1969 that no human fertilizer be used in designated urban cities and towns in each province. The Ministry of Health and Social Affairs promulgated the order to be enforced 1 March 1969. The areas totaled 55, including the capital metropolitan city and Busan special city. At the same time, the Korean Association for Parasites Eradication established a supply center for parasite-free vegetables in several cities. One center has to be established in each district of the special city, more than three in each provincial capital, and at least one in each city and *Gun* (county level). Parasite-free vegetables have been designated as those not cultivated with human night soil. In this way, the fundamental way of improving the environment must be by sanitary control of human night soil, especially to prevent soil-transmitted helminthic infection. Since no treatment plant is provided, except in a part of the metropolitan city of Seoul and some other main cities as regards human night soil in

Korea, it may not be an exaggeration to say that all human excreta are utilized as fertilizer without any satisfactory management.

In principle, night soil should be collected and emptied into an open water drain, buried in the earth, composted with town refuse, or treated by itself or jointly with sewage. In Korea, as in most developing countries, the privy has been almost completely neglected as regards sanitation. The conventional types have been the surface privy and the pit privy, except in limited areas of urban districts. The feces are deposited in crockery or in cement-lined pits. Urine is permitted to leak away or evaporate. The privies of this kind are often poorly built with ill-fitting doors and cracks between the boards. Insects and animals play an important role in spreading the pathogenic agents. Given this situation it is understandable that 35 of 45 species of human parasites in Korea are related to the digestive system. To eliminate these parasites, flush-toilet system construction, night soil disposal plants, and sufficient amounts of chemical fertilizers are the basic requirements. As intermediary means, several new devices have been tried as pilot projects in Korea since 1959.

b) Urine-feces separation system: Urine can be utilized as a fertilizer at all times, since no pathogenic agent is present in the urine under normal condition. Feces are treated first with chemicals: sodium nitrite, 0.1% – 0.2%, calcium superphosphate 4% – 8% for 2 days, or else feces are mixed with grass matter for composting. In the conventional privy, the *Ascaris* eggs survive as long as 2 – 8 months. But when night soil is treated by chemicals or by composting, the pathogens are destroyed within 3 days. Night soil is placed in the center of a heap of grass to destroy pathogenic organisms by high temperature during composting [257].

c) Methane gas-producing tank system: The primary concern of this system has been to furnish methane gas, a fuel source. A pilot study has been carried out on the resistance of parasite eggs in the main tank. The results up to date suggest that eggs of *Ascaris* and hookworm are not destroyed in the main tank [255]. It is recommended that live eggs be treated with sodium nitrite in an effluent tank before using them as fertilizer.

d) Multicompartment tank system: Three- or five-compartment systems were tested in limited areas without success. Several devices mentioned above have both advantages and disadvantages, although no method is considered far superior to the conventional privy system in reducing fecal pollution. All are theoretically reasonable, but the practicability is still under question for reasons of expense and insufficient understanding of the people concerned. At this stage, the government is encouraging the methane gas-producing tank system through the Bureau of Rural Development, Ministry of Agriculture and Forestry.

8. Food Sanitation

As long as night soil is used as fertilizer, most vegetable products are subject to the spread of various pathogenic agents. Currently, makers of chemically produced food-stuffs have also been prosecuted for selling excessive amounts of toxic chemicals. Increased use of pesticides causes pollution of grain and fruit. Under such environmental conditions all the nutrients of animal and plant life surrounding us may turn into a hinderance rather than a help. For this reason, food sanitation should be practiced very carefully not only by individuals but on a nationwide basis. Actually, food sanitation affairs are handled by the Ministry of Health and Social Affairs in accordance with Article 23 of the Law of Government Organization. According to the regulation of the organization, food-handling and entertaining businesses are to be controlled by the Bureau of Sanitation. Regulations, Ministry orders, and decrees promulgated since 1950 have been set on soft drinks and prepared food, restaurants, the sale of meat, ice cakes, and cake manufacture, etc. Since 1954 sanitarians throughout the country have been engaged in the following inspection activities: condition of sanitation facilities, food for military forces, and other related matters. However, no matter show strict the law is, the more important thing is the comprehensive understanding of the people and improvement of environmental sanitation.

9. Countermeasures Against Public Nuisances

Some pollutants can be easily eliminated by effective and faithful application of technology. Usually, it requires the setting up of a large-size factory or a recycling system in some cases.

Within the framework of the government's anti-pollution plans water supply projects have been carried out in areas with no previous water supply services. Until the end of 1971, simple water supply service was provided for 463 village settlements throughout the country. In the coming 10 years, 8880 settlements will be provided with water supply services. And this project will contribute to lessening the incidence of various kinds of water-borne diseases in the countryside.

More and more night soil discharged from the city which used to be used as fertilizer is being replaced by chemical fertilizers, and less night soil goes back to the countryside. Instead it is discharged into the river, causing serious water pollution. Since 1968, night soil tanks have been set up around the city at more than 100 places for treatment purposes. This project will be expanded in the future. A sewage treatment plant was established in Seoul in 1970, a first for Korea. This project plans to treat the sewage of one-fourth of the population of Seoul, to reduce the BOD, and to discharge the treated water into the Han River. The facilitiy was completed by 1976. Thus, the river water near the Bogwangdong area was treated, after reducing the BOD level from 29 ppm to 5 ppm so as to clean the water flowing into Han River.

Sewage treatment facilities in Seoul will be expanded to treat all the sewage of Seoul in the future, and these facilities will be extended nationwide. For example,

Jeonju, capital city of Jeonra Bug Do, completed a sewage treatment plant in October 1976 with a daily capacity of 100 kiloliters of human excrement.

A long-range plan provides for the development of a river basin area for the four large rivers. By 1981 in an effort to utilize 28% of the total flow of the rivers in Korea, 14 dams for hydraulic power plants and industrial water, 10 dams for public water supply, and 13 dams for irrigation will be built in the form of individual or multipurpose dams. These projects aim not only at the expanded use of river water but also at leveling off the average amount of river flow at a certain level so as to increase the water supply in times of drought and to dilute the pollution of the water, thus contributing to improved water quality.

The control air pollution, exhaust gas from automobiles will be kept below 1.5% emission of carbon monoxide in new cars and below 5.5% in old cars. Soot will be checked under "Ringleman" degree. Obsolete cars will gradually be replaced by new cars made more accessible through subsidy offers.

One means of solving the noise problem caused by automobiles will be subway construction in the metropolitan area. The first subway line was completed and began operation in Seoul in August 1974. Control of noise, vibration, waste, and effluent discharged by medium-small industries in Seoul will be tackled by relocation of the factories to the suburbs. However, such a costly project will be impossible to implement without the financial support of the government.

The oil refinery industry is not cooperating in the desulfurization of low-quality fuel. Neither has the use of lead-free gasoline been adopted, nor has any other measure been taken to reduce the 70 kg lead emitted daily from automobiles in Seoul city.

In conclusion, environmental pollution in Korea has been increasing for two decades, keeping pace with industrial growth. Unless some check is made on pollutants, Korea will face a serious crisis in the near future.

IV. Animal Factors

1. Domestic Animals

Excluding large scale livestock and poultry farms, domestic animals are usually raised by individual households without resorting to enclosures. For example, a yard scattered with vegetable matter serves as a pigsty for hogs, garbage serves as food. In such an environment a domestic animal may develop into a paratenic host, an intermediate host, or a carrier of various pathogens. An example of the distribution of Salmonella among animals in Korea [259] is provided by the 25 strains of Salmonella isolated from the feces of 1200 animals comprising 200 each of rats, swine, cattle, chicken, dogs, goats, and sheep. Three strains of Salmonella were identified from 400 chicken egg shells, and of the Salmonella strains isolated, about half (14 strains) were S. typhimurium. Of these seven strains were *S. pullorum,* three were group E organism, and two each were *S. paratyphi C* and group D organism. The rate of isolation from feces of animals averaged 2.1%, the highest being from chickens (5.0%). Another report by Park [157] indicates that human-specific parasite eggs are relatively common in the fecal matter of domestic animals in Korea. From a total 183 samples from pigs, dogs, chickens, and rats 44 samples (24%) contained parasites specific to humans: *Ascaris lumbricoides, Trichocephalus trichiurus, Clonorchis sinensis,* and *Taenia sp.* Pig manure in Jeju Do contained ascaris and whipworm eggs as well as other nonpathogenic types of coccidia. Specifically, 93 *ascaris* eggs, not distinguished from swine species, were found per gram manure and 34.4% of them were alive [260]. All the above reports suggest that the control of domestic animals is also an important means of blocking the channel of infection to humans.

Some wild and domestic animals may be considered reservoirs for some endemic diseases. However, no record or morbidity survey of wild animals is available in the literature. Of domestic animals the dog is known to be the most common reservoir of *Clonorchis* and *Paragonimus* in Korea.

Cattle may serve as reservoirs of *Fasciola hepatica,* but only two cases of human infection with this parasite have been recorded in Korea to date (see Section CIX. 1). However excellent the control measures, they will be in vain if the control measures for the reservoir hosts are neglected. Strict surveillance of the suspected animals is as important as testing human beings.

2. Rodents

Since most houses in Korea are made of wood and earth, damage caused by rats is considerably greater than in other countries. A variety of rodents recorded in the literature as extant in Korea are listed below (Chun, 1975) [3]:

1) *Microtus mandarinus kishidai*
2) *Microtus fortis fords* pellenus Thomas
3) *Microtus fortis uliginosus* Jhonson
4) *Clethrionomys rufocanus regulus* (Thomas, 1907)
5) *Cricetulus triton nestor* Thomas
6) *Cricetulus barabensis* fumatus, 1909
7) *Apodemus agrarius coreae* Thomas
8) *Apodemus agrarius pallescens* Jhonson and Jones, 1955
9) *Apodemus agrarius chejuensis* Jhonson and Jones
10) *Apodemus sylvaticus draco* (Barro-Hamilton 1700)
11) *Apodemus speious peninsulae* (Thomas, 1906)
12) *Micromys minutus ussuricus* Barret-Hamilton, 1899
13) *Micromys minutus hertigi* Jhonson and Jones
14) *Rattus rattus* (Linnaeus, 1753)
15) *Rattus rattus* alexandrinus (Geoffroy, 1803)
16) *Rattus norvegicus* Berkenhout
17) *Rattus norvegicus* longicuadus Mori, 1937
18) *Rattus norvegicus* caraco (Pallas, 1788)
19) *Mus musculus yamshinai* Kuroda, 1931
20) *Mus musculus utsuryonis* Mori, 1938.

Of these 20 species, the Norwegian rat, familiarly known as the house rat, can be a factor in carbon monoxide poisoning from coal bricks (Yeon-tan-coal mash). The rat probably digs a tunnel in the wall of the room which permits the carbon monoxide to enter. Additionally, rats can serve as carriers of pathogenic agents from latrines or sewage, but salmonellosis and rat-bite disease are considered to be directly transmitted. They may sometimes contribute to an unexpected outbreak of an epidemic of plague by carrying pathogenic organisms through vessels or air-cargo from the endemic areas. In 1972 a survey on rats and rat fleas was performed in the coastal areas of Korea, Incheon, Gunsan, Yeosu, Busan, and other areas [81] (see Section C. II. 2). In general, the index showed a decline in the summer months which again increased in the fall season. By species, however, *M. anisus* increased in the spring season and *X. cheopis* in the fall.

Rodent control is one of the important tasks in preventing various diseases that prevail in countries with poor standards of sanitation. Campaigns utilizing rodenticides have been conducted each year for several decades, but the rat population has not appeared to decline significantly, remaining at an estimated threefold the total human population. Success can only be achieved by regular well-planned campaigns that are faithfully carried out at least twice a year.

3. Arthropod Vectors

Blood-sucking insects, ticks, and mites are the important known arthropods that transmit pathogens biologically. Some of the important diseases in Korea related to these arthropods are malaria, filariasis, epidemic typhus, Japanese encephalitis, recurrent fever. However, scientific investigations of arthropods have not been very exact in the past [3, 261].

a) Mosquitoes: The following 42 species of mosquitoes have been recorded in Korea [3, 262, 266]:

1) *Anopheles (A.) koreicus* Watanabe, 1918
2) *An. (A.) lindsayi japonicus* Yamada, 1918
3) *An. (A.) pullus* Yamada, 1937
4) *An. (A.) sinensis* Wiedemann, 1928
5) *An. (A.) sineroides* Yamada, 1925
6) *An. (A.) yatsushiroensis* Miyazaki, 1951
7) *Toxorhynchites (T.) christophi* (Portschinsky, 1884)
8) *Tripteroides (T.) bambusa* Yamada, 1917
9) *Mansonia (C.) ochracea* Theobald, 1903
10) *M. (M.) uniformis* (Theobald, 1901)
11) *Heizmannia lii* Wu, 1936
12) *Aedes (Ochelerotatus) dorsalis* (Meigen, 1830)
13) *Ae. (Finlaya) hatorii,* Yamada, 1921
14) *Ae. (F.) japonicus* (Theobald, 1901)
15) *Ae. (F.) koreicus* (Edwards, 1917)
16) *Ae. (F.) kobayashii* Nakata, 1956
17) *Ae. (F.) nipponicus* Lacasse and Yamaguti, 1948
18) *Ae. (F.) oreophilus* (Edwards, 1916)
19) *Ae. (F.) seoulensis* Yamada, 1921
20) *Ae. (F.) togoi* (Theobald, 1907)

21) *Ae. (Stegomyia) albopictus* (Skuse, 1894)
22) *Ae. (S.) Chemulpoensis* Yamada, 1921
23) *Ae. (S.) flavopictus* Yamada, 1921
24) *Ae. (S.) galloisi* Yamada, 1921
25) *Ae. (Aedimorphus) vexans nipponii*
26) *Ae. (neomelaniconion) lineatopennis* (Ludlow, 1905)
27) *Ae. (Aedes) esoensis* Yamada, 1921
28) *Armigeres (A.) subalbatus* (Coquilett, 1898)
29) *Culex (Lutzia) fuscanus* Wiedemann, 1820
30) *C. (L.) vorax* Edwards, 1921
31) *C. (Neoculex) hayashii* Yamada, 1917
32) *C. (Neo.) rubensis* Sasa and Takahasi, 1948
33) *C. (Culiciomyia) kyotoensis* Yamaguti and Lacasse, 1952
34) *C. (Culex) bitaeniorhynchus* Giles, 1901
35) *C. (C.) jacksoni* Edwards, 1934
36) *C. (C.) mimeticus* Noe, 1899
37) *C. (C.) orientalis* Edwards, 1921
38) *C. (C.) pipiens pallens* Coquillett, 1898
39) *C. (C.) sinensis* Theobald, 1908
40) *C. (C.) tritaeniorhynchus summorosus* Dyar, 1920
41) *C. (C.) vaagans* Wiedemann, 1828
42) *C. (C.) whitmorei* (Giles, 1904)

Although the recent massive application of insecticides seems to have reduced the mosquito population, mosquito-borne diseases are still prevalent which show an endemic pattern. *Plasmodium vivax* infection exists mostly in the northwestern part of southern Korea, and *Anopheles sinensis* and *A. yatsushiroensis* are the recognized vectors for the protozoa. *Aedes togoi* is suspected as the most probable vector of *Brugia malayi* on Jejo Do island, and *A. sinensis* and *A. yatsushiroensis* are suspected to be vectors of filaria in the Gyeongsang Bug Do area. The main vector of Japanese encephalitis is *Culex tritaenio-rhynchus*, but both *Aedes vexans* and *Culex pipiens* have also been recognized as probable vectors [262].

b) Fleas: To date, 33 species of fleas have been recorded in Korea. Of this number, *Xenopsylla cheopis* is distributed throughout the land, especially along the coastal areas [81]. The flea indexes of rats caught from areas surrounding the main ports in Korea were 0.35 – 1.72 in January through March, 2.02 – 2.59 in April through June, 0.65 in July, and 1.58 – 2.84 from August into the deep winter months. Since this species is a known vector for plague, which is prevalent in neighboring Manchuria and southeastern Asia, and sometimes for murine typhus, control measures must be undertaken. The elimination of fleas depends on the eradication of rats. Insecticide resistance poses another problem. Mortality rates of *X. cheopis* when exposed to various concentrations of DDT for 1 h were 43.8% in 4.0% concentration, 30.4% in 2.0% concentration, and 13.6% in 1.0% concentration with 24-h exposure to DDT, the mortality rates soared to 100% in 4.0% concentration, 91.5% in 2.0% concentration, 73.5% in 1.0% concentration, and 37.5% in 0.5% concentration.

c) Lice: As in other parts of the world, *Pediculus humanus corporis, P. h. capitis,* and *Phthirus pubis* also exist in Korea. In the past, possibly before World War II, the body louse played an important role in Korea as the vector of epidemic typhus and recurrent fever, both of which are now extinct. In spite of the fact, that lice were reported during the Korean War in 1950 – 1953 to be resistant to DDT, their number has been greatly diminished by massive and repeated applications of insecticides along with the improvement of sanitary conditions. In ordinary communities they are almost completely extirpated. Only pubic louse infestation is occasionally seen in dermatology clinics. Ten species of lice inhabit rodents, five species of the genus *Polyplax* and five species of *Haplopleura.*

d) Other insects: Two species of bedbugs, which transmit typhoid fever, have been found in Korea, *Cimex lectularius* and *C. hemiptera,* but these insects have also been almost completly extirpated from the normal community. Twenty-four species of biting midges or culicoides have been reported in Korea, but none has proven to be a biologic transmitter of human pathogens. However, their biting habits may hinder the sleep of humans and domestic animals.

Fourteen species of black flies (*Simulium flies*) are recorded in Korea. The black fly bite produces focal erythema or papules. Since *Onchocerca gibsoni* have infected a number of cattle in Korea, some of the species may serve as vectors of filaria [264, 265]. About 23 species of horse and deer flies (*Tabanidae*) have been found in the land [3]. Except for biting humans and animals, no report on their biologic role as disease transmitters, as is the case in Africa, has yet appeared. The housefly and its allies are the most common hexapods in Korea, consisting of 24 species of *Muscidae*, 15 species of *Calliphoridae*, and 39 species of *Sarcophagidae* [267]. As in other parts of the world, flies play various roles in transmitting typhoid, salmonellosis, dysentery, cholera, tuberculosis, and some other diseases. Myiasis is not uncommon. *Psychodidae* (sand flies) have not yet been found in Korea.

Of the insects of less medical importance are two species of *Coleoptera*, some species of *Hymenoptera*, and 12 species of *Lepidoptera*. No report has been made on the systematic pathogenicity of these arthropoda, but during the summer *Euproctis flava*, belonging to *Lymantriidae* of *Lepidoptera*, causes dermatitis from its toxic hairs.

e) Ticks and mites: About 22 species of ticks have been reported; the major ones one *Argas, Ixodes, Boophilus, Rhipicephalus, Haemophysalis,* and *Dermacentor.* Even though ticks transmit various pathogenic agents, no scientific report has yet been recorded in Korea, except for some descriptions of Russian spring-summer encephalitis occurring in North Korea [3]. Nevertheless, the mechanical damage from tick infestation is not negligible. For example, a nationalpasture in Jeju Do was destroyed due to the prevalence of *Boophilus.* Medically important mites are *Trombiculidae* (Ewing, 1944) and *Laelaptidae* (Berlese, 1892), of which about 64 species have been reported in Korea. Of these are 22 species of *Trombiculid* mites, the species that transmits scrub typhus is known to be included. Traub [268] suspected some mites of transmitting *Rickettsia tsutsugamushi,* the agent of scrub typhus.

f) Crustaceans: As the intermediate host of fish tapeworm and *Dracunculidae, Cyclops,* and *Diaptomus* are widely distributed. Of the *Decapoda* occurring in Korea, *Eriocheir japonicus, E. sinensis, Sesarma dehaani, Helice tridens de Haan, Cambaroides similis, C. douricus,* and *Macrobrachium (Palaemon) nipponensis* are incriminated of being secondary intermediate hosts of *Paragonimus, Eriocheir species* and *Cambaroides similis* being the most popularly known and widespread. The former prefer to inhabit ditches in flat lands, and the latter inhabit clean streams in mountainous areas.

g) Others: Millipedes and *Chilopodes* are the commonly distributed *Myriapoda. Chiracanthium japonicum, Ch. lascivum,* and *Ch. cutitha* are recorded as *Arachnida.*

V. Nutrition

1. Korean Food Habits

A nation's food pattern is characterized by its climate, natural features, and traditions. Cooking methods may continuously change with the current of thought and the various social conditions. Food patterns, which are also influenced by economic conditions, contribute to the changing eating customs, quality of foods, and variety of cooking methods. The Korean food pattern of three meals a day has changed very little [16].

Korea grows a variety of cereals since its temperature range, and the fertile land optimal for cereals-planting. The first agricultural products in Korea were grass and millet. Around 500 BC, paddy-field rice plants (aquatic rice) appeared first, and since that time, the food pattern with rice as a staple food has become traditional. Rice, with its excellent flavor and nutritive and caloric value, is considered as the most important cereal in Korea (see Map 5).

A food consumption survey performed by the Office of Rural Development with the aid of FAO, WHO, and UNICEF in 1968 showed the fish, meat, and egg intake to be 26 g per day per adult, considerably lower than the recommended daily allowance, 120 g. A dietary pattern survey conducted by the National Chemistry Laboratory in 1946 and 1948 showed Korean eating habits after liberation. The total food intake was 1038 – 1180 g, of which 97% – 98% consisted of vegetable food and 611 to

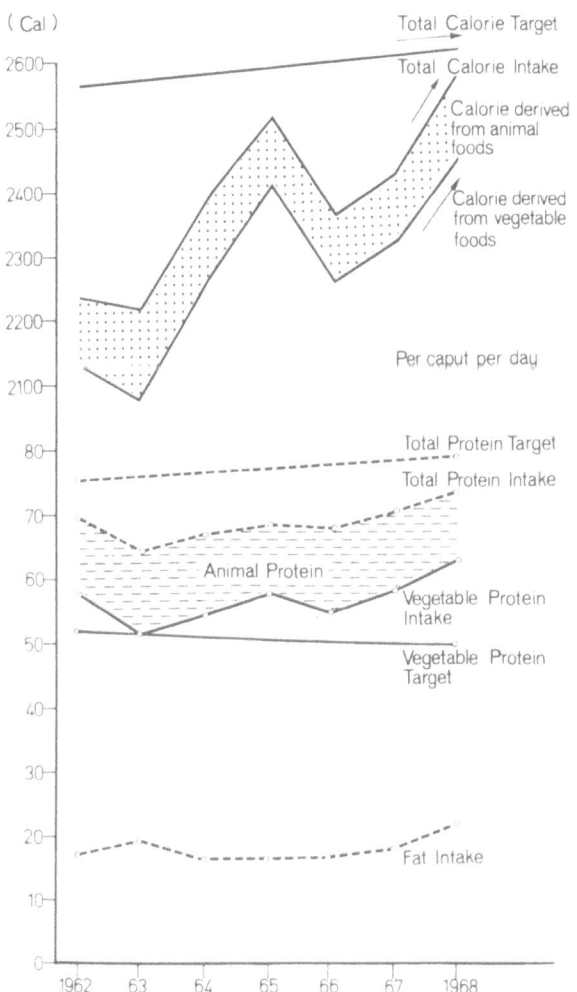

Figure 15. Quantitative levels of nutrition in Korea
(Source: UN FAO/Source: Food Consumption Table, K-FAC.
K. Y. Lee and H. S. Kim, 1974 [16])

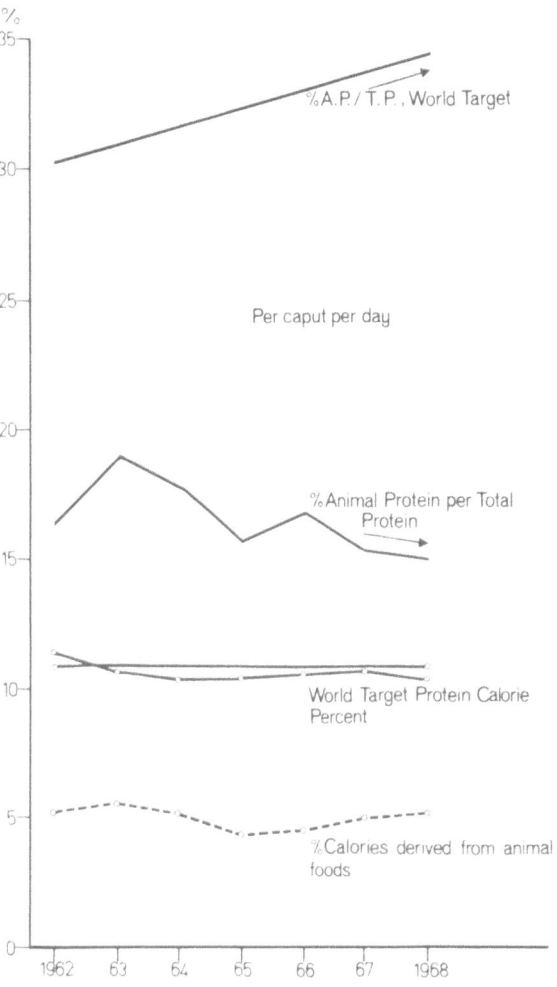

Figure 16. Qualitative levels of nutrition in Korea
(Source: UN FAO/Source: Food Consumption Table, K-FAC.
K. Y. Lee and H. S. Kim, 1974 [16])

674 g was cereal. According to a similar survey performed in 1961 – 1962, plant food intake averaged 842 – 1352 g, but cereal consumption, only 470 – 605 g. These two surveys revealed that for 14 – 15 years no critical changes occurred in Korean food habits or the kinds of foods used in main or side dishes [16, 269, 270].

Although agriculture is gradually declining in importance in the Korean economy, it is still the principal industry of Korea, with 40% – 50% of the people living by farming. Therefore, the problems of rice self-sufficiency rate, rice production, and foreign rice imports are treated seriously in the Korean economy.

Before liberation, Korea was one of the few countries self-sufficient in food stuffs; in 1942 in excess of rice was exported to Japan. But with the increasing influx of refugees, growing insecurity of living, and unstable prices, the food problem became unsolvable on a self-sufficient basis. Foreign grain totaling 800 000 million tons was imported for the first time in 1946 – 1948. There is still a serious imbalance between grain production and consumption. Because the increasing rate of rice yield still met only one-third of the rice demand, it was inevitable that foreign grains be imported. The Korean government has spent ever more dollars to buy foreign grain; over 150 million dollars was spent in 1969.

Food shortages, foreign grain imports, imbalance of grain supply and demand, and the difference between production costs and selling prices have created many serious problems. Added to this situation are the growing numbers of farmers who are leaving the land and moving into the cities.

Korean farmers grew their rice by traditional methods until the end of the nineteenth century. Immense changes in political and social situations have greatly affected agricultural policy, e.g., plant breeding has been improved and distributing centers of the open-port type have been turned into national cooperative federations.

In 1966 a national propaganda campaign was launched to reduce grain consumption. At that time the government prohibited the use of rice in alcoholic beverages and glutinous rice jelly. Boiled rice had to be mixed with miscellaneous cereals in all restaurants and inns. In 1970 the Korean government enforced a high price policy for rice for two main reasons, to institute a price innovation and to compensate for food shortages. A decision was also made to import grains from abroad, especially wheat flour. As a link in the chain of recommendations for mixed cereal consumption, rice foods were prohibted on Wednesdays and Saturdays from 11:00 – 5:00 pm. These austere measures continued until 1 December 1976.

2. Nutrient Intake

a) By area:

Farming area: A nutrition survey was conducted in a farming area located in Digam Ri, Suhcheon Gun, Chungcheong Nam Do, in February 1973 [16]. An approximate measuring method was used to evaluate the kinds of foods and nutrient intake of 40 households during a 3-day period. The results obtained are as follows (Table 33).

The average calorie intake was 2641 for men and 1952 for women; both intakes were lower than the recommended dietary daily allowance (RDA) revised in 1973. The total calorie intake was composed of carbohydrates, 84%, protein, 11.5%, and fat, 4.5%. The average protein intake was 67 g, also lower than the RDA. Protein usually consisted of staple foods and grains. The food pattern lacked variety since most of the foods were grains, starchy foods, and vegetables. The amount of meat in the daily meal was very small, and most of the fish was salted, so that the protein intake was very limited. The vegetable protein (bean protein is good in its essential amino acids) was usually eaten as a seasoning, soy sauce or soy paste, but this contains a very high level

Table 33. Average Nutrients Intake Per Adult Per Day in Farm, Fishery, Mountain, and Urban Area {16}

Area	Sex / Nutrients	Calorie	Protein	Fat	CHO	Ca	Fe	Vitamin A	Vitamin B₁	Vitamin B₂	Niacin	Vitamin C
		Cal	g	g	g	mg	mg	IU	mg	mg	mg	mg
Farm	Average	2296	67	13	482	382	14.5	1060	0.89	0.59	14.8	31
	M	2641	77	15	554	439	16.7	1219	1.00	0.68	17.0	36
	F	1952	57	11	410	325	12.4	901	0.74	0.50	12.6	27
Fishery	Average	2050	53	8	427	312	8.4	2043	0.87	0.64	14.9	27
	M	2358	61	9	506	359	10.0	2349	0.99	0.73	17.1	31
	F	1743	45	7	374	215	7.1	1736	0.74	0.54	12.6	23
Mountain	Average	2491	68	15	473	446	11.1	3010	1.1	0.87	17.4	60
	M	2864	78	17	544	513	12.8	3461	1.27	1.00	20.0	69
	F	2117	57	12	402	379	9.4	2558	0.94	0.74	14.8	51
Urban	Average	2228	79	42	383	579	13.1	3669	1.33	1.36	17.4	43
	M	2600	87	49	452	614	13.6	3568	1.55	1.55	20.2	49
	F	1856	70	31	324	543	12.7	3769	1.1	1.16	14.5	37
Average	Average	2266	66	19	441	430	11.8	2445	1.04	0.87	16.1	40
	M	2606	76	22	507	494	13.6	2812	1.19	1.0	19.5	46
	F	1926	56	16	375	365	10.0	2079	0.08	0.74	13.7	34

of sodium chloride, thus limiting the amount eaten. Protein, especially good quality protein, an essential component for body composition and its maintenance, was urgently needed.

The average fat intake was so limited that only 4.5% of the total calorie intake came from fat. Major sources of fat are of vegetable origin, e.g., sesame oil. The average carbohydrate intake was 482 g or 84.5% of the total calorie intake. This was due to the large amount of grain eaten, which results in nutritional imbalance.

Calcium (382 mg) intake was lower than RDA, because there was no milk or milk products in their daily food, and it is very difficult to meet the calcium requirement without milk or milk products. Most of calcium intake was obtained from seaweeds, fish, and shellfish. The iron intake (14.53 mg) was relatively adequate.

The other nutrients, vitamin A, vitamin B complex, and vitamin C were lower than RDA. Deficiency in vitamin B complex intake especially affects the carbohydrate metabolism, which is a serious problem. The vitamin C intake was low because the survey was carried out in the winter season.

Fishing area: The nutrition survey was conducted in a fishing area located in Chang-gu Ri, Dasa Ri, Suhcheon Gun, Chungcheong Nam Do in February 1973 [16]. An approximate measuring method was used in evaluating the kinds of foods and nutrient intake for 80 households during a 3-day period (Table 33). Average calorie intake was 2052; 2358 for men, 1743 for women, both of which were lower than the RDA. The total calorie intake was composed of carbohydrates, 86%, protein, 10.3%, and fat, 3.5%.

The average protein intake was 53 g, also lower than the RDA. Although the people live by fishing, they do not eat an adequate amount of fish and shells, so they lack animal protein in their total protein intake. Average fat intake totaled only 8 g, the lowest among the three areas, farming, fishing, and mountainous areas. Carbohydrate intake was 450 g or 86% of the total calorie intake. This ratio of carbohydrate intake to total calorie intake was the highest of all the areas.

Calcium intake 312 mg, was lower than the RDA, most of it coming from seaweeds, fish, and shellfish. Iron intake was 8.4 mg, also lower than the RDA (11.5 mg). Iron intake in women was especially low, only 7.1 mg, thus leading to iron deficiency anemia.

Vitamin A intake (2043 IU), vitamin B_1 (0.87 mg), vitamin B_2 (0.6 mg), niacin (14.8 mg) were all lower than the RDA, as was the amount of vitamin C intake.

Mountainous area: Nutrition surveys were conducted in mountainous areas located in Sundong Ri, *Suhcheon Gun* (Chungcheong Nam Do), and Auchun Ri, Gaebug Myeon (Chungcheong Bug Do) in February 1973. An approximate measuring method was used in Sundong Ri, and the precise weighing method was used in Auchun Ri to evaluate the kinds of foods and nutrient intake for 40 and 24 households, respectively, during a 3-day period (Table 33) [16].

The average calorie intake for an adult was 2491; 2864 for males and 2117 for females, both lower than the RDA. The total calorie intake was composed of carbohydrates, 82%, protein, 12%, and fat, 6%. The ratio of fat to the total calorie intake was higher than in the farming and fishing areas.

The average protein intake per day was 68 g, relatively close to the RDA, but in fact most of it came from grain, which is lacking in essential amino acids. Fat intake was 15 g obtained from grain, beans, vegetable oil, and as seasoning. Carbohydrate intake was 473 g or 82% of the total calorie intake on the average.

Calcium intake was 446 mg, lower than the RDA, and iron intake was 11.1 mg for men, 9.4 mg for women.

Vitamin A intake (3009 IU), vitamin B_1 (1.11 mg), vitamin B_2 (0.87 mg), niacin (17.4 mg), and vitamin C (60 mg) were all lower than the RDA, but vitamins A and C were relatively adequate.

Urban area (Seoul): The nutrition survey was also conducted in an urban (Seoul) area in February 1973. Approximate measuring method was used in evaluating the kinds of foods and nutrient intake for 120 households during the 3-day period (Table 33).

The average calorie intake was 2228 for adult; 2600 for men, 1856 for women, lower than the RDA, which is 3000 for males, 2200 for females. The total calorie intake was composed of carbohydrates, 68.8%, protein, 14.1%, fat, 10.1%, and components which were near to the recommended requirement; carbohydrates, 60%, protein, 15%, and fat, 25%.

Protein intake was 78.6 g on the average. Their dietary contents had more variety than in any other area, and the amounts of meat, fish, and beans in their daily meals were relatively high. Although there was a difference between the high economic class and the low economic level of household in the urban area, urban dwellers ate protein of good quality and in larger amounts.

The average fat intake was 42 g or 17.1% of the total calorie intake because they ate adequate amounts of meat, fish, milk, milk products, and vegetable oil. A difference in the fat intake was observed between the high and low economic class.

The average intake of carbohydrates was 383 g, 68.8% of the total calorie intake, lower than in other countries. Although they cannot attain the ideal level of 60% carbohydrates, the kind of foods they eat are more varied than in other countries, therefore the carbohydrate intake was relatively low.

Calcium intake was 579 mg, close to the RDA. This is explained by the recent increase in milk and milk product consumption in the urban area. The calcium intake also varied according to the economic level of the household. The iron intake was 13.1 mg, an average of 13.6 mg for male and 12.7 mg for females, which was up to the RDA.

The other nutrients-vitamins were also relatively sufficient (see Table 34).

Table 34. Nutrients Intake by Income Level (per adult per day) {270} [a]

Areas \ Sex	Nutrients	Calorie	Protein	Fat	CHO	Ca	Fe	Vitamin A	Vitamin B$_1$	Vitamin B$_2$	Niacin	Vitamin C
		Cal	g	g	g	mg	mg	IU	mg	mg	mg	mg
Mansion	M	2670	115	63	410	590	29.7	3711	1.39	1.31	25.1	49
	F	1974	85	46	303	436	21.9	2743	1.03	0.97	18.5	36
KIST	M	2496	106	45	419	522	12.1	3403	1.16	1.23	20.0	45
	F	1845	78	33	310	386	9.0	2515	0.86	0.91	14.8	33
Mun-hwacheon	M	2566	90	50	487	518	15.0	3482	1.24	1.16	25.1	40
	F	1897	67	37	360	383	11.1	2574	0.92	0.86	18.6	29
Buk-ahyun	M	2 193	81	41	435	409	13.2	3208	1.22	1.21	17.5	38
	F	1 621	60	30	322	302	9.7	2371	0.90	0.89	13.0	28
Middle Class	M	2742	81	45	469	740	8.9	3829	1.90	1.70	18.1	56
	F	1874	71	31	324	690	12.4	4805	1.25	1.24	13.0	41
Business and profes-sional	Professor	2818	132	54	450	611	15.9	3600	1.87	1.76	29.7	91
	Writer	2981	116	43	412	673	15.2	4273	1.41	1.34	21.2	100
	Business man	2976	152	73	415	704	20.0	4185	2.30	2.79	24.8	91
	Average	2825	133	57	425	663	17.0	4019	1.86	1.95	25.2	94
Average	M	2600	87	49	452	614	13.6	3568	1.55	1.55	20.2	49
	F	1856	70	31	324	543	12.7	3769	1.10	1.16	14.5	37
	Average	2228	79	42	383	579	13.1	3669	1.33	1.36	17.4	43

[a] Mansion, rich upper class; KIST (Korea Institute of Science and Technology), high income level; Munhwacheon, various income levels; Bukahyun, various income levels.

Table 35. Comparison of Nutrients Intake between Average and Middle Class (per adult per day) {270}

Kinds \ Sex	Nutrients	Calorie	Protein	Fat	CHO	Ca	Fe	Vitamin A	Vitamin B$_1$	Vitamin B$_2$	Niacin	Vitamin C
		Cal	g	g	g	mg	mg	IU	mg	mg	mg	mg
Middle class	M	2742	81	45	469	740	8.9	3829	1.9	1.7	18	59
	F	1874	71	31	324	690	12.4	4805	1.25	1.24	13	41
Average	M	2606	76	22	507	494	13.6	2812	1.19	1.0	19	46
	F	1926	11	8	375	365	10	2078	0.88	0.74	14	34
R.D.A	M	3000	80	40	580	600	10	2000 (6000)	1.5	1.8	20	70
	F	2200	70	33	406	600	13	2000 (6000)	1.3	1.3	15	60

Table 36. Nutrient Intake by Age in Urban and Rural Areas (per adult per day) {16}

Age \ Area	Nutrients	Calorie	Protein	Fat	CHO	Ca	Fe	Vitamin A	Vitamin B$_1$	Vitamin B$_2$	Niacin	Vitamin C
		Cal	g	g	g	mg	mg	IU	mg	mg	mg	mg
School age	Urban	1688	67	41	263	400	9	2391	0.8	0.8	11.0	41
	Rural	1447	19	8	326	253	5.7	1279	0.66	0.53	10.98	34
Adoles-cent	Urban	2280	73	37	396	720	10	4186	1.59	1.45	15.7	46
	Rural	2585	99	20	508	441	15.8	2380	1.06	1.79	18.5	42
Adult	Urban	2158	80	43	381	440	15.2	2952	1.09	1.24	19.1	37
	Rural	2294	64	12	481	371	11.9	2099	1.01	0.75	17.9	36

b) By income levels: The samples were selected from various kind of apartments, income levels, and business and professional workers who could manage relatively comfortable livelihoods (Table 34) [270, 271].

High income level: Intakes of protein, fat, iron, and niacin were higher than the RDA, but vitamin A, riboflavin, thiamine, and vitamin C intakes were low. Chyu [270] analyzed various reports during the period 1972 to 1974 and found that the majority of the total population consumes much less fat than is recommended, and the calorie intake is also lower than the average. Well-to-do groups, a minority of the total population, consume far above the recommended calories together with protein and fat [43].

Middle income level: The nutrient intake was low compared to that of the high income level. However, for calcium, thiamine, and riboflavin, their intake was very much higher than that of the high income level. The above results point to an unbalanced diet and lack of consideration for the nutritive values of food.

Low income level: The total calorie intake was lower than that of other income levels, but as regards the composition of calories, the ration of carbohydrate, protein, and fat was almost the same as the total Korean average.

Business and professional workers: The nutrient intake of professionals and businessmen — professors, writers, and businessmen — was relatively high in comparison with the RDA. The nutrient intake of businessmen was the highest of all. This is explained by the fact that the amount of nutrient intake depends on the economic level, and the samples were selected in the Seoul area, especially from the high educational level.

The total daily intake of protein and fat of Seoulites averaged 79 g and 42 g, respectively. These average intakes exceeded that of the other areas due to the many nutritive advantages afforded by living in the Seoul area. Here the obvious differences of nutrient intake depend on income levels.

There were differences in the intake of vitamins and minerals between high and low income levels, but not between high and middle income levels. According to the studies, the income levels of professors, businessmen, and writers were classified above the upper-middle class. In the Hangang Mansion apartment, occupied by the rich upper class, nutrient intake was very much higher.

In summary, the amount of nutrient intake is a little higher than the total average. Calorie, protein, fat, calcium, vitamins A, B_1, B_2, niacin, and vitamin C intake were high, while carbohydrates, iron, niacin intake were somewhat low.

The most critical difference was seen in the fat intake amount and vitamin A intake. Middle class people in the Seoul area consumed twice as much fat as the national average. In contrast, the carbohydrate intake of the middle class was very low, but the vitamin intake also exceeded the national average (Table 35).

c) By age difference: The nutrient intake of rural children is generally lower than that in the Seoul area. Even though the children in the Seoul area attend private schools and belong to a relatively high economic level, there are very critical differences in protein, fat, carbohydrate, calcium, iron, and vitamin A intake. Children in the Seoul area ingest 3.5 times as much protein, 5 times as much fat, and twice as much vitamin A as those in rural areas (Table 36).

The total calorie count in the Seoul area consists of carbohydrates, 63.0%, fat, 21.1%, protein, 25.9%, whereas the respective values in rural areas are 89.8%, 4.9%, 5.3% [16].

University and college students in the Seoul area and 18 – 22 year old men and women in the rural area were selected for this survey. The amount of nutrient intake in the rural areas is relatively high in comparison with that of the Seoul area [16]. Calories, protein, carbohydrate, iron, vitamin B_2, and niacin intake were high in the rural areas, but fat, calcium, vitamins A, B_1, and C intake were very low. Especially fat, carbohydrates, and calcium intake differed sharply in the various survey areas.

The ratios of carbohydrate, fat, and protein intake to the total calorie intake were 72.5%, 14.6%, and 12.9%, respectively, in the Seoul area, but 77.8%, 7%, and 15.3% in the rural areas, i.e., fat intake is very high and carbohydrate intake is very low in the rural areas. Of the minerals, iron intake is very high in Seoul, but calcium, vitamin A, and vitamin B_1 intake are very low.

In general, almost all kinds of nutrient intake in adults are lower than the RDA [16].

3. Korean Food Problems

The per capita cultivated land in the whole world amounts to 0.4 ha. In the case of South Korea with a population of 35 million, the land area is about 2.3 million ha; per capita land area is about 0.08 ha or only one-fifth of the world average.

Of the total amount of grain consumed in a year (10 million tons), 4 million tons rice, 2 million tons barley, 1 million tons miscellaneous grains are produced in Korea. The other 30% (3 million tons grain) is imported.

In 1 year, 2 million tons wheat and 1 million tons other miscellaneous grains were imported from foreign countries. Therefore, the Korean government has instituted a food policy plan to achieve self-support in rice by 1976 and to substitute for other deficient foods by importing wheat and corn. The population is estimated to exceed 50 million by the year 2000.

An important role in this plan is accorded to potatoes, used as a main dish in western countries, they are also produced in Korea, but only as a seasonal side dish or snack. Consequently potato consumption is very small, and potato yield is only 670 thousand million tons from 150 thousand ha land. Sweet potatoes are considered a very important farm product in Korea, which has plenty of hills and low mountains for the potato cultivation. Thus, if the method of cooking potatoes is effectively developed, potatoes may contribute to the self-support food plan in Korea.

4. Nutritional Problems

There are several nutritional problems in Korea (Table 37).

Due to their meal content as was described above, Koreans ingest an excessive amount of grains but have a vitamin and mineral deficiency. Especially vitamin A deficiency has prevailed in farming and fishery areas (Table 37).

Table 37. Caloric Composition of Food in Three Countries (Unit: %) {16}

Country	Carbo-hydrate	Fat	Protein	(Animal Vegetables)
United States	64	19	19	(29 : 71)
Japan	60 – 01	7 – 8	12	(60 : 40)
Korea	80 – 90	4 – 6	8 – 10	(3 : 97)

Because of their excessive carbohydrate intake, Koreans tend to prefer salty side dishes. They use a lot of foods preserved in salt because they are easy to store for long periods of time.

Lee [16, 269, 270] has reported that Koreans ingest 20 – 30 g table salt per day and that this dietary habit is already established by 6 years of age. Before discussing whether this excessive table salt intake comes from an unbalanced diet of grains or not, we must seriously consider the problem of table salt in order to improve Korean eating habits and to reduce grain consumption.

Not only is the sodium chloride in table salt of concern, but sodium itself can raise a more serious problem. Therefore, the solution to the problem of sodium content in food must be approached from the viewpoint of improving Korean eating habits.

The problem of essential amino acid imbalance, resulting from excessive vegetable food intake and insufficient animal food intake, has been observed in nutrition surveys to vary with seasonal changes and income level differences.

Fat and oil intake of Korean children and adolescents was the lowest of all Pacific Ocean countries. This situation raised the problem of fat intake deficiency in farm and fishery areas, where the fat intake averaged only 2 – 6 g per day per adult. The widespread view that the quantity of food, not its nutritive value, is most important lies at the root of this problem (Figures 15 and 16) [16]. If one kind of food comprises 70% – 90% of the total food intake, a functional defect occurs in the body, because one kind of food cannot provide all the various nutrients required.

The above nutritional problems of Korea may also exist in many neighboring countries where excessive amounts of rice and other grains are eaten.

5. Dietary Improvement

a) A food policy: A nation's food habits is determined by many factors, political, social, and economic structures, especially those of agriculture. The content of food

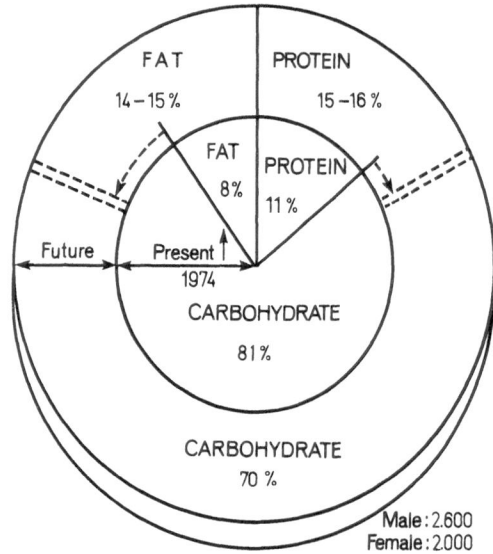

Figure 17. Korean recommended dietary allowances, energy (Cal/day/adult)
(Source: K. Y. Lee and H. S. Kim [16])

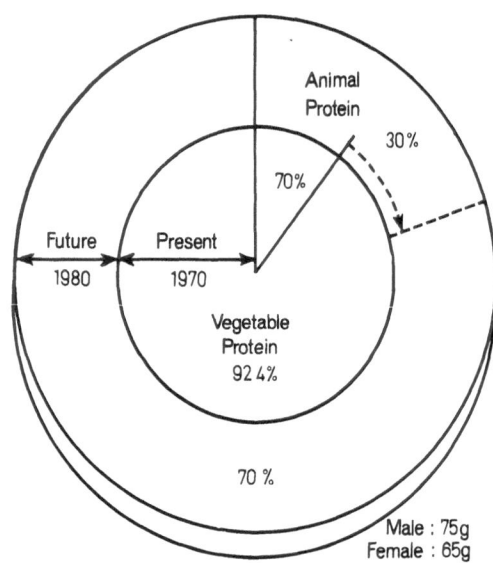

Figure 18. Korean recommended dietary allowances, protein (g/day/adult)
(Source: K. Y. Lee and H. S. Kim [16])

demanded varies with time. The nutritional problems of a nation are closely related to its people's physical development. Scientifically analyzed data, i.e., death rate, morbidity rate, and malnutrition are identical not only with hygienic problems of food and nutritional plans, but also with the essential factors of a country's economic development. A clear understanding of the differences in regional dietary patterns has been considered in setting up a policy for food and food production.

Food shortages, which should be solved on the basis of a nutrition policy, still remain a very serious problem. The solution to the food shortage problem requires the establishment of very thorough countermeasures. Therefore, while coping with the fluctuations in food demand, we must establish short-term or long-term plans to bring

Table 38. The Food Intake in Present and Tentative Plans (per day per adult) {16}

Foods groups	Present					% of present foot intake to the tentative plan	Tentative plan				
	Amount (g)	Cal	CHO (g)	Protein (g)	Fat (g)		Amount (g)	Cal	CHO (g)	Protein (g)	Fat (g)
Cereals	558.5	1899	435	31.5	2.3	159.6	350	1190	272.6	22.7	1.4
Beans	109.2	448	27	45.2	19.2	218.4	50	205	10.8	20.6	8.8
Green vegetables	102	35	5	2.6	0.7	51	200	68	9.8	5.2	1.4
Kimchies	260	49	3	5.2	1.5	130	200	38	2.6	4	1.2
Meats	36.7	48	–	7.4	2.1	73.4	50	66.5	0.1	10.0	2.9
Fishes	52	42	–	0.9	0.9	86.7	60	48	0	10.5	0.5
Seaweeds	9	19	4	0.9	–	90	10	21.6	4.2	1.0	0.1
Potatoes	20	14	3	0.5	0.1	40	50	36	7.4	1.2	0.2
Milk and milk products	17	10	1	0.5	0.6	14.2	120	70.8	5.4	3.5	3.9
Fruits	7.3	4	1	–	–	7.3	100	52	12.1	0.3	0.3
Oils	6.7	60			6.7	33.5	20	180	0		20
Seeds	–						10	59.4	1.7	2.0	5.0
Sugar	–						20	79.6			
		2628	480	94.8 (8.8)	33.8			2114.9	326.8	81.1 (25) [a]	45.8

[a] () Animal Protein.

about an improvement in the food and nutritional status of Korea and to rationalize food production (see Map 5).

There are many research institutes and investigators in Korea, but their functions still remain at a primitive level. The government must establish a fundamental research institute as a kind of national body to solve various kinds of problems pertaining to a nutrition policy. One of the most important points is to ensure that each ministry has an organic relationship with the other; i.e., through cooperation of the Ministry of Agriculture with other related Ministries, we can expect more effective methods of solving the nutritional problems of Korea.

b) The necessity of establishing a menu bank: Although we are well acquainted with our nutrient demands, scientific and concrete data on the amount of food production and food consumption have not yet been proposed. It is necessary to devise a menu suited to each class and each profession. The nation must be able to propose menus for households, and official institutes must be able to serve food in the ideal quantities set by a menu bank.

c) A tentative plan for Korean RDA: According to the total average in nutrient intake, adult men ingest 2606 Cal and women, 1928 Cal. This total calorie intake is composed of carbohydrates, 81%, fat, 8%, and protein, 11% of total protein. The animal protein intake percentage is only 7.6%.

The recommended allowances of present and future calorie and protein intake, are the result from various viewpoints, the nutrient intake survey, animal experiments, and the presently operating Korean Recommended Dietary Allowances (RDA) (Figures 17 and 18).

The present Korean R.D.A. recommends 3000 Cal for men and 2200 Cal for women. But Lee and Kim [16] prepared tentative plan reducing calorie allowances to 2600 Cal for men, 2000 for women to prevent obesity.

The tentative plan of food intake allowances recommends reducing the present grain intake and increasing the meat, milk, and milk products intake. The following table shows a comparison of the present food intake situation to the tentative plan for a recommended food intake amount (Table 38).

In summary the nutritional status of Korean people is not characterized by serious symptoms of kwashiorkor like other malnourished regions. However, nutritional deficiency symptoms are still prevalent in communities with unbalanced diets. Inspite of the great improvement made in Korean eating habits, it cannot be denied that a very deep gulf in the amount and kind of food intake still remains between the urban and rural people. A high prevalence in the rural areas of angular stomatitis due to vitamin B deficiency may be one example of this discrepancy.

E. Conclusion: Korea and Its Diseases – A Geomedical View

Diseases in Korea may be divided roughly into communicable and noncommunicable, and both are closely related to the climate, geographic conditions, the people's habits and ways of living, and other factors. Vector-borne diseases are in Korea prevalent in the summer months. The warm, rainy season is favorable for the breeding of mosquitoes; a good example of a mosquito-borne infection is Japanese encephalitis. Infections of *Brugia malayi* and plasmodia may also occur during the summer months, because the mosquitoes hibernate during the winter season. *Brugia malayi* is limited geographically mainly to Jeju Do and Gyeongsang Bug Do, but even in the same province, the endemic pattern of *Brugia malayi* differs. It is found in the southern coastal area of Jeju Do, but not in the north of the island. In the north-western mountain valleys of Gyeongsang Bug Do, there are foci of *Brugia malayi* but not in the remainder of that province. The vectors vary with the geographic characteristics of the landscapes, which was prevalent in the whole land until 1945, is now restricted to the north-western part of Gyeongsang Bug Do. Even though a considerable effort was made by the government to eliminate the disease from the area, it is still prevalent. Reluctance among people raising silk worms in the area to use insecticides may favor the maintenance of the mosquito vector population *Anopheles sinensis*.

In general, diseases related to insects and the soil are prevalent in the warm months. They are enteric infectious and parasitic diseases. Only louse-borne epidemic typhus is more prevalent in the cold months. *Enterobius vermicularis* and *Mycobacterium tuberculosis* infections are related to dust and a crowded environment. Living in small rooms during the long winter months may provide more chances for infection. Hookworm distribution is less common in the plain area where the soil is heavy. Heavy soil such as clay and silty soil is not favorable to maintenance of the larvae of parasites outside their host. This was proven in the Honam plain. Distribution of the intermediate host does play a role in determining the endemicity of certain parasites. For example, parago-nimiasis is more prevalent in mountainous areas inhabited by the first intermediate host, *Semisulcosphira* snails; and *Clonorchis sinensis* is common in the plain areas where the *Parafossarulus* snails, the first intermediate host, propagate. The former prefers a sandy, clean stream, and the latter a muddy soiled-stream or paddy.

In an unsanitary environment, domestic animals act as carriers, intermediate hosts, reservoirs or causes of anthropozoonotic diseases. Salmonellosis, tapeworm infection, and trematode infections are good examples. They are in some way connected with food habits. As a matter of fact, a large part of the communicable and parasitic diseases in Koreans are caused by the ingestion or improperly prepared foods. Eating habits are also related to customs, traditions, and ignorance. Shamanistic background and poor knowledge about the modes of transmission may contribute to this in other ways. Some diseases such as venereal diseases are concerned with ethical matters, too. In this way every disease has its own background related to habit, culture, soil, water, climate, and other geomedical factors of the environment.

Communicable diseases are still serious health problems in Korea, although the general trend reveals a decline due to strenuous control activities. In contrast, noncommunicable degenerative diseases have been increasing in recent years, e.g., hypertension, diabetes, malignant tumors, mental diseases, and others. Industrial development, urbanization, and tensions in the changing world may be responsible for the increase of these, the so-called civilization diseases. Carbon monoxide poisoning due to inadequate utilization of pressed coal should be

eliminated by replacing the fuel source or by safe construction of the hot-floor (ondol). Low intake of animal protein and fat causes nutritional imbalance, especially in persons of low economic status. Encouragement of stock farm industry may help adjust such an imbalance.

In considering the etiology of disease, two groups of factors should be taken into account: etiologic hygienic factors and sociocultural factors. The former may be determined empirically by finding the specific pathogenic agent, but the latter is rather complex and must be considered from the viewpoints of both socioeconomic culture and geo-environment. In this summary, only the latter will be dealt with.

Before western medicine was introduced, medical problems were controlled by the traditional concepts of disease, shamanism and oriental medicine. Folk medicine is still closely adhered to by the people, especially among rural and socioeconomically poor communities. A survey in rural towns indicate that the greatest proportion, about 32.4%, receive their initial medical care from pharmacies, followed by folk medicine (27.0%), and lastly hospitals or clinics (20.4%). Acupuncture is a valued native medicine and is still valued empirically by Koreans.

Since shamanism is nothing besides superstition, all behavior relating to this philosophy has misled the people adversely, leaving the nation in an underdeveloped state. Oriental medicine (herb medicine) has characteristically put more weight on empiricism; however, oriental medicine has more or less been deeply influenced by shamanistic psychology, so that the analysis of diseases has not been conducted in a scientific way. Even though a theory of constitutional medicine (Sasang Eui-Hak) has been introduced by Je Ma Lee already in 1901, the anatomic knowledge of the human body was rather illusory at that time. For such reasons, oriental medicine has remained stagnant until recently. This new development is seen in the curricula of the College of Oriental Medicine, Kyeung Hee University, which have been appreciably modified toward a more scientific viewpoint, consequently producing more scientific-minded graduates. The methodologies in clinical, preventive, and basic oriental medicine are expected to coincide eventually with the so-called western style, thus inducing the people to take a more reasonable direction away from the erroneous traditions of the past.

Maldistribution of health personnel is turning into another health problem. Because metropolitan growth has severely aggravated the problem, shortage of medical personnel in the rural or remote areas may result in certain diseases remaining endemic and acting as foci in the spread of disease to other areas.

Health personnel are concentrated mostly in the large cities. The ratio of medical doctors to the population on the average is roughly 1 : 1999, but about 60% of the medical doctors are concentrated in Seoul and Busan. By city or province, Seoul city shows the highest proportion 1 : 1140, but Chungcheong Bug Do province shows 1 : 13 550 (see Map 4). This is strong evidence that the distribution of medical personnel is highly distorted in favor of the urban over the rural areas. Along with health personnel, medical facilities also show maldistribution. In 1974, of a total of 11 228 facilities, over 30% were in Seoul and about 10% in Busan (see Map 4).

Maternal and child health (MCH) services were started as a part of the overall health program in 1945. Since then, activities have been carried on successfully. Infant mortality in the rural communities during the years 1944 to 1948 was 171.4 per thousand, but decreased to 30.2 in 1961 – 1968. MCH services include nutritional services, immunization, periodic physical examination, health education, and public health nursing services. Together with MCH services, school health programs have been carried out as a part of government programs and by several related voluntary groups. The government carries out health activities more effectively with the cooperation of voluntary organizations.

The present crude death rate in Korea is about 6.8 per thousand; it is expected to decline gradually to about 5.4 in 1985. As regards the status of sickness and death, no reliable data cover the whole land. Some reports indicate that morbidities per 1000 population in some rural areas were 255 in the group 60 years old and over, 247 to 240 in those 40 – 59, and 211 in the 0 – 4 year old group, but the lowest was 75 in the 10 – 19 year old group. Overall statistics imply that the causes of death are changing from infectious to noninfectious diseases, as industrialization, urbanization, and modernization go forward. It is predicted that nosologic features will change in accordance with the development of the country.

High medical expenditures also hinder effective utilization of medical facilities by the people, especially the low income class comprising the larger part of the population. To cover such handicaps, a medical insurance system may be one conceivable method of approach. Before 1981, the final year of the Fourth National Economic Development Plan, the national per capita income is expected to exceed U.S. $ 1000. The Korean government will wait until then before expanding the medical insurance system on a large scale, although it was started as a demonstration project in 1965. This kind of social security system is necessary, not only for the equal distribution of medical services, but also for the preservation of healthy workers for the growth and expansion of industry. But the voluntary medical insurance movement, which was begun by devoted community leaders such as doctors, social workers, businessmen, and pastors who donated their services has met with a lot of difficulties because the government provides no comprehensive health planning nor administrative and financial support.

Environmental factors in disease must not be overlooked. Rapid industrialization has contributed to more air pollution, noise nuisance, water contamination, soil pollution, and traffic accidents. A recent increase in the incidence of noncommunicable diseases and even some communicable diseases may be largely due to these environmental factors. All wastes, including human excrement, garbage, and other materials, should be treated before they come into contact with living organisms or food matter. Because these treatment plans are still in a rudimentary stage, contagious diseases and zoonoses are still prevalent in the land. Various kinds of rodents and insects are propagated in the unsanitary environment and spread diseases.

Another unexpected problem is the ecologic changes of some biologic and mechanical transmitters of diseases. For example, many dams have either been constructed or are under construction as a part of the national economic development plan. New systems of water utilization may afford new sorts of snail- and vector-borne diseases and establish new endemic foci. Indiscriminate use of pesticides for agricultural purposes may contaminate water resources and soil, then break down the balance of the free-living stages of the pathogenic organisms. Consequently, pollution of water and soil may change the patterns of morbidity.

The contents of the foregoing monograph suggest the following needed improvements: abolition of shamanistic behavior — improvement of the environment — development of a health care system based on the realistic needs of the people.

Annex I. Republic of Korea — Administrative Districts (Dec. 31. 1974)

Special city or province	Area (sq. km)	Rate (%)	Gu	Shi (si)	Gun	Dong (legal)	Eub	Myeon
				South Korea [a]				
Whole Country	98 757.69	100.0	28	33	138	1984	122	1348
City of Seoul	627.86	0.6	11	–	–	468	–	–
Busan city	375.09	0.4	6	–	–	173	–	–
Gyeonggi Do	11 068.72	11.2	4	6	18	235	14	181
Gangweon Do	16 784.68	17.0	–	4	15	99	15	98
Chungcheong Bug Do	7 436.64	7.5	–	2	10	64	8	96
Chungcheong Nam Do	8 752.22	8.9	–	2	15	83	21	158
Jeonra Bug Do	8 057.68	8.2	–	3	13	95	8	153
Jeonra Nam Do	12 074.94	12.2	2	4	22	199	17	214
Gyeongsang Bug Do	19 802.01	20.1	5	5	24	236	22	229
Gyeongsang Nam Do	11 957.87	12.1	–	6	19	307	14	209
Jeju Do	1 819.98	1.8	–	1	2	25	3	10

[a] The figures indicate areas under the jurisdiction of the Republic of Korea as of December 31, 1973 surveyed on the basis of land register.
  Source: Korea Statistical Yearbook, 1975, Bureau of Statistics, Economic Planning Board, Republic of Korea

Annex II. Myeon Population without Medical Personnel, Republic of Korea [a]

	No. of towns and villages			Population in Myeon and Eub	Without physician		Without dentist		Without herb-doctor		None of the three	
	Gu · City	Gun	Eub · Myeon		No. of Myeon	Population	No. of Myeon	Population	No. of Myeon	Population	No. of Myeon	Population
1965	45	139	1 466	19 380 347	649	7 503 267	1 343	16 031 745	887	10 303 849	504	5 925 982
1968	45	139	1 466	19 355 995	645	6 858 433	1 337	15 865 030	956	11 017 993	508	5 320 632
1971	45	140	1 473	18 513 282	536	5 068 207	1 333	14 853 316	1 064	9 924 469	483	4 517 933
1974	62	139	1 459	18 262 204	602	5 625 095	1 321	14 608 231	1 115	12 104 652	536	4 954 489
City of Seoul	11	–	–	–	–	–	–	–	–	–	–	–
Busan City	7	–	–	–	–	–	–	–	–	–	–	–
Gyeonggi Do	10	18	195	2 292 210	39	278 777	170	1 654 727	117	1 018 046	31	249 821
Gangweon Do	4	16	107	1 431 400	37	296 452	100	1 179 399	76	796 280	27	198 647
Chungcheong Bug Do	2	10	104	1 255 192	48	450 137	96	1 038 815	69	699 495	38	356 366
Chungcheong Nam Do	2	15	179	2 375 268	65	638 356	156	1 768 615	123	1 363 517	58	587 381
Jeonra Bug Do	3	13	157	1 939 700	101	934 233	155	1 735 923	143	1 556 326	98	908 243
Jeonra Nam Do	6	22	231	3 126 809	147	1 620 304	215	2 620 966	201	2 600 776	138	1 458 577
Gyeongsang Bug Do	10	24	250	3 220 725	93	836 259	222	2 487 013	190	2 251 130	80	676 881
Gyeongsang Nam Do	6	19	223	2 348 135	70	541 825	196	1 937 301	185	1 633 610	64	486 114
Jeju Do	1	2	13	272 765	2	32 459	11	185 472	11	185 472	2	32 459

[a] Population in 1974 based on the Report on Resident Population Survey as of Oct. 1. 1973

Explanations to Figure 11 (page 37)

1. Brain.
2. Throat.
3. Lungs.
4. Pericardium.
5. Heart.
6. Spleen.
7. Diaphragm.
8. "Oil Sac," omentum?
9. Stomach.
10. Neck of spleen. Ligament?
11. Neck of stomach. Cardia?
12. Neck of liver. Inf. vena cava?
13. Internal anus. Pylorus?
14. Liver.
15. Small intestines.
16. Inside of large intestines.
17. Large intestines.
18. Gall bladder.

19. Kidney.
20. Original source of urine. Renal vessels?
21. Urethral sphincter.
22. Bladder.
23. Straight intestines.
24. Center of breast.
25. Navel.
26. Inside face of navel.
27. Urinary meatus.
28. Sphincter.
29. End of large intestine.
30. Anus.
31. Seven parts of heart.
32. Three parts of spinal column.
  I. Thoracic Portion.
  II. Abdominal Portion.
  III. Pelvic Portion.

References

A. Textbooks and Monographs

1. Central Meteorological Office, Rep. of Korea: Climatic atlas of Korea, 1931–1960, 1962
2. Central Meteorological Office, Seoul, Korea: Climate of Korea Korean Weather Service. 1964. pp. 4–6
3. Chun, C. H.: Outline of acute infectious diseases in Korea. Choe-Sin Eui-Hak-Sa Co., 1975
4. Chung, T. H.: Korean flora I, II. Kyo-Yook-Sa Pub. Co., 1972
5. Dong-A printing Co.: Hankook Dae-Gwan (Korea big view, 20th century) 1974
6. Economic Planning Board, Rep. of Korea: Korea Statistical Yearbook, 1975
7. Fact-Finding and Coordinating Committee (December 30, 1961) R. O. K.: Report on Korean Health, ROK-WHO-AID Health Evaluation and Planning Programs, 1961
8. Forest Research Institute: An annotated Checklist of the birds of Korea (author: Prof. Won, P. O.), 1969
9. Hur, J., Park, Y. S.: A study on medical care expenditure (Monograph). J. of Public Health, Seoul National University, 1973
10. Institute of National Geology: Mineral Resources of Korea. Lee, J. H. (ed.). No. 3, 1975
11. Institute of Plant Environment, Office of Rural Development, Suweon, Korea: Official soil series description, I and II, 1971
12. Kim, H. S.: Illustrated Encyclopaedia of Fauna and Flora of Korea, 14, Anoplura. Samwha Pub., 1973
13. Korean Overseas Information Service: Wildlife and Flowering Plants (authors: Dr. Won, P. O., Dr. Lee, Y. N.), 1973
14. Kwang Myong Pub. Co.: Korea-past and Present, 1972
15. Lee, Je Ma: Sa Sang-Eui Hak Won Ron. (1836 A.D.). It is renamed as "Dong-Eui Soo-Se Bo-Won", Published from Soo-Moon Pub. Co., 1973
16. Lee, Ki Yull, Kim, He Sook: A scientific search for the improvement of Korean diet. March, 1972–June, 1974 Published Yonsei University and Ewha Women's University, 1974
17. Ministry of Education, Republic of Korea: Statistical yearbook of education. Ministry of Education, 1975
18. Ministry of Health and Social Affairs, Republic of Korea: Report on Korean Health. ROK-WHO-AID health evaluation and planning program, 1961
19. Ministry of Health and Social Affairs, Republic of Korea: Yearbook of public health and social statistics, 1974
20. Moon, O. R., Hong, J. W.: Health services outcome data. J. Family Plan. Stud., KIFP 3, 135–224 (1976)
21. Suh, Cheong-Soo, Pak Chun-Kun: Aspects of Korean culture. Soo-do Women's Teachers College Press, Seoul, Korea 1974
22. Office of Rural Development (ROK): Reconnaissance Soil Map, 1971, pp. 1–9
23. Jin Dan Hak Hoe: Korean History – Modern Age – 1st period. Eulyoo Pub. Co., 1966
24. Office of Rural Development (ROK): Official soil series description 1–2, 1975

B. General

25. McCune, Shannon: Korea's Heritage, A Regional and Social Geography. Tokyo: Tuttle, 1956
26. Sohn Pow-Key: Early Korean Printing. J. Amer. Oriental Studies 79, 101–28 (1959)
27. Sohn Pow-Key et al.: The History of Korea (Korean National Commission, UNESCO), 1970
28. Sohn Pow-Key: Early Korean Typography (The Pochin chai, Seoul), 1971
29. Sohn Pow-Key: Early Korean Printing. Der gegenwärtige Stand der Gutenberg-Forschung. 1, 217–31 (1972)
30. Sohn Pow-Key: The Upper palaeolithic Habitation, Sockchang-ni, Korea (Yonsei University Museum), 1973
31. Sohn Pow-Key: L'Analyse palynologique et l'Environment paleolithique de la Corée. Etudes Franco-Coreennes, 1, 9–32 (1974)
32. Sohn, Pow-Key: Lee Cultures Palaeolithique. Revue de Corée 6-1. 4–17 (1974)
33. Sample, L. L.: Tongsan-dong, A Contribution to Korean Neolithic Culture History. Arctic Anthropology II. No. 2., 1975
34. Sohn Pow-Key: Early Palaeolithic Cultures of Sokchang-ni, Korea. Early Palaeolithic in South and East Asia, 1976
35. Lee, S. Y.: Various blood types in Korean and related clinical problems. Yonsei Med. J. 1, 40–44 (1960)
36. Lee, S. Y.: Blood types of Korean. Official J. Research Inst. Med. Science, Korea 1 (10), 711–720 (1969)
37. Lee, S. Y.: Further analysis of Korean blood types. Act. Crim. Japan 33 (3), 117–127 (1967)
38 a. Yang, M. Y.: Chemical analyzes of hot springs in Korea. Geology and Ore Deposit 13, 42–49, 1973
38 b. Yang, M. Y.: Contents of Li, Sr. in hot springs in Korea. Geology and Ore Deposit 13, 50–52 (1973)
39. Lee, Joung Hwan: Mineral resources of Korea. Geological and Mineral Institute of Korea, Part 1, 1975, pp. 1–22
40. Kim, Ok Joon: Geology and tectonics of South Korea. United Nations ESCAP, CCOP Technical Bulletin, 8, 17–34 (1974)
41. Kim, Ok Joon: Mineral resources of Korea. Circum-pacific Energy and Min. Resources No. 25, 1976
42. Lah, K. Y.: Studies on venomous snake bites in Korea. Trop. Med. News (Korea), No. 62, 1975
43. Soh, C. T., Lee, K. T., Cho, K. M., et al.: Prevalence of Clonorchiasis and Metagonimiasis along rivers in Jeonra Nam Do, Korea. Yonsei Rep. Trop. Med., 7 (1), 3–16 (1976)
44. Soh, C. T., Lee, K. T., Ahn, Y. G.: Shrimp as intermediate host of Paragonimus westermani in Korea. J. Parasitol. 52 (5), 1034–1035 (1966)
45. Government of the Republic of Korea: The third 5 year "Economic Development Plan", 1972–1976, 1976
46. Kwon, T. H., Lee, H. Y. et al.: The population of Korea – The Population and Development Studies Center at Seoul National University, 1975
47. Planned Parenthood Federation of Korea: Annual report of family planning 13, 1975
48. Cho, L. J.: The demographic situation in the "Republic of Korea" prepared for presentation at the annual meeting of the "Population Association of America", New Orleans, April, 1973
49. Im, T. B.: Population projection for the Republic of Korea 1960–1980, Monthly Statistics of Korea (Seoul), 5 (11–12), 5–47 (1963)
50. Lee, Y. C., Kim, K. S., Yun, D. J.: Studies on birth, death and cause of death of the inhabitants of Korean rural, mountainous and island areas, part I, II, III. J. Rural Health 3 (1), 75–118 (1970)
51. Choe, E. H., Park, J. S.: Some findings from the special demographic survey, Seoul. Population and Development Studies

Center, Seoul National University, and Economic Planning Board Bureau of Statistics (April), 1969

52. Cho, L. J., Ratherford, R. D.: Comparative analysis of recent fertility trends in East Asia. Paper prepared for the 1973 General Conference of the International Union for the Scientific Study of Population (IUSSP), Liege, Belgium, August 27–Sept. 1, 1973

53. Lee, D. W.: Derivation of life table functions from the recent Korean census. Unpublished M. A. thesis, London School of Hygiene and Tropical Medicine, London University, 1972

54. Brass, W.: The demography of tropical Africa. Princeton University Press, 1968

55. Kim, I. S. et al.: Mortality patterns of a rural village in Korea. Kor. Cent. J. Med. *31* (2), 177–189 (1976)

56. Borman, N. H.: The history of ancient Korean medicine Yonsei Med. J. 7, 103–115 (1966)

57. Lee, W. C.: Medical education and practice in Korea. J. Med. Educat. *45*, 283–292 (1970)

58. Rice, R. G.: Medical education in Korea, needs and opportunities. *14*, 91–108 (1973)

59. Jung, M. H., Hong, J. W.: A review of the reported research data on maternal and child health in Korea. Kor. J. Pub. Health *11* (2), 328–340 (1974)

60. Sabin, A. B.: Japanese B. encephalitis in American soldiers in Korea. Amer. J. Hyg. *46*, 356 (1974)

61. Ree, H. I. et al.: Studies on over-wintering of Culex tritaeniorhynchus Giles in the Republic of Korea. Cah. O.R.S.T.O.M., ser Ent. med. et parasitol. *14* (2), 105–109 (1976)

62. Mifune, J.: Transmission of Japanese encephalitis virus to susceptible pigs by mosquitoes, Culex tritaeniorhynchus after experimental hibernation. End. Dis. Bull. Nagasaki Univ. 7 (3), 178–191 (1965)

63. Hayash, K.: Ecological studies on Japanese encephalitis virus. Reports of Investigation in Nagasaki area, Japan in 1969–1971. Trop. Med. *15*, 212–224 (1973)

64. Wada, Y.: Mosquitoes overwintering in Izu peninsula and Mt. Nakogiri, Jap. J. Sanit. Zool. *19* (1), 82–83 (1968)

65. Mathis H. L., Jolivet, P. H. A.: Seasonal and yearby population variations of adult Culex tritaeniorhynchus in Korea, WHO/VBC/74 499, 1974, pp. 1–7

66. Kim, S. W., Chun, C. H. et al.: Changing patterns of epidemiology and death-causes on Epidemic hemorrhagic fever in Korea. Kor. J. Inf. Dis., 8 (1), 58–74 (1976)

67. Lee, H. W., Lee, P. W.: Demonstration of causative antigen and antibodies. The Kor. J. Intern. Med. *19:* 371–383 (1976)

68. Paul, J. R., MeLure, W. W.: Epidemic hemorrhagic fever attack rates among United Nations troops during Korean war. Amer. J. Hyg. *68*, 126–139 (1958)

69. Gauld, R. L., Craig, J. P.: Epidemic hemorrhagic fever. Amer. J. Hyg. *59*, 32–38 (1954)

70. Seo, J. S.: Rubella among Koreans-Hemagglutination test. Rep. from Nat. Health Inst. *9*, 119–124 (1972)

71. Park, H. J.: Epidemic of mumps in a primary school. Kor. J. Pediatrics No. 1, 1949.

72. Kang, D. Y.: Case of infectious mononucleosis. Hyundae Eui-Hak *3*, 73–77 (1965)

73. Lee, S. B.: A case of infectious mononucleosis. J. of Kore. Med. Assoc. *10*, 36–44 (1967)

74. Kim, C. K. et al.: A case of infectious mononucleosis. J. Pediatrics (Korea) *11*, 441–446 (1968)

75. Chun, C. H. et al.: Present status of Murine typhus in Korea. J. Kor. Med. Assoc. 7, 267 (1964)

76. Kim, S. T.: Report on outbreak of Epidemic typhus. Kor. J. Microbiol. *1*, 4 (1958)

77. Chun, C. H., Chung, H. Y.: Complement fixing antibodies among Epidemic typhus cases in Korea. J. Kor. Med. Assoc. *5*, No. 2, 38 (1962)

78. Kim, D. H.: Epidemiological study on Typhoid fever in Korea. J. Kor. Med. Assoc. 8, 848–856 (1965)

79. Chang, H. N. et al.: Vibrio parahemolyticus isolated from a typical food poisoning. Korea Centr. J. Med. *17* (4), 1–5 (1969)

80. Chun, D. K.: Vibrio parahaemolyticus as the cause of a food poisoning outbreak in Korea. Trop. Med. News (Korea) No. 25–26 (1971)

81. Anh, Y. K., Soh, C. T.: Flea fauna of rodents in coastal region of Korea. Yonsei Rep. Trop. Med. *4* (1), 41–49 (1973)

82. Lee, S. Y.: Epidemiological survey of whooping cough in Seoul area. Chong-Hap Eui-Hak *8*, 1073 (1963)

83. Chung, H. Y., Chun, C. H.: Clinical observation of Meningococcal meningitis. Kor. J. Intern. Med. *2* (1) (1949)

84. Yoshida: Study on Diphtheria in Korea. Mansen-No Ikai, Nos. 154–157, 1934

85. Han, S. S.: Study on the types of Diphtheria bacilli isolated in Seoul. J. Exp. Med. *24*, 952 (1940)

86. Lee, Y. G.: Study on the types of Diphtheria bacillus. First Ann. Meet. of Microbiol. (Korea), Nov. 26, 1954

87. Hong, C. H.: Effect of Sulfamethoxazole-Trimethoprin on bacillary dysentery. J. Pediat. (Korea) *16*, 137–143 (1973)

88. Hong, P. W., Scott, K. M.: A review of pulmonary tuberculosis in Korea and at Severance Hospital. Yonsei Med. J. *2*, 80–89 (1961)

89. Eui-Hyup-Sin-Bo, No. 976: Improvement on supply of health personnels, 1976

90. Kim, S. J.: Tuberculosis and its national control programme in Korea. J. Kor. Med. Assoc. *19* (8), 604–611 (1976)

91. WHO: Seventh Report, WHO expert committee on tuberculosis. WHO Technical Report Series *95*, 13 (1969)

92. Yun, S. W., Yang, J. M.: Study of rural tuberculosis (2). J. Rural Health *2* (1), 1–12 (1953)

93. Lew, J.: Leprosy in Korea. Yonsei Med. J. *1*, 77–93 (1960)

94. Lew, J.: The present status on leprosy in Korea. Yonsei Report on Trop. Med. *5* (1), 156–157 (1974)

95. Kim, K. H., Lew, J.: A study of estimation of total number of leprosy patients in Korea. Kor. Lepr. Bullet. *6* (1), 1–23 (1969)

96. Chung, I. S., Lew J.: Clinical epidemiology of ambulatory leprosy patients. Kor. Lepr. Bullet. *8*, 97–114 (1972)

97. Cho, S. K.: Study on the epidemiology of leprosy and the classification aiming social rehabilitation. J. Kor. Lep. Bullet. *4* (1967)

98. Kim, Y. S. and Lew, J.: Clinical epidemiology of ambulatory leprosy patients in Korea. Kor. Lepr. Bullet., *5* (1): 79–94, 1968

99. Kim, Y. P.: The broad aspect of present medical mycology and mycotic diseases in Korea. J. Kor. Med. Assoc., *18* (11): 931–939, 1975

100. Suh, S. B.: Flora of aerial fungi in Korea. J. Kor. Med. Assoc., *18* (11): 948–952, 1975

101. Lew, J., Kim, J. D.: Recent trends of syphilis incidence in Korea. Yonsei Med. J. *9* (1), 74–80 (1968)

102. Huh, J. et al.: Cost benefit analysis of tuberculosis control program in Korea. Tuberculosis and Respiratory Diseases, *2* (1), 19–24 (1974)

103. Beerman, H.: Syphilis. Arch. Intern. Med. *109*, 323–344 (1962)

104. Yang, H. D., Ye, H. H.: Non-treponemal test to the prostitutes before and after penicillin treatment. Chong-Hap. Eui-Hak, *12* (2), 139–148 (1967)

105. Ruge, H. G. S., Fromm, G.: Serological findings in leprosy. WHO Bull. *23*, 793–802 (1960)

106. Edmundson, W. F.: A clinico-serologic study of leprosy I. Internat. J. Leprosy *22*, 440–449 (1954)

107. Idsoe, O., Guthe, T.: The rise and fall of the treponematoses I. Brit. J. Vener Dis. *43*, 227–243 (1967)

108. Lyo, K. G.: Sociomedical investigation of venereal disease on Korean. Med. Digest. *3*, 235–240 (1961)

109. Kim, J. D., Lew, J.: Syphilis among the prostitutes and a new test plan for the serologic diagnosis of syphilis. J. Kor. Med. Assoc. 6 (11), 1143–1152 (1963)

110. Woo, J. S. et al.: Prevalence of syphilis and findings of the urine sediments among ROK army personnel. Chong-Hap-Eui-Hak, *12* (3), 219–228 (1967)

111. Kim, H. J., Lew, J.: Unpublished. Abstract in the proceed of Kor. Society of Microbiol., 1967

112. Park, D. I. et al.: Trends in the prevalence of the early syphilis in Korea. Centr. Med. (Korea) *9* (2), 617–629 (1965)

113. Kim, Y. L. et al.: Serological test of syphilis in Korean young healthy groups. Kor. J. Dermatol. *4* (1), 1–6 (1965)

114. Kim, J. D. et al.: Some STS findings among prostitutes in Wonju area. Chong-Hap-Eui-Hak *15* (2), 149–152 (1968)

115. Rhee, K. S.: A comparative study of serological reactions for syphilis using reiter protein antigen and cardiolipin antigen. Kor. Centr. J. Med. 7, 765 (1964)

116. Choi, I. H., Suh, H. S.: Serologic tests for syphilis to the prostitutes in Busan. Kor. Med. J. 10, 593 (1965)

117. Moon, K. H.: Studies on present conditions and venereal diseases of prostitutes. J. Kyungpook Univ. 8, 24 (1964)

118. Lew, J., Kim, J. D.: Recent trends of syphilis incidence in Korea. Yonsei Med. J. 9 (1), 74–79 (1968)

119. Yang, H. D. et al.: Serologic tests for syphilis to the prostitutes in Busan in 1965. J. Pusan Med. Coll. 6 (1), 95 (1966)

120. Yang, H. D. et al.: Non-treponemal test (VDRL, Kolmer, W. T.) to the prostitutes before and after penicillin treatment. Kor. Med. J. 12 (2), 139 (1967)

121. Kim, C. K. et al.: Serologic tests for syphilis to the prostitutes in Pusan in 1967–1968. J. Kor. Soc. Microbiol. 42 (1), 9 (1969)

122. Wang, C. S. et al.: An epidemiological study on the present status of venereal diseases among prostitutes in areas surrounding army bases in Korea. Kor. J. urology 17 (2), 39–68 (1976)

123. Kim, J. W.: Serological tests of syphilis in Koreans. Collected thesis, Catholic Med. College, No. 10, 1966

124. Miki, S.: The history of Korean Medicine and of diseases in Korea. Osaka prefecture (Japan), 1955

125. Hyung, S. H., Lee, K. M.: A hydrophobia case due to bite of wolf. J. Veter Med. (Korean) 3, 2–8 (1962)

126. Mills, R. G.: A contribution to the nosography of northern Korea. China Med. J. 25, 277–293 (1911)

127. Suzuki, K.: On Ascaris. Chosen lgakkai zassi No. 1, 1911

128. Soh, C. T.: Control of soil-transmitted helminths in Korea, Yonsei Reports on Trop. Med. 4 (1), 102–125 (1973)

129. Jhee, H. T.: A survey of the incidence of oral protozoan infection. Kor. Choong Ang Med. J. 2 (2), 153–156 (1962)

130. Kim, N. W., Cho, K. M.: Prevalence of oral protozoa in Korean soldiers. Yonsei Rep. Trop. Med. 4 (1), 59 – 64 (1974)

131. Kessel, J. F.: A preliminary report on the incidence of human intestinal protozoan infection in Seoul, Korea. China Med. J. 39, 975–982 (1925)

132. Choy, D.: Examination of intestinal protozoa and helminths in Koreans. Chosen Igakkai Zassi, No. 66, 1926

133. Soh, C. T. et al.: Incidence of parasites in Seoul area based on an examination of the Severance Hospital outpatients. Yonsei Med. J. 2, 31–41 (1961)

134. Kim, C. H. et al.: Prevalence of intestinal parasite in Korea. Yonsei Rep. Trop. Med. 2 (1), 30–43 (1971)

135. Brooke, M. M. et al.: Intestinal parasite survey of Korean Prisoner of war camp. U.S. armed Force Med. J. 7 (5), 708–714 (1956)

136. Cho, K. M. et al.: Incidence of Entamoeba histolytica and hepatomegaly in Jeju Do. Kor. New Med. J. 7, (5), 605–612 (1967)

137. Soh, C. T. et al.: Parasites examination of roadside soils in Seoul. Han-kuk Eui-Hak 2 (2), 81–83 (1959)

138. Kim, O. C.: Entamoeba histolytica survey in Jeju Do. Chong-Hap Eui-Hak 12 (2), 935–945 (1967)

139. Min, H. K.: Intestinal protozoa of residents in Jeonra Bug Do. Kor. J. Parasitol. 10, 8–21 (1972)

140. Rim, Y. C. et al.: Size of E. histolytica, and the pathogenecity. Catholic Med. College Collect. paper, No. 9, 109–115 (1965)

141. Choi, Y. M., Kang, D. Y.: A case of Balantidium coli infection in Ehwa Women's Med. College. Eui-Hyup Shinbo, No. 705, Jan. 14, 1974

142. Soh, C. T.: Latent infection by Toxoplasma gondii in Korea. Yonsei Med. J. 1, 52–54 (1960)

143. Soh, C. T. et al.: Serological observation of Toxoplasma antibody among neurologically and physically deficient groups in the Seoul area of Korea. Yonsei Rep. Trop. Med. 6 (1), 23–30 (1975)

144. Chyu, I.: Report of 3 cases of Isospora belli infection. Kor. J. Parasitol. 4 (1), 64 (1966)

145. Paik, Y. H., Tsai, F. C.: A note on the epidemiology of Korean vivax malaria. Choi-sin Eui-Hak 6 (2), 189–195 (1963)

146. Paik, Y. H., Van der Gugten, A. C.: Evaluation report on the results of the passive case detection conducted in the Korea Malaria pre-eradication programme during the period 1960–1965. Kor. J. Parasitol. 4 (1), 1–9 (1966)

147. Park, S. H.: The first observation of Isosporiasis in Korea. Abst. 1st Kor. Soc. Parasitol., 5 (1959)

148. Seo, B. S., Rim, H. J.: Falciparum and quartan malaria among adicts in Seoul. Thesis collection, Seoul National Univ. 8, 213–220 (1959)

149. Lim, S. K.: Pneumonitis due to Pneumocystis carinii infection. Medical Digest (Korea) 2, 285–294 (1960)

150. Kim, S. E.: Case report of Pneumocystic carinii infection among Korean infants. J. Kor. Med. Ass. 4 (10), 40 – 52 (1961)

151. Song, Y. K. et al.: Pneumocystis carinii pneumonitis, its 20 cases and the radiological examination. Choi-Shin Eui-Hak 5 (3), 201–207 (1962)

152. Chung, K. W. et al.: Examination of parasites by concentration method in a orphanage. Collect. Thesis. Catholic Med. College, No. 7, 245–252 (1963)

153. Soh, C. T.: Human parasites in Korea. Yonsei Med. J. 7, 93–103 (1966)

154. Shin, H. S., Kim, Y. J.: Trichomonas vaginalis in Korean women. J. Kor. Med. Assoc. 2 (1), 282–283 (1957)

155. Chung, P. R. et al.: Prevalence of Trichomonas vaginalis and Candida in the vagina of the prostitutes at Yeosu and Kunsan, Korea. J. Rural Health 3 (1), 283–288 (1969)

156. Seo, B. S., Rim, H. G. et al.: Study on the status of helminthis infections in Koreans. Kor. J. Parasitol. 7, 53–70 (1969)

157. Park, Y. S.: A study of the role of coprophagous animals in the transmission of human parasites. Choi-Sin Eui-Hak 8, 65–84 (1965)

158. Kim, K. S.: Parasites incidence of rural residents. Medical Digest (Korea) 4, 1707 (1962)

159. Choi, D. I., Lee, S.: Incidence of parasites found in vegetables collected from markets and vegetable gardens in Taegu area. Kor. J. Parasitol. 10, 44–51 (1972)

160. Soh, C. T.: The effects of food-preservative substances on the development and survival of intestinal helminths eggs and larvae. Amer. J. Trop. Med. and Hyg. 9 (1), 1–10 (1960)

161. Komiya, Y.: For ascaris zero-pro achievement. Jap. Assoc. of Parasite Control. 1962, p. 28

162. Park, B. J.: Epidemiological study on oxyuris infection in Koreans. Chong-Hap Eui-Hak 10 (1), 57–72 (1965)

163. Seo, B. S., Rim, H. J.: Epidemiological studies on Enterobius vermicularis in Korea. Seoul J. Med. 4 (1), 23–27 (1963)

164. Lee, K. T., et al.: Epidemiological studies on Enterobius vermicularis in Jeju Do, Korea. Kor. Med. J. 12 (2), 31–38 (1967)

165. Soh, C. T. et al.: Hookworm distribution in Korea, according to the species. J. Kor. Med. 4 (5), 49–51 (1966)

166. Choi, D. W. et al.: Treatment of hookworm infection with Furfurol. Kor. J. Parasitol. 10, 22–26 (1972)

167. Kim, S. W.: Soil-composition on prevalence of ascaris and hookworm. Modern Med. (Korean) 2 (1), 37–57 (1965)

168. Soh, C. T.: The effects of natural food-preservative substances on the development and survival of intestinal helminth eggs and larvae I, II. Amer. J. Trop. Med. Hyg. 9, 1–10 (1960)

169. Soh, C. T.: The distribution and persistence of hookworm larvae in the tissues of mice in relation to the species and to routes of inoculation. J. parasitol. 44 (5), 515–519 (1958)

170. Beaver, Paul C.: Observation on the epidemiology of ascariasis in a region of high hookworm endemicity. J. Parasitol. 38, 445–453 (1952)

171. Kim, D. C. et al.: Mass treatment for the control of ascariasis and hookworm infection in inhabitants in Koyang Gun. Report from National Institute for Health, 7, (1970)

172. Soh, C. T. et al.: Resistance of free-living stage of soil-transmitted parasites to pesticides. Yonsei Rep. Trop. Med. 6 (1), 3–13 (1975)

173. Mori, H. S.: Experiment on Anguilla intestinalis which causes diarrhoe in west-northern part of Korea. Chosen Igakkai Zassi, No. 13, 1964

174. Soh, C. T.: Ascites due to Strongyloides stercoralis infection. J. Kor. Med. Assoc. 2 (1), 91 (1954)

175. Lee, B. H., Min, D. Y.: Clinical studies on Trichostrongylus infections in Korea. Yonsei Rep. Trop. Med. 5 (1), 130–135 (1974)

176. Soh, C. T. et al.: Clinical manifestation of Brugia malayi infection in Korea. Kor. J. Parasitol 4 (2), 1–8 (1966)

177. Seo, B. S. et al.: The epidemiological studies on the filariasis in Korea II. Kor. J. Parasitol. 6 (3), 132–141 (1968)

178. Lee, K. T.: Malayan filariasis (1). Rep. National Institute of Health (Korea) 4 (1), 107–111 (1961)

179. Senoo, T., Lincicome, D. R.: The presence of Malayan filariasis in Korea. Trans. Roy. Soc. Trop. Med. Hyg., 45 (2), 269–273 (1951)

180. Hwang, C. H. et al.: A report on elephantiasis and microfilariasis found in Yongju-Gun, Gyeongsang Bug Do, Korea in 1963. Kor. Central. J. Med. 9, 491 (1965)

181. Bun, I. J.: On elephantiasis in southern chosen and in aetiology. Tokyo Iji Shin-ji, 3178, 1940.

182. Paik, Y. H. et al.: Epidemiological survey on filariasis in Nonsan area. Kor. Med. J. 2, 1175 (1957)

183. Im, K. I. et al.: A human infection with Thelazia sp. in Korea. Yonsei Rep. Trop. Med. 5 (1), 137–139 (1974)

184. Kobayashi, H.: Structure of metacercaria of Paragonimus. Jap. J. Zoology 31, 364–365 (1919)

185. Seo, B. S., Kon, B. Y.: Studies on the lung fluke, Paragonimus iloksuenensis Chen 1940. Seoul, Med. J. 12, 13–43 (1971)

186. Kobayashi, H.: Lung fluke disease in Chosen. Mitt. Med. Acad. Zu Keijo 9, 3–4 (1926)

187. Walton, B. C., Chyu, I.: Clonorchiasis and Paragonimiasis in Korea. WHO Bulletin 21 (6), 721–726 (1959)

188. Kim, H. J.: Study on pesticides on the life cycle of Paragonimus westermani. Collected thesis of Yonsei Univ. Medical College 2 (2), 106–121 (1969)

189. Kawamura, R.: Experimental infection of Paragonimus westermani to domestic animals. Annual Report from Serological Laboratory for Animal Epidemics 4, 7–14 (1916)

190. Soh, C. T.: Paragonimiasis in Korea. Yonsei Med. J. 3 (1), 79–84 (1962)

191. Kang, S. Y. et al.: Study on the Paragonimiasis in Ko-heung, Jeonra Nam Do. Kor. J. Parasitol. 2 (3), 53–58 (1964)

192. Chyu, I., Lee, J. S.: Metacercariae of Paragonimus in Eriocheir sp. in Jeju Do. Collect thesis of Catholic Med. College 10, 289–301 (1965)

193. Soh, C. T. et al.: A new second intermediate host of Genus Paragonimus. Kor. J. Parasitol. 2 (1), 35–40 (1964)

194. Ahn, Y. G. et al.: Paragonimus survey in Nam-hae, Gyeongsang Nam Do. Modern Med. (Korea) 19 (11), 1117–1121 (1966)

195. Oh, S. J.: Cerebral Paragonimiasis. Trans. Amer. Neurolog. Assoc., pp. 275–277, 1967

196. Kim, S. K., Walker, A. E.: Cerebral Paragonimiasis. Acta Psych. et Neuro. Scandin., suppl., 153, 36: 1 – 84 (1961)

197. Kim, J. S.: Mass chemotherapy in the control of paragonimiasis. Kor. J. Parasitol. 7 (1), 15–24 (1969)

198. Hunter, G. W. et al.: Parasitological studies in the Far East. J. Parasitol. 35 (supple), 41 (1949)

199. Kim, D. C.: Ecological studies of Clonorchiasis Sinensis Yonsei Rep. Trop. Med. 5 (1), 3–44 (1974)

200. Soh, C. T.: Clonorchiasis in Korea. Proceed, 4th SEA Seminar on Parasitol. and Trop. Med. 1 (1), 219–229 (1970)

201. Kim, D. C.: Intermediate host of Clonorchis sinensis in Koyang-Gun. Progr. and Abst. 10th Annual Meet of Parasitol. (Korea), 1968, pp. 8

202. Soh, C. T., Im, K. I.: Clinical manifestation of Clonorchiasis in Korea. Yonsei Rep. Trop. Med. 1 (1), 68–81 (1970)

203. Im, K. I., Koh, T. Y.: One case of Dicrocoelidae infection. Kor. J. Parasitol. 9, 58–60 (1971)

204. Yeo, T. O., Seo, B. S.: Study on Metagonimus yokogawai in Korea III. Seoul Med. J. 12, 259–267 (1971)

205. Kobayashi, H.: On scientific name of Metagonimus yokogawai, Chosen Igakkai Zassi, 55, 1925

206. Chyu, J. K.: On metacercaria of Pygidiopsis summus. personal communication, 1969

207. Ludlow, A. I.: Inguinal hernia – ova of Schistosoma japonicum in hernia sac. China Med. J. 38, 829–832 (1924)

208. Hara, C., Himeno, K.: Report on intestinal parasites eggs in Gyeongsang Nam Do. Chosen Igakkai Zassi, 48, 1924

209. Kim, C. H. et al.: The occurrence of sparganum in the tissues of frogs in Korea. Hyundae-Eui Hak 6 (2), 191–195 (1967)

210. Weinstein, P. P. et al.: Sparganosis in Korea. Amer. J. Trop. Med. Hyg. 3 (1), 112–128 (1954)

211. Cho, S. Y. et al.: Two cases of sparganosis caused by survival training in Army. J. Kor. Med. Assoc. 17 (5), 367–371 (1974)

212. Chang, K., Sung, W. Y.: Sparganum found from Natrix sp. snake and its significance as the infection source. Thesis of Catholic Med. College 10, 267–273 (1966)

213. Cho, K. M. et al.: Taenia infections in Jeju Do. Modern Med. (Korea) 7 (4), 455–461 (1967)

214. Lee, K. T. et al.: Taenia infections in Jeonra Bug Do. Kor. J. Parasitol. 4 (1), 39–45 (1966)

215. Nakanishi, J.: Statistical observation of cysticerci in Korean cattle. J. Jap. Veter. Med. 5 (3) (1926)

216. Choi, W. Y. et al.: The first case of human infection with tapeworms of the genus Mesocestoides in Korea. Kor. J. Parasitol. 5 (2), 21–25 (1967)

217. Chun, C. H.: Zoonotic contageous disease in Korea. Yonsei Rep. Trop. Med. 6 (1), 67–68 (1975)

218. Lee, C. S., Lee, J. K.: Case report on human infection of Hymenolepis diminuta. Kor. J. Parasitol. 4 (2), 41–44 (1966)

219. Chyu, I. et al.: 3 cases of Hymenolepis diminuta infection. Abst. 6th Annual meet. Parasitol (Korea), p. 28, 1964

220. Chyu I. et al.: Urinary test, infection with Trichomonas vaginalis in man. Kor. Cent. J. Med. 26 (3), 325–330 (1974)

221. Sohn, E. S.: Epidemiology of hypertension in Koreans. J. Kor. Med. Assoc. 16 (11), 897–905 (1973)

222. Kim, K. S. et al.: An observation on hypertension among Korean rural people. J. Rural Health 3 (1), 235–249 (1970)

223. Kim, E. J. et al.: Epidemiological study of Diabetes mellitus in Koreans. Seoul Med. J. 11, 25–30 (1970)

224. Kee, C. S. et al.: Epidemiological study on 378 cases of Diabetes mellitus cases. Kor. J. Int. Med. 13, 551 (1970)

225. Kim, E. J.: Epidemiological studies on Diabetes mellitus in Korea. Diabetes Mellitus in Asia Proc. of a Symp., Kobe, Japan, 1970

226. Kim, E. J. et al.: Epidemiological study on Diabetes mellitus among rural people in Korea. J. Kor. Diabetes Assoc. 1, 17–24 (1972)

227. Hwang, J. W., Kim, E. J.: Epidemiological studies on Diabetes mellitus in Korea. J. Kor. Diabetes Assoc. 2 (1), 27–32 (1974)

228. Shim, B. S.: Inherited metabolic diseases. J. Kor. Med. Assoc. 16 (7), 13–14 (1973)

229. Nam, K. H.: The prevalence of mental disorders in a rural area, Korea. Kor. J. Pub. Health. 10, (1), 33–38 (1973)

230. Cho, K. S.: Occupational diseases and industrial accidents in Korea. J. Kor. Med. Assoc. 18 (10), 837–842 (1975)

231. Chung, K. C.: Occurrence and diagnosis of lead poisoning. J. Kor. Med. Assoc. 17 (5), 319–324 (1974)

232. Cha, C. H.: Metallic poisoning. J. Kor. Med. Assoc. 17 (5), 351–357 (1974)

233. Kim, D. J., Park, S. H.: Aplastic anemia due to chronic exposure to benzene. J. Kor. Med. Assoc. 17 (6), 342–346 (1974)

234. Kim, S. I. et al.: A study on carbon monoxide concentration in Kitchens of houses in Seoul, Korea. Kor. J. Pub. Health 10 (1), 95–96 (1973)

235. Hong, S. S. et al.: Venerupin shellfish poisoning in Koje island, Korea. Asian J. Med. 8, 385–388 (1972)

236. Hong, S. S. et al.: Studies on venerupin shellfish poison in Korea. Kor. J. Gastroentero. 7 (3), 39–55 (1975)

237. Lee, M. H. et al.: Clinical investigation and treatment of thyroid diseases with radioactive iodine. Kor. J. Intern. Med. 5 (3), 157–177 (1962)

238. Min, B. S.: Radioiodine therapy of hpyerthyroidism. Kor. J. Nucl. Med. 9 (1), 15–19 (1975)

239. Chun, C. H.: Changing patterns of epidemiology and death causes on epidemic hemorrhagic fever in Korea. Kamyeon 8 (1), 25–41 (1976)

240. Kim, K. I.: The mythical factors in mental illness of Korean. Kor. J. Neuropsych. 7, 35–43 (1968)

241. Kim, K. I.: Psychoanalytic study on the primitive dream interpretation in Korea (1). Kor. J. Neuropsych. 8 (2), 15–27 (1969)

242. Kim, K. I.: New religions in Korea. Kor. Neuropsych. Assoc. 11 (1), 31–36 (1972)

243. Kim, K. I.: Traditional concept of disease in Korea. Kor. Journ. (Kor. Nat. Commission for UNESCO) *13* (1), 12–19 (1973)

244. Kim, K. I.: Shamanist healing ceremonies in Korea. Kor. J. (Korean National Commission for UNESCO) *13,* (1973)

245. Kim, K. I.: Culture and mental illness in Korea. Kor. J. (Kor. Nat. Commission for UNESCO) *14* (2): 1974

246. Kim, K. I., Kim, T. G.: The psychological function of shaman's dreams during initiation process. Kor. J. Neuropsych. 1, *9* (1), 47–56 (1970)

247. Chun, S. Y.: Folk culture in Korea. Kor. Cult. Series No. 4. Intnatl. Cult. Found, Seoul, 1972

248. Korea Scientific Book Publishing Co: Medical yearbook of Korea, 1975

249. Choi, D. K.: Medical care delivery system in Korea. Eui-Sa Sin-Moon No. 1436, Sept. 1976

250. Yang, J. M.: Study on the medical care in Korea. Thesis for Doctorate Degree to Yonsei University Graduate School, 1960

251. Lee, C. O., Park, Y. S., Huh, J.: A survey on medical care expenditure on rural area. Kor. J. Publ. Health *11* (1), 154–161 (1974)

252. Cho, D. B., Park, Y. S., Huh, J.: A survey on medical care expenditure in urban population. Trop. Med. News (Korea), No. 63, 1975

253. Im, K. I. et al.: Prevalence of intestinal parasites in ROKA soldiers. Kor. J. Parasitol. *10,* 1–7 (1972)

254. Cha, C. W.: Problems of pesticide usage in rural area. Kor. J. Rural Med. *1* (1), 22–28 (1976)

255. Kwon, S. P.: National Report of Korea – Rapid economic development calls for more social investment. An Internatl. Forum under the auspices of the Friedrich Ebert Stiftung, Tokyo. Nov. 25, 1973

256. Kim, C. S.: Health education of Korean students. Collected thesis No. 4, Sook-Myung Women's Univ., 1964, p. 27

257. Soh, C. T. et al.: Treatment of human nightsoil for prevention of Parasite infection. Kor. J. Int. Med. *8* (3), 157–167 (1965)

258. Soh, et al.: Elimination of ascaris and hookworm infection by mass treatment (1). Modern Medicine (Korea) *1* (3), 41–48 (1968)

259. Tak, R., Chun, D. K.: Distribution of salmonella among animals in Korea. Chung-Ang Eui-Hak *20* (3), 259–263 (1971)

260. Kim, C. H. et al.: Examination of parasite eggs and their survival rate in pig-manure heap in Jeju Do. Modern. Med. (Korea) *2* (2), 167–169 (1965)

261. Chow, C. Y.: Arthropods of public health importance in Korea. Kor. J. Entomol. *3,* 31–54 (1973)

262. Lee, H. W. et al.: Japanese encephalitis virus isolation from mosquitos of Korea. J. Kor. Med. Assoc. *12,* 429–440 (1969)

263. Issiki, O.: Studies on bovine Onchocerciasis in Korea, I. Jap. J. Veter. Sci. *25,* 375–385 (1963)

264. Issiki, O.: Studies on bovine Onchocerciasis III–IV. Jap. J. Veter. Sci. *26* (5–6), 259–297 (1964)

265. Lee, K. W.: The culicidae. In "Illustrated Encyclopedia of Diptera in Korea. Seoul Sam Hwa Publ. Co., 1971, pp. 679–757

266. Park, S. H.: The muscidae, Calliphoridae and Sarcophagidae. In: "Illustrated Encyclopedia of Diptera in Korea, Seoul Sam Hwa Publ. 1969, pp. 949–1042

267. Traub, R. et al.: Potential vector and reservoirs of Hemorrhagic fever in Korea. Amer. J. Hyg. *59* (3), 291–305 (1954)

268. Lee, K. Y.: Food of Koreans. pp. 26–27, 1975

269. Lee, K. Y.: A study on the food intake and nutritional status of elementary schoolchildren and their family in urban and rural Korea 1967–1968, Yonsei University, 1968

270. Chyu, J. S.: Clinical nutrition. J. Clinical Research *10,* 126–127 (1976)

271. Kim, S. P.: A study on the levels of nutrition in Korea. Kor. J. Pub. Health *10* (1), 138–144 (1973)

272. Hashikura, M.: One case of Dracunculus medinensis infection in Korea. Chosun Igakkai Zassi, 69, 1926

273. Nakada, K.: One case of Thelasia callipaeda infection in Korea. Chosun Igakkai Zassi, 24 (6), 1934

274. Kobayashi, H.: The first intermediate host of Paragonimus westermani II. Chosun Igakkai Zassi, 27, 1919

275. Kim, D. C. et al.: Distribution of human helminths (Koyang, Gyeong-gi Do). Rep. National Inst. Health *5,* 154–155 (1968)

276. Ree, H. I. et al.: Study on natural infection of Plasmodium vivax in Anopheles sinensis in Korea. Kor. J. Parasitol. *5* (1), 3–4 (1967)

277. Cho, S. Y. et al.: A case of human Fascioliasis in Korea. Kor. J. Parasitol. *14* (2), 147–152 (1976)

278. Chang, C. H. et al.: Heat flow in Korea. Rep. Geophys. and Geochem. Exploration geology survey of Korea *4* (1), 30–37 (1970)

279. Chang, C. H.: Outline of the geology of Korea. Geolog. Survey of Korea. Bullet. No. 1, 1956

280. Kim, O. J.: The stratigraphy and geological structure of the metamorphic complex in the northwestern area of the Kyong-gi massif. J. Kor. Inst. Mining Geol., 6 (4) (1973)

281. Reedman, A. J., Um, S. H.: The geology of Korea. Geol. and Min. Inst. Korea, 1975

282. Kim, I. S., Lee, D. W.: Recent mortality trends in Korea. Kor. J. Prev. Med. *2* (1), 61–76 (1969)

283. Kim, I. S. et al.: Mortality patterns of a rural village in Korea. Kor Centr J Med. *31* (2), 177–189 (1976)

284. Kim, I. S.: Study of medical care in health subcenter Kor. J. Prev. Med. *9* (1), 109–116 (1976)

Illustrations

Historical Background

Plate 1. Namdae-Mun (*Soong-ye Mun* or south gate) Seoul. Built in the beginning of the Yi dynasty (14th century) (National treasure No. 1)

Plate 2. Chomsongdae observatory. This stone tower, located in the capital of the ancient Silla dynasty, Gyeongju, is considered the oldest astrodome or gnomon in the world.
Built in 647 A. D., it is 9.1 m in height

Plate 3. Guard posts. The male post on the *right* represents the domain under heaven, and the female post, left, is said to rule under the earth. Formerly the posts were found on roadsides near the village entrance

Plate 4. An example of classic Korean architecture

Plate 5. A folk farmers' dance. At harvest time peasants dance in groups in celebration of a good crop

Plate 6. Traditional Korean cabinet and chest hold clothes and other belongings. The floors pictured here, called *Ondol,* are oil-paper covered earthen surfaces under which heat passes through flues

Plate 7. An elderly rural resident

Urban and Rural Views

Plate 8. Seoul, the capital of Korea

Plate 9. Busan, the largest harbor of Korea

Plate 10. A rural village

Plate 11. A coastal village

Plate 12. A rural village in Jeju Do, the only island province in Korea. Southern-most Mt. Hanra, 1 950 m, is seen in the distance. About 1 700 species of subtropical, temperate, and arctic plants are found and distributed over the island. The sea level climate is semi-tropical

Way of Life

Plate 13. Washing is often done at a stream

135

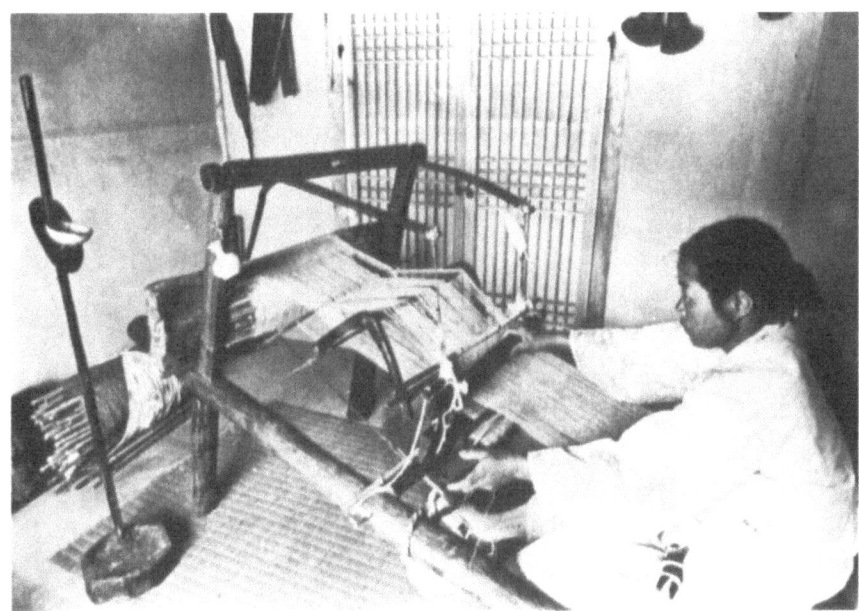

Plate 14. A woman weaving

Plate 15. Women divers off Jeju island

Plate 16. Village life around a water wheel. A woman is carrying a water jar

Plate 17. A village food market

Plate 19. Earthen jars (*Jangdokdae*). *Kimchi* (pickled vegetables), soy sauce and hot pepper sauce are stored in earthen jars

Plate 18. Preparing *Kimchi* (pickled vegetables)

Agriculture

Plate 20. Man-powered irrigation

Plate 22. Rice harvesting with minitractor: farmers are beginning to employ modern technical know-how to increase productivity

Plate 21. Ploughing

Plate 23. A wooden carrying device called an "A" frame (*Ji-ge*)

Plate 24. Rural scenes: threshing

Plate 25. Dairy farming is being encouraged in Korea to make use of grazing land that would otherwise serve no productive purpose, as well as to supplement the diet of Koreans and provide raw material for export. For this purpose, pure-bred cattle is imported

Industry and Transportation

Plate 26. A 428-km highway connects the port city of Busan with Seoul. The upstream section of the Geum river is seen

Plate 28. Ulsan industrial complex

Plate 27. Paldang dam. A low head, 11.5 m, value style installment for hydroelectricity. Completed: 24 May 1974; Capacity: 80 000 kw; Location: middle stream of the Han *gang* (river)

Plate 29. Girls assemble delicate electronic equipment at a factory

Plate 30. A *Tong-il* (unification) road that stops at the border line between South Korea and North Korea

Folk Medicine

Plate 31. Panax ginseng

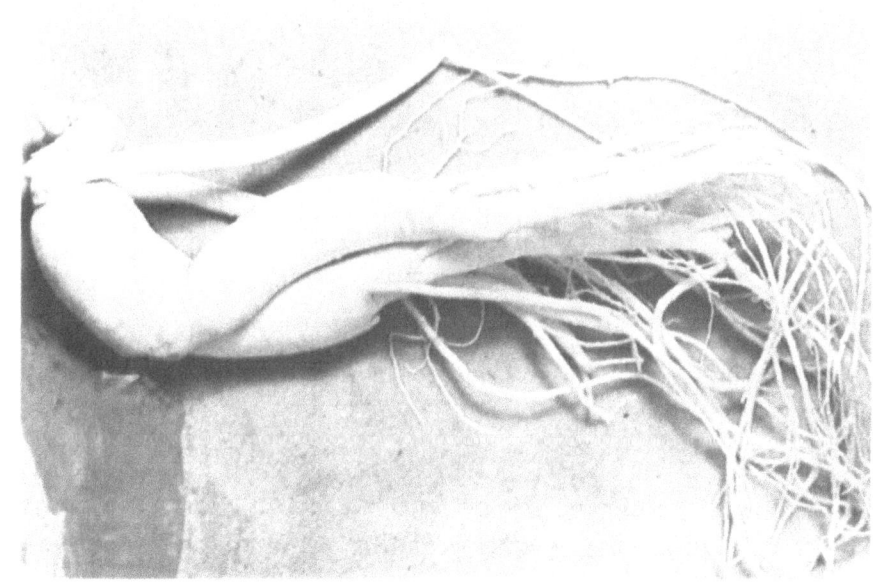

Plate 32. Panax ginseng; the root. From ancient times valued as a panacea not only in Korea but throughout the Orient, it has been used as an anticancer drug, to effect good muscular tone, to control blood pressure, to alleviate fatigue, to maintain stable body temperature, etc.

Plate 33. Ginseng garden. The ginseng root of 6-year-old plants is usually highly valued

Pate 34. Pinus densiflora is the most common species of pine trees in Korea

Plate 35. Pulsatillla carnua (Thumb) spreng Var. Koreana (Nakai) (Y. Lee); folk name: *Halmi-ggot.* Its root is used as a herb to treat malaria

Plate 36. Paeonia japonica (Makino). The root is used as a herb to treat menstrual irregularity

Plate 37. Chrysanthemum zawadskii Herb ssp. acutilo. Abounding in hilly areas, particularly in northern and central Korea, the leaf and the stem of this plant are used for fumigation

Plate 38. Songhwang-dang. A guide post with the function of protecting a village is situated hillside at its entrance. Passers-by attempt to prevent evil spirits from entering by throwing stones at the spot, thus ridding the village of evil, injury, and all kinds of hazards. The spot is usually situated beside an old tree considered sacred

Plate 39. Fresh water, flanked by incense and candles, is offered to the mountain goddess while a Shaman prays for good health and good fortune

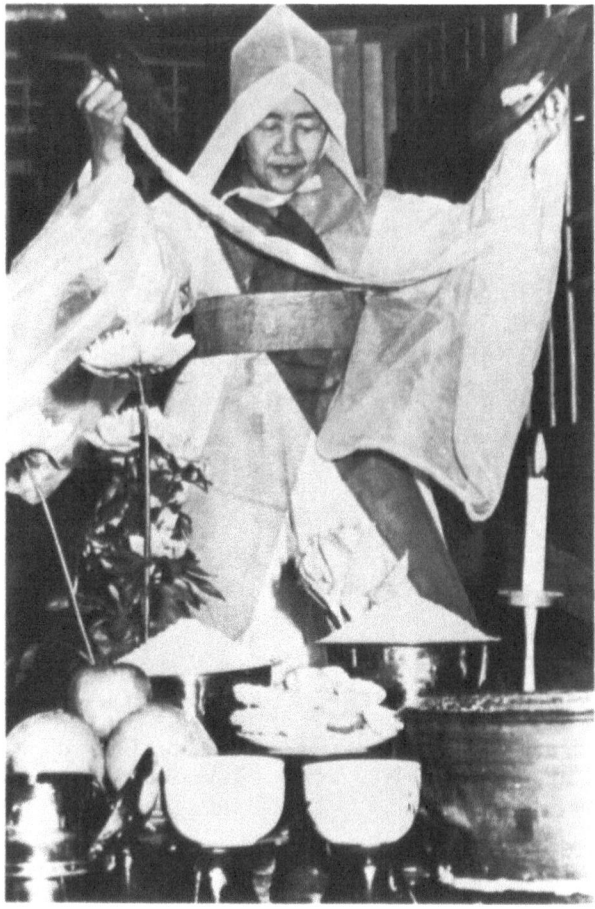

Plate 40. A shaman in a trance offers prayer to drive out the evil spirits that bring diseases to the people

Plate 41. Traditional herb medicine. Herbs in the original state are kept dry in paper bags (*above*), while prepared ones are stored in individual drawers ready for use. The kettle is used to boil the herbs

Public Health

Plate 42. Free examinations are offered along the streets during Anti-Parasite Campaign Week, held twice annually

Plate 43. Korean babies during health examination

Plate 44. Mass examination of parasites in a local laboratory of
the Korean Association for Parasite Control

Plate 45.
Yonsei University Medical Center. Founded in 1884, it was originally named *Kwang Hei Won,* the
first hospital of "modern medicine" in Korea

Medizinische Länderkunde/
Geomedical Monograph Series

Beiträge zur geographischen Medizin/Regional Studies in Geographical Medicine. Schriftenreihe der/Series of Monographs of the Heidelberger Akademie der Wissenschaften, Mathematisch-naturwissenschaftliche Klasse. Begründet von/Founded by E. Rodenwaldt. Herausgeber/Editor: H. J. Jusatz

Volume 1:

Libyen – Libya

Eine geographisch-medizinische Landeskunde. A Geomedical Monograph. Von/By H. Kanter.
For the English Translation: J. A. Hellen, I. F. Hellen
1967. 70 figures, 17 maps. XVI, 188 pages. (German/English)
ISBN 3-540-03925-2

Volume 2:

Afghanistan

Eine geographisch-medizinische Landeskunde. A Geomedical Monograph. Von/By L. Fischer.
For the English Translation: J. A. Hellen, I. F. Hellen.
1968. 16 plates, 15 figures, 10 maps. XII, 168 pages (German/English)
ISBN 3-540-04266-0

Volume 3:

Äthiopien–Ethiopia

Eine geographisch-medizinische Landeskunde. A Geomedical Monograph. Von/By K. F. Schaller.
Mit einem geographischen Beitrag von/With a Geographic Contribution by W. Kuls. For the English Translation: J. A. Hellen, I. F. Hellen
1972. 64 photos, 34 figures, 7 maps. XV, 180 pages (German/English)
ISBN 3-540-05829-X

Volume 4:

Kuwait

Urban and Medical Ecology. A Geomedical Study. By G. I. Ffrench, A. G. Hill.
1971. 61 figures, 26 text-figures, 56 tables, 3 maps. XIII, 124 pages.
ISBN 3-540-05384-0

Volume 5:

Kenya

A Geomedical Monograph. By H. J. Diesfeld, H. K. Hecklau.
1978. 60 photos, 17 figures, 9 map-plates, 55 tables. XII, 134 pages.
ISBN 3-540-08729-X

Volume 6:

Korea

A Geomedical Monograph of the Republic of Korea.
By Chin Thack Soh
With Cartographical Contributions of E. Dege.
1980. 45 photos, 18 figures, 5 map-plates, 38 tables. XV, 146 pages.
ISBN 3-540-09128-9

Springer-Verlag
Berlin
Heidelberg

Additional information of this book

(Korea;978-3-642-67137-1) is provided:

http://Extras.Springer.com